When Maps Become the World

# When Maps Become the World

Rasmus Grønfeldt Winther

*The University of Chicago Press*  CHICAGO AND LONDON

The University of Chicago Press, Chicago 60637
The University of Chicago Press, Ltd., London
© 2020 by Rasmus Grønfeldt Winther
All rights reserved. No part of this book may be used or reproduced in any manner whatsoever without written permission, except in the case of brief quotations in critical articles and reviews. For more information, contact the University of Chicago Press, 1427 E. 60th St., Chicago, IL 60637.
Published 2020
Printed in the United States of America

29  28  27  26  25  24  23  22  21  20     1  2  3  4  5

ISBN-13: 978-0-226-66967-0 (cloth)
ISBN-13: 978-0-226-67472-8 (paper)
ISBN-13: 978-0-226-67486-5 (e-book)
DOI: https://doi.org/10.7208/chicago/9780226674865.001.0001

Library of Congress Cataloging-in-Publication Data

Names: Winther, Rasmus, author.
Title: When maps become the world / Rasmus Grønfeldt Winther.
Description: Chicago ; London : University of Chicago Press, 2020. | Includes bibliographical references and index.
Identifiers: LCCN 2019046233 | ISBN 9780226669670 (cloth) | ISBN 9780226674728 (paperback) | ISBN 9780226674865 (ebook)
Subjects: LCSH: Cartography—Philosophy. | Knowledge, Theory of. | Abstraction. | Reification. | Mental representation. | Science—Philosophy.
Classification: LCC GA102.3 .W56 2020 | DDC 526.01—dc23
LC record available at https://lccn.loc.gov/2019046233

♾ This paper meets the requirements of ANSI/NISO Z39.48-1992 (Permanence of Paper).

*Til Grethe Grønfeldt Winther*

# Contents

*Preface* xi

1. Introduction: Why Maps? *1*

    A History and Philosophy of Map Thinking *4*
    *The Nature of Map Thinking—Elements of Map Thinking—
    Deep Mapping—Five Hundred Years of Western Mapping*

    Maps Today *9*
    *Cartography Meets GIS—A Definition Based on Representation—
    Characterizations Based on Process and Function*

    Three Maps *15*
    *Waldseemüller's Map—Guaman Poma's Countermap—
    Van Sant's Ultimate Map?*

    Conclusion *24*

    ## PART 1: PHILOSOPHY

2. *Theory* Is to *World* as *Map* Is to *Territory* *29*

    Analogy *30*
    *Three Types of Analogy—Critical Cautions*

    The Map Analogy *35*
    *A Typology of Map Analogies—Uses of the Map Analogy in
    Humanistic Inquiry*

    Assumption Archaeology *52*

    Conclusion *56*

3. **From Abstraction to Ontologizing**  59

   The Abstraction-Ontologizing Account  60

   Abstraction  61
   - *Abstraction Stage I: Calibration of Units and Coordinates — Abstraction Stage II: Data Collection and Management — Abstraction Stage III: Generalization*

   Ontologizing  81
   - *Ontologizing 0: Representation Testing — Ontologizing I: Changing the World — Ontologizing II: Understanding the World — Ontologizing III: Classroom Communication*

   Conclusion  87

4. **Long Live Contextual Objectivity!**  89

   Pernicious Reification  90

   Contextual Objectivity  95
   - *Conformation — The Essential Indexical*

   A History of the Mercator Projection I: Gerardus Mercator  99
   - *Mercator's Critique of Earlier Projections — Mercator's New Purpose: Navigation — Mercator's Clear Presentation of Latitude and Longitude — Mercator's Awareness of Alternative Projections*

   A History of the Mercator Projection II: Post Mercator  102

   Integration Platforms  107
   - *A Beyond-Mercator Integration Platform: Blocking Pernicious Reification and Seeking Contextual Objectivity — Philosophical Aspects of Integration Platforms*

   Conclusion  116

5. **Projecting Maps into Our Worlds**  119

   Two Canonical Philosophical Accounts of Representation: Isomorphism and Similarity  120
   - *The Isomorphism Account — The Similarity Account*

   The Multiple Representations Account  127
   - *Ontologizing — Merely-Seeing-As — Pluralistic Ontologizing — Climate Change and Multiple Representations*

   Conclusion  141

## PART 2: SCIENCE

6. **Mapping Space**  *149*

    Extreme-Scale Maps in Cosmology  *150*
    > *The Universe's Baby Portrait—The Universe Growing Up (and Outward)—Cosmic-Scale Maps and the Abstraction-Ontologizing Account*

    Literal Cartographic Maps in Geology  *159*

    State-Space Maps in Physics and Physical Chemistry  *165*

    Analogous Maps in Mathematics  *169*

    Conclusion  *174*

7. **Mapping Ourselves**  *177*

    Migration Maps  *180*
    > *Arrowized Assumptions—Arrowized Maps—Countermapping Migration*

    Brain Maps  *187*
    > *Decompositional Assumptions—Phrenological Maps—The Somatosensory and Motor Homunculi—Functional Magnetic Resonance Imaging (fMRI)—Countermapping the Brain*

    Statistical Causal Maps  *199*
    > *Linear Model Assumptions—Correlation and Causation—"Genetic" and "Environmental" Diseases—Path Diagrams as Statistical Causal Maps—When Causal Maps Become the World*

    Conclusion  *208*

8. **Mapping Genetics**  *211*

    Building a Mapping-Genetics Integration Platform  *213*
    > *Assumptions—Terminology—Map Types*

    The Linear Genetic Map  *215*
    > *Linear Genetic Maps of Phenotypic Linkage—Linear Genetic Maps of Nucleotides—Assumptions of the Linear Genetic Map*

    The Gene Expression Map  *224*

    The Genotype-Phenotype Map  *225*

    The Literal Cartographic Genetic Map  *227*

    The Comparative Genetic Map  *229*

The Adaptive Landscape Map  *231*

An Analogous Genetic Map: The Tree of Life  *234*
   *Darwin's Hypothesis—Contemporary Phylogenies*

Future Extensions: Mapping Genetics as a Paradigmatic Integration Platform  *239*

9. **Map Thinking Science and Philosophy**  *243*

   Existence, World Making, and Responsibility  *244*

   Map Thinking Scientific Methodology  *249*

   Map Thinking Philosophical Methodology  *251*
      *Assumption Archaeology—Tracking Ethics and Power—Imagining "What If . . . ?"*

   An Invitation to Dream  *254*

*Appendix: Cognitive Map Exercise  257*
*References  261*
*Index  303*

*Color illustrations follow page 126.*

# Preface

It is no accident that I wrote this book about maps. Maps are reflections of geography. And in my life I have felt lost within my own geography: born in a Danish household; raised in Caracas, Venezuela, with annual summers spent in Denmark; and having lived extensively in Venezuela, Mexico, Denmark, the United States of America, and France. Family is now bedrocked in Denmark, and close friendships are scattered across all continents, making traveling enjoyable—and inevitable. More homes await me. I suspect this is a set of experiences and feelings shared by many readers of this book.

From my childhood fascination with science to a teenage desire to understand the complexity of human relations, I have continually sought out a broad mix of natural and social sciences as well as the humanities. I still feel perhaps most comfortable in studying corals and ants, slightly less so in learning about planets and galaxies, and even less so in reading about human psychology, human governance, economics, and culture. But examples from all these areas pepper this book.

To make sense of and order my interests, experiences, and hopes—past, present, and future—I turned to the professional study of maps in my thirties. A map is a symbolic depiction emphasizing relationships between elements in space creating a lens through which you and I view the world. The organization of sentiments, knowledge, and adventures on a grid of symbolism became my motherland.

Perhaps because of my nomadic life, my strong attraction to all scientific theories across the disciplines, and my simultaneous and mutually constitutive obsessions with philosophy *and* cartography, I suspect that I am the right person to write a book that, I also feel, needed to be written. When I was a graduate student, a close friend offered me advice in the form of a quote he attributed to the computer scientist Edsger W. Dijkstra: "Do only what only you can do." This book is my attempt to make that contribution. Hopefully it will inspire you in ways you might not have imagined, and I look forward to hearing from you about your response to the book—but also about where I can improve its thesis and structure, and how I can work with you to build a better world.

## Icon Key

Time is scant; energies are limited. While I recommend chapters 1 and 9 to every reader, certain parts of the book could be skipped, depending on your domain of expertise, your curiosity, and your stamina. The chapters of part 1 ("Philosophy") and their philosophically challenging sections are classified as either "philosophical deep diving" ( 🤿 ) or "philosophical snorkeling" ( 🤿 ). Non-philosophically inclined readers might consider skimming or even skipping philosophical deep-diving sections, perhaps to return to them later. However, please try to process the philosophical snorkeling sections, because they are necessary to the book's argument. Finally, philosophically easygoing chapters or sections are labeled "easy reading" ( 🛶 ). A general reader could perhaps locate the main lessons of part 1 in chapters 2 and 4; the cartographically minded might wish to focus especially on the middle chapters, 3 and 4; and the philosophically inclined might study the bookend philosophy chapters, 2 and 5. As for part 2 ("Science"), all might benefit from chapters 6, 7, and 8 because of the parade of examples they offer.

In many respects, even this simple description of the book's organization is a kind of map, perhaps a semantic or conceptual map—and so your journey into how maps become our world has already begun. To those of us who feel drawn to mapping, and to those who have always felt their world is navigationally rich, I invite you to follow me though the vast and truly staggering influence that maps of many types can have on one's self and the world.

## Acknowledgments

Following geographic themes, these acknowledgments organize individuals and institutions spatially and temporally. With apologies to many, and following simplification and selection abstraction protocols from cartography, the only people mentioned are those strongly involved with the book or its archaeology.

I wish to pluck out the following for special acknowledgments: My father, Aage Bisgaard Winther, who supported this project in more ways than can be explained. *Tak far*. Helen Longino and Ian Hacking for providing nonstop encouragement; Lucas McGranahan for always finding *le mot juste*; and Ann Lipson, Heidi Svenningsen Kajita, and Cathrine Winther Jørgensen for challenging, and believing in, me.

*When Maps Become the World* was born in the "New World," in Latin America. I remain grateful to my parents, Grethe Grønfeldt Winther and Aage Bisgaard Winther, as well as my sister, Rikke Winther, for creating a loving and fascinating home there, and later in France and Denmark. In Caracas, Venezuela, like so many other children do, I pored over any and every map and atlas I could get my hands on. At the Colegio Internacional de Caracas, Linda Mishkin always challenged her students to think critically and creatively. Thank you to Mr. Lan, Ms. Lombardi, Ms. Meyer, and Mr. Berrendero for helping guide my deep and developing interests in, respectively, physics, biology, math, and Spanish literature. And much gratitude to Natasha Deganello, Jan den Hartog, Yair Klein, and Cristina Sitja. In Mexico City, Mexico, the first embryonic lines of this book emerged while I was given wonderful working conditions as a postdoc and subsequent assistant professor at the Instituto de Investigaciones Filosóficas, Universidad Nacional Autónoma de México. During my four years at UNAM, particularly Sergio Martínez, Nattie Golubov, Carlos López Beltrán, Alba Ruibal, Fermín Fulda, Siobhan F. Guerrero McManus, Carlos Pereda, Roy Brand, Kira Barrera, Octavio Valadez Blanco, Axel Barceló, Natalia Carrillo, and Guillermo Hurtado were constant sources of intellectual inspiration and wisdom, as this actual book started taking shape. *Gracias a todos ustedes.*

It was only much later (2009–2011, 2016–2017) that I was able to live in Copenhagen for extended periods rather than just make relatively short visits. The Center for the Philosophy of Nature and Science Studies at the Niels Bohr Institute provided a strong physics-based context in which to try out ideas. Jácome Armas and Claus Emmeche continue as dialogue partners. Jeremy Klein, now in the US, offered guidance. Kim Sneppen (Center for Models of Life) and Thomas Heimburg (Membrane Biophysics Group) provided office space and discussion regarding network theory and thermodynamics, respectively, as well as sociology of science on the ground. Johanna Seibt at Aarhus University and Frederik Stjernfelt at Aalborg University provided constructive commentary. Ongoing interactions with architects, artists, and writers in Nørrebro (my "home away from home") are much appreciated. *Tak især til* Peter Adolphsen, Heidi Beckmann, Liam Bialach, Christian Frankel, Mikkel Willum Johansen, Gitte Harbo Pedersen, Lene Harbo Pedersen, Jes Høgsberg Lind, Marie Lumholtz, Christina Okai Mejborn, Wolfgang Mostert, Søren Mørk, Line Richter, Heather Spears, Danni Tell, Sia Signe Sander, Janne Breinholt Bak, Heidi Svenningsen Kajita, *og*

Cathrine Winther Jørgensen, *samt* Harbo Bar, Bevars, Brus, *og* Original Coffee. *Det har været belærende, berigende, hyggeligt og... sjovt.*

In the US, whether at Stanford University, Indiana University, or University of California, Santa Cruz, I have been lucky to interact with a wonderful variety of impressive academics. At Stanford University, Brian Cantwell Smith, Hasok Chang, John Dupré, Marcus Feldman, Peter Galison, Peter Godfrey-Smith, Deborah Gordon, and Debra Satz provided guidance during my undergraduate and master's days. Matthew Brockwell, Mark Brown, Michelle Friedland, Benj Hellie, Jonathan Kaplan, Sandhya Kilaru, Josh Knobe, and Siddhartha Mukherjee were highly supportive peers. Thank you to Amir Najmi, Hervé Kieffel, and Simon Dickie for ongoing friendship and advice since those early days. We only live once, indeed. Returning to Stanford both during dissertation writing (2002/03) and for a sabbatical year as visiting scholar (2015/16; office space provided by Helen Longino and Noah Rosenberg), I met some of the other members of the "Stanford school" of philosophy of science. Thank you to Helen Longino, Tom Ryckman, Nancy Cartwright, and Solomon Feferman—your friendship is much appreciated. During the sabbatical, Silvia De Toffoli, Thomas Icard, Arezoo Islami, and R. A. Briggs in particular provided valuable feedback.

The period in which I earned a PhD in history and philosophy of science at Indiana University was graced by world-class professors across the biology, history and philosophy of science, mathematics, and cognitive science departments: Michael J. Wade, Elisabeth A. Lloyd, Frederick Churchill, Michael Friedman, Curtis Lively, Rudolf Raff, and Victor Goodman. Supportive friends during my IU time and beyond include Janice Alers-Garcia, Nicole Howard, Narisara Murray, and Michael S. Young.

Living in San Francisco and the San Francisco Bay Area offers an opportunity to stumble upon a variety of remarkable people. Some of them have also been generous with their time in reacting to the book, and in helping make San Francisco one of several homes. Thank you to Mark Detweiler, Marcia Baum, Michael "Doc" Edge, Sarah Dihmes, Joseph Hendry, Claire Heidi Hafner, Alicia Claire Kletter, Ryan Giordano, Trudy Singzon, Anthony Hunt, Franses Simonovich, Barbara Wein, Michael Whitehurst, Jack and Gay Reineck, Jack Dangermond, Josiah McElheny, and Bernie Sanders. Gratitude to Craftsman and Wolves, Dandelion, Tartine, and California Wine Merchant for distributed office space, and for first-rate coffee, chocolate, pastries, and occasional glasses of wine, a number of them on the house (thanks in particular to Hannah Kram, Aili Constable, Joe Keefe, Elisabeth Balmer, Sarah Gagnon, Julían Gervasi, Ian Roorda, Laurence Lai, and Deb O'Flynn). Mette Smølz Skau provided support for five years in San Francisco, before deciding to return home. Thank you to Bøllemis, Minimis, and Ubuntu, and Asger and Getrud, for being exemplary members of the next generations.

Lectures on the book at various universities across the globe provided excellent critique: Cambridge University, Copenhagen University, University of Kassel, National Tsing Hua University, University of South Carolina, Stanford University, Université de Bordeaux, Université Paris 1 Panthéon-Sorbonne, University of Toronto, UNAM, and Aarhus University. Other people afforded thoughtful feedback, including Rolena Adorno, Hanne Andersen, Judith Baker, Oliver Baker, Ulrik Bohnstedt Christensen, Nicholas Chrisman, Robert Cioffi, Hélène Courtois, Floren C. de Teresa, Nathaniel Deutsch, Sébastien Dutreuil, Kirk Fitzhugh, Elihu Gerson, Sara Green, Michael Hawrylycz, Philippe Huneman, Alistair Isaac, Douglas Karpa, Evan Kletter, Robert Kohler, Richard Lewontin, Natalie Lo Stoiber, Susan Oyama, Daniel Pech, Russell Poldrack, Anja Schubkegel, Alina Shron, Alok Srivastava, Mauricio Suárez, Bas van Fraassen, Denis Walsh, Michael Weisberg, Eugene Whyte-Earnshaw, and Eric Winsberg. Denis Wood, Sergio Martínez, Karl Niklas, Fermín Fulda, Hasok Chang, Marco del Seta, Tom Ryckman, and Ann Lipson tirelessly commented on significant parts of the text; their sharp critiques, encyclopedic knowledge, calls to action, and unconditional underlying inspiration are much appreciated. Helen, Marco, Sergio, and Tom provided constant encouragement. The Harvard Map Collection offered assistance in locating certain maps. Pedro Espinosa Ruiz, Isis Espinoza, and the library at Filosóficas, UNAM, provided bibliographic support.

At University of California, Santa Cruz, I am grateful to Ann Lipson, Ed Lipson, William Ladusaw, Scott Lokey, Richard Otte, Anne Callahan, David Hoy, Faye Crosby, Tyler Stovall, and George Blumenthal, all of whom provided support. As research assistants, the following UCSC students offered critique: Quentin Coudray, Cory Knudson, Jan Mihal, Lia Salaverry, Amy Coffin, and Anna Zaret.

That a book is a communal effort is perhaps common knowledge, but until you embark on the adventure, it is impossible to know how much of a combined effort it actually is. Karen Merikangas Darling at the University of Chicago Press always answered emails informatively, knew how to prod and critique in the right kinds of ways, and always helped me along, especially when it was most needed. Thank you to Susannah Marie Engstrom, Johanna Rosenbohm, and Christine Schwab, also at the University of Chicago Press, for editorial and production support on the ground, and to Mats Wedin for cartographic design, Helmut Filacchione for creating the index, and Laura Laine for kindly locating countless permissions. Three developmental editors deserve special mention for helping open up the prose: Gloria Sturzenacker and Marilyn Freedman (the entire manuscript), and Jennie Dusheck (part 2). Art discussions with Heidi Svenningsen Kajita, Larisa DePalma, Joe Keefe, and Pablo Carlos Budassi were delightful. Claire Heidi Hafner and Mette Bannergaard Johansen provided project management assistance.

Each of us pays the ultimate price for being given the gift of world making, and on October 5, 2011, Grethe Grønfeldt Winther, my mother, died after a three-year struggle with cancer. The death of my mother changed my life deeply, and though her physical being was no longer present during the writing of this book, she nonetheless contributed energetically to its creation. Perhaps the yearning to write books, have children, or embark on voyages of discovery across the oceans or into space is an attempt to escape the eternal anonymity of a death without legacy. My mother will continue to live on in my heart and mind, and now yours, in part because her essence is woven through the stories filling the pages of *When Maps Become the World*.

FIGURE 1.1. In 1507, Waldseemüller designed and completed an enormous map: 1.32 by 2.36 meters, in twelve sheets. The title, in English, is roughly *A Universal Cosmography According to the Tradition of Ptolemy and the Discoveries of Amerigo Vespucci and Others*. (Geography and Map Division, Library of Congress.)

# 1
# Introduction: Why Maps?

Maps help writers build new worlds, and then open these worlds to the reader. Consider the map of Middle Earth in J. R. R. Tolkien's *The Lord of the Rings*; Robert Louis Stevenson's map of Treasure Island; or R2-D2's and BB-8's complementary holographic maps in *Star Wars: The Force Awakens*. Ocean charts, paper maps, digital maps, and other cartographic objects are as intuitive and familiar as they are disregarded and forgotten, despite being fundamental to finding our way in the world.

Ponder, for instance, J. L. Borges's "On Exactitude in Science" (1946), with its map as big as an entire empire, which highlights the practicality—even necessity—of reducing scale and abstracting symbolism. While Borges's very short story will be familiar to many, it is worth bearing in mind:

> In that Empire, the Art of Cartography attained such Perfection that the map of a single Province occupied the entirety of a City, and the map of the Empire, the entirety of a Province. In time, those Unconscionable Maps no longer satisfied, and the Cartographers Guilds struck a Map of the Empire whose size was that of the Empire, and which coincided point for point with it. The following Generations, who were not so fond of the Study of Cartography as their Forebears had been, saw that that vast Map was Useless, and not without some Pitilessness was it, that they delivered it up to the Inclemencies of Sun and Winters. In the Deserts of the West, still today, there are Tattered Ruins of that Map, inhabited by Animals and Beggars; in all the Land there is no other Relic of the Disciplines of Geography. —Suárez Miranda, *Viajes de varones prudentes*, Libro IV, Cap. XLV, Lérida, 1658[1]

---

1. Borges 1999, 325. With humor and irony, Eco (1994) playfully questions the very possibility of a one-to-one map.

A map the same size as the territory, a one-to-one scaled map, that also copies the entire terrain will fail to subtract irrelevant and noisy features. Due to its magnitude, level of detail, and sheer practical unwieldiness, Borges's map of the empire is useless in guiding your eye, mind, and feet through a particular neighborhood or city. A perfectly realistic map is a useless representation.

In Italo Calvino's work of fiction *Invisible Cities* (1974; *Le città invisibili*, 1972), Marco Polo recounts to the Mongol emperor, Kublai Khan, the many cities he has visited in both dreams and reality. In the chaotic and smelly city of Eudoxia, according to Polo, there is a symmetrical, patterned carpet.[2] Each place in the carpet purportedly corresponds to a place in the city, and each inhabitant reads into the carpet her or his unique perspective. An oracle had told that either the city or the carpet is a map of the entire universe, the other merely an approximate reflection; but neither the inhabitants nor we, the readers, know which is the universe map and which is the reflection. A map represents the territory, and structural analogies between the carpet and the city—and between them and the universe writ large—allow for novel and informative representation relations to emerge.

Consider how Ursula Le Guin's pathbreaking Earthsea series starts with a map that she herself drew (plate 1), as she reports in an oral interview:

> I had written a couple of short stories, very light short stories that took place on these islands where there were wizards and dragons. And when I sort of began thinking about a book, these islands grew in, just boom, it sort of, this is a whole archipelago of islands, and now I draw the map. And so the first thing I did for the book was the map. And I've always used that map all six books.[3]

A map organizes the entire narrative, both for the author in the process of composition and for the reader in the process of comprehension. No story without a map.

Writing in *Tom Sawyer Abroad*, Mark Twain humorously satirizes the conflation of map and territory as Huck Finn and Tom Sawyer fly over the American Midwest in a balloon:

> [Huck:] "I know by the color. We're right over Illinois yet. And you can see for yourself that Indiana ain't in sight."

2. Calvino 1974, 96–97.
3. Simon & Schuster Books 2012.

[Tom:] "I wonder what's the matter with you, Huck. You know by the *color*?"

"Yes, of course I do."

"What's the color got to do with it?"

"It's got everything to do with it. Illinois is green, Indiana is pink. You show me any pink down here, if you can. No sir; it's green."

"Indiana *pink*? Why, what a lie!"

"It ain't no lie; I've seen it on the map, and it's pink."

You never see a person so aggravated and disgusted. He says:

"Well, if I was such a numbskull as you, Huck Finn, I would jump over. Seen it on the map! Huck Finn, did you reckon the States was the same color out-of-doors as they are on the map?"

"Tom Sawyer, what's a map for? Ain't it to learn you facts?"[4]

Huck's sincere questions are striking. Perhaps we can forgive him for interpreting the map as the territory. But what if such errors become systemic, shared by many individuals? And what if such errors lead to harm or even death?

Borges's map of the empire, Calvino's carpet of Eudoxia, Le Guin's map of Earthsea, and Mark Twain's pink Indiana exemplify themes from *When Maps Become the World*: maps are abstractions discarding detail, focusing only on essential features of the territory. What is essential depends on one's purposes. Moreover, we map because we have a deep spatial human cognitive and social capacity, and need. We orient ourselves in different kinds of spaces. Maps therefore represent their territories in spatial, multiple, and creative ways. But we also sometimes perniciously conflate and confuse map and territory. In order to realize that a map is not the territory, we can, for instance, consider multiple points of view on—multiple maps of—the same territory.

Literature occasionally goes further and explores the meaning and significance of maps and mapping for science and philosophy. For the philosopher—as for the self-aware writer, artist, creator, explorer, and responsible citizen—this realization allows the map to serve as an apt analogy for theory and model; as metaphor for how we know; and even as illustration of magic, mystery, and the beyond. In a sentence, this book is about the power and limitations of maps and mapping, including those ambitious and interconnected maps that we call scientific theories.

To lay the cartographic groundwork for our journey, this chapter explores my concept of "map thinking," first theoretically, with sections on elements of map thinking and definitions of our central notion, "map,"

4. Twain 1894, chap. 3.

and then through three actual cartographic objects. It concludes with some thoughts on digital mapping.

## A History and Philosophy of Map Thinking

Visualize a simple paper street map of your favorite city. It can guide you to the museum, to the nearest park, or to a new grocery store. It speaks volumes. Conversely, a bad map serves no purpose but frustration and a disoriented loss of time and direction. Maps can be wrong, or may not fulfill your purposes. But let us take a minute to focus on that street map that guides you properly without incident. Who designed and produced your useful map? According to which data, techniques, and conventions was it made? How did the mapmakers simplify and abstract the teeming city? A well-designed street map is an effective tool, not just a lifeless visual archetype.[5]

In our mundane map use, we may not care about such questions, since we wish only to get from point A to point B in the most timely or easiest fashion possible. But I want to invite you to become a map thinker—someone who seeks to answer these questions and more. Do maps entrance and enchant you? Did you pore over an atlas for hours at home as a child, traveling continent to continent in your imagination? Let us traverse a new territory in which we investigate the hidden power of mapping within realms of being and action.

### THE NATURE OF MAP THINKING

*Map thinking* refers to philosophical reflection concerning what standard geographic maps are and how they are made and used. The purpose of such contemplation is to explore the promises and limits of representations—cartographic and beyond. Representations in general are not just visual, but also linguistic, physical, and mathematical; they proliferate across the sciences, the arts, and everyday life. You might be more comfortable with precise mathematical representations than with visual ones. Your friend might prefer poetry, literature, or other linguistic representations. Important lessons about senses other than visual can be learned from cartographic representations.

Map thinking massages the imagination; excavates hidden assumptions; challenges and synthesizes dualisms; and invites us to reflect on

---

5. Mette Smølz Skau provided constructive feedback.

space and time—including the future. In this book, map thinking is used to develop an analogy from representation in the domain of cartography to representation in the domain of scientific theory.

ELEMENTS OF MAP THINKING

Map thinking involves four elements: cartography, mapping, map studies, and map analogizing.

*Cartography.* Cartography is the study and practice of "philosophical and theoretical bases, principles, and rules for maps and mapping procedures."[6] From the early twentieth century onward, cartography has involved the self-reflexive investigation of principles and rules of mapmaking and map use. These theoretical investigations further impel the practice of mapping and form the basis for training students in making and using maps. One aim of this book is to use cartography's nuanced analyses of abstraction and representation, map design, and map politics to illuminate scientific practices.

*Mapping.* The actual making and using of maps can be thought of in two ways: deep mapping and modern mapping. Deep mapping is probably as old as *Homo sapiens*, perhaps older. It entails the yearning to draw pictures of geographic, cosmic, and local space, and to visualize and communicate about journeys, dreams, and memories. Modern Western mapping traditions began around 1500 CE and refer to the mapmaking and map use that emerged from European colonialism, nation-state building, and imperialism. Both of these senses of mapping matter to me, and shall be explored presently.

*Map studies.* From around the 1970s onward, the history, philosophy, and social studies of maps, mapping, and cartography became a discipline.[7] Map studies provide insight into the nature, structure, history, and purpose of maps and mapping, thereby expanding map thinking.

---

6. Muehrcke 1972, 1; see also Robinson and Sale 1969, esp. chap. 1, "The Art and Science of Cartography," 1–18. The origin of cartography can usefully be dated to the publication of one or more scholarly books between 1921 and 1967 (e.g., Eckert 1921; Raisz 1938; Harley 1962; Imhof 2013; and Bertin [1983] 2011); to the founding of the International Cartographic Association (ICA) in Bern, Switzerland in 1959; or to the institutional appointment of particularly important figures, such as the "dean" of (North) American cartography, Arthur Robinson, to the Geography Department at University of Wisconsin–Madison in 1947 (Cook 2005; Harley 1987; Wolter 1975).

7. See, for instance, Robinson and Petchenik 1976; Wood 1992a; Harley 2001c.

Although I appeal to map studies in this chapter, I draw more heavily on them in chapters 3, 4, and 5.

*Map analogizing.* From at least the eighteenth century onward, professionals across many fields of knowledge have used maps and mapping as grand, powerful analogies of knowledge and representation. For instance, in his "Preliminary Discourse" (1751) to the European Enlightenment epoch–making *Encyclopédie* that he coedited with Denis Diderot until 1759, Jean le Rond d'Alembert wrote: "The encyclopedic arrangement of our knowledge . . . is a kind of world map which is to show the principal countries, their position . . . [and] the road that leads directly from one to the other. . . . individual, highly detailed . . . maps will be the different articles of the *Encyclopedia* and the Tree or Systematic Chart will be its world map." Although we will encounter elements of map analogizing in this chapter, the map analogy is the explicit focus of chapter 2.[8]

While I prefer the term *map thinking*, a synonym might be *cartology*.[9] Cartology is the study and practice not only of the philosophy of mapmaking and map use, but of *anything* having to do with maps and mapping: concretely making and using them; the historical, social, and political context of maps; the analogies, metaphors, art and personal stories associated with mapping, as it moves into so many other areas of human experience. Map thinking and cartology are quite general.

## DEEP MAPPING

*Mapping* is older and broader than either cartography or the contemporary geographic information system (GIS). When and where did mapmaking and map use originate? Cognitively speaking, *deep mapping* is a

---

8. Incidentally, this categorization of elements of map thinking resonates with categories regarding science. Analogous to deep mapping, rudiments of experimenting and theorizing are at least as old as our species, and are practiced by other species; analogous to modern Western mapping, Western science was arguably formally born with the establishment of the Royal Society of London in 1660, and of the Académie des sciences in Paris in 1666; analogous to map studies, the history, philosophy, and sociology of science is primarily a twentieth- and twenty-first-century phenomenon. One disanalogy here is that the reflection on the "philosophical and theoretical bases" for science are at least as old as Ancient Greek philosophy, much older than cartography. Claus Emmeche provided constructive feedback.

9. I am not here using the term in the specific, information-theoretic sense of Ratajski 1972.

strong and highly intuitive representational impulse. The British geographer, cartographer, and map historian John Brian Harley makes the point perfectly at the beginning of his *History of Cartography*: "There has probably always been a mapping impulse in human consciousness, and the mapping experience—involving the cognitive mapping of space—undoubtedly existed long before the physical artifacts we now call maps."[10]

*Space thinking* remains a key cognitive desire and challenge throughout our lives. To think space is to create, imagine, and control said space through a spatial representation—a map. We manage space and our spaces when we can map them, and then use our map or maps to guide us through space, and help us understand and act in our worlds, whether personal, geographic, or scientific. The American cartographers Arthur H. Robinson and Barbara Petchenik conclude their book thus: "The reason for the common use of mapping as a metaphor for knowing or communicating . . . has finally become clear: the concept of spatial relatedness which is of concern in mapping and which indeed is the reason for the very existence of cartography, is a quality without which it is difficult or impossible for the human mind to apprehend anything."[11]

We do not want to just know what is beyond that hill or across that river or on the next city block; we also want a tool with which to remember these things and how they are connected, and an easy way to talk about all of this with other people. Maps and mapping give us these tools, helping us write a script to communicate with others. *Our individual, cognitive deep yearning to map space and the world, and to share and communicate our maps, are as ancient as any of our other general representational practices.*[12]

## FIVE HUNDRED YEARS OF WESTERN MAPPING

In Western culture, mapmaking and map use are roughly five hundred years old. In his studies of the history of probability, Ian Hacking makes a similar claim about the emergence of concepts and practices related to probability: "We do not ask how *some* concept of probability became

---

10. Harley 1987, 1. For a discussion of space from the heartland of classic cartography, see Robinson and Petchenik 1976, chap. 5, "The Conception of Space."

11. Robinson and Petchenik 1976, 123.

12. Spatially structured maplike representations have been found in Paleolithic caves; see Utrilla et al. 2009. Or consider the Bedolina Map from circa 1500 BCE (see Turconi 1997). On prehistoric spatial and mapping cognitive and cultural drives, see Wynn 1989; Liebenberg 1990; Burke 2012; Utrilla et al. 2009.

possible. Rather we need to understand a quite specific event that occurred around 1660: the emergence of *our* concept of probability." Hacking explores the "preconditions for the emergence" of this concept and how such logical and temporal preconditions regulate and "determine" the meanings, conditions of application, and the "very nature" of probability as a concept and set of methodological and analytical practices in institutional contexts.[13]

As with the emergence of probability, so with the emergence of mapping. Martin Waldseemüller's 1507 map, to be explored below, is a prime candidate for the birth moment of the modern Western tradition of mapping. The number, uses, and variety of maps exploded in the sixteenth century. Scholars have proposed a variety of theses to explain this: the Renaissance fascination with antiquity, including a strong interest in Ptolemy's *Geography*; the Scientific Revolution and its impulses toward experiment, measurement, and quantification; artistic developments of perspective and realism, especially in the Low Countries, France, and various Italian city-states and regional states; the commissioning of "estate plans" by landowners; political nation building; and the unquenchable thirst for maps of distant lands, peoples, and resources by European colonizers and imperialists.[14] Focusing on large-scale maps, the historian David Buisseret and his collaborators found the answer to "just when did monarchs and ministers in various [European] countries begin to perceive that maps could be useful in government?"[15] to be roughly the end of the sixteenth century. The interdisciplinary variety of such theses reflects how versatile in application map thinking can be.

It is important to note that maps were drawn and used centuries earlier, for example, in Athens, Babylon, China, and Rome;[16] but these are not the kinds of maps playing a role in contemporary Western, or globally Westernized, life. The case of Japan is particularly interesting because mapping emerged as part of complex nation building around 1600, roughly simultaneous with Western mapping.[17] However, this Japanese tradition is independent of Europe, despite the similar timeline.

---

13. Hacking 1975, 9.
14. Buisseret 1992, 1–4, esp. 1–2; Conley 1996, 1–2.
15. Buisseret 1992, 2.
16. For a partial history of mapping across a large swath of global cultures, contemporary and historical, see Dilke 1985, 1987; Harley 1987; Berry 2006; Raaflaub and Talbert 2010; Winther 2014d, 2019a.
17. In Japan, the samurai served as cartographers and scholars (Berry 2006, chap. 3). Interestingly, for Berry the central mapmaking challenge of "reduc[ing] space to generic attributes" is "inherently ideological rather than mechanical" (60).

## Maps Today

Is there a single useful definition of what a map is, especially today? Might the challenges we face when defining our central concept raise worries or insights about the nature of knowledge? Exploring nonstandard meanings of *map* moves us away from rigid and closed geographic definitions, and into more flexible conceptual territory, as we shall see. Maps are a broadly applicable tool for understanding human thinking and doing.

### CARTOGRAPHY MEETS GIS

The map is not a stable representational object. Cartography was fundamentally transformed in the 1990s by GIS. With GIS's ability to capture, store, manipulate, analyze, and depict data, the modern, digital, computational map was born.

In her history of GIS, Nadine Schuurman describes the switch in approach. The digitally displayed map is "the tip of the iceberg"; "the body of the iceberg... is the database" of facts about the place represented. "In order to simplify the map, it is necessary to simplify the database from which it is produced. Model-based generalization [in GIS] focuses on reducing the detail in the database while cartographic generalization simplifies the display."[18] Cartography is primarily concerned with generalizing data into a map. GIS is as interested in the complex database as in the display.

In an earlier paper, Schuurman reminds us to take seriously the philosophical implications of GIS: "GIS is frequently concerned with prediction rather than explanation that requires identification of structural and causal mechanisms, hallmarks of realism. Nevertheless, GIS scholars are far from agreeing epistemologically. Indeed, they approach the modeling of space from every conceivable angle which may speak to the nature of space as much as their investigative skills. It is safe to say, however, that the tradition of philosophical inquiry is weaker in the GIS community than in social geography."[19] Space is indeed multifaceted, and a philo-

---

18. Schuurman, 2004, 47. "'Model,' in this case, refers to the database and 'cartographic' to its display" (46). Schuurman further notes, "Though this differentiation may seem self-evident, it took two decades of generalization research for it to be recognized and articulated" (48). She cites Brassel and Weibel 1988 in that regard.

19. Schuurman 1999, 75.

sophical analysis of GIS could benefit philosophy, GIS, and cartography alike.[20]

Comparing GIS and traditional cartography, the American philosophical geographer Helen Couclelis has expressed great concern that GIS would make pen-and-paper cartography obsolete: "Map symbolism evolved over the centuries through extensive trial and error, guided by the need to minimize the distance between the cognitive and the graphical representations of the geographic world," and "the brash newcomer still has a great deal to learn from the old master."[21]

Whatever the fate of the map within GIS, and whatever the fate of GIS, a rich conceptual framework in cartography has survived. I will use that framework to develop heuristic tools and analyze the sciences. It is to this older, more classic cartographic understanding of *map* that we now turn.

## A DEFINITION BASED ON REPRESENTATION

An unadorned concept of the map can be found in the geographer John Andrews's article "What Was a Map? The Lexicographers Reply," which is a tour de force exploring 321 definitions of our term published between 1649 and 1996. According to Andrews, a "group of map definitions" overpowers all other meanings. The essence of these definitions is that a map is "*a representation . . . in a plane . . . of all or part of the earth's surface.*"[22] This stark definition suggests one possible, representational definition of *map*:

> A map is a visual representation, on a surface or a plane, of Earth's spatial properties or processes.

An important term in Andrews's constellation of map definitions is *representation*, which he tells us is semantically close to words such as *picture, projection, description, model,* and *diagram*. Andrews argues that we must qualify the word *representation* with some emphasis on the visual features of a map and suggests *graphic* as one of the more popular options.[23] Andrews also points out that *surface* is an expression with multiple meanings. For example, we can construe a surface experientially, as

---

20. I take steps in this direction in Winther 2015b, 2019b.
21. Couclelis 1992, 71, 75.
22. Andrews 1996, 1; emphasis added.
23. Andrews 1996, 2.

something that "breaks the fall of an object dropped from a considerable height"; alternatively, a more abstract and mathematical treatment of surface is as "a purely conceptual form perhaps synonymous with 'geoid,'"[24] which is a curved, Earth-like surface, which we shall learn about in chapter 3. Indeed, both *representation* and *surface* are historically situated terms.[25]

Barbara Petchenik and Arthur Robinson also provide plainly representational definitions, arguing that "*representation* and *space* are the two critical elements" of "a general definition of a map."[26] They render another simple definition with necessary and sufficient conditions: "*A map is a graphic representation of the milieu.*"[27] They characterize the term *graphic* as a "visual construct," whereby the map is itself a "space" in which "marks" and their relative positions are "assigned meanings."[28] Second, "to represent" means "to stand for, symbolize, depict, portray, present clearly to the mind, describe, and so on, and *seems to occasion no problem in meaning.*"[29] Finally, the space of a map "normally refers to the three-dimensional field of our experience," and the "spatial framework" as a feature of the mapmaker's "cognitive endowment" is "essential."[30] The word *milieu* is taken to best describe the representational target of a map: *space*.[31] Furthermore, a "mapper" is someone "who actively conceives of spatial relationships in the milieu, whether those relationships involve the hills and valleys of the topographer, the familiar nodes and

---

24. Andrews 1996, 2–3. On the geoid, see chapter 3.

25. Andrews divides the history of these definitions into four sequential yet overlapping periods: the scientific map, the popular map, the cartographer's map, and the philosopher's map. He places the coalescence of cartography in the mid-twentieth century. The last period appeals to the work of "philosophers who took the generalization of the map concept as far as it could go" (Andrews 1996, 6). A variety of "committee definitions," all highly representational, are also presented. For instance, the International Cartographic Association's 1973 definition is: "A representation normally to scale and on a flat medium of a selection of material or abstract features on, or in relation to, the surface of the earth or of a celestial body" (quoted in Andrews 1996, 8).

26. Robinson and Petchenik 1976, 15.

27. Robinson and Petchenik 1976, 16; emphasis added.

28. Robinson and Petchenik 1976, 16.

29. Robinson and Petchenik 1976, 16; emphasis added. The different accounts of representation explored in chapter 5 belies their optimism.

30. Robinson and Petchenik 1976, 15, 17.

31. "Territory" connotes "proprietorship"; "environment" has "aspatial ecological overtones"; and "place" and "area" are, Robinson and Petchenik argue, "quite impersonal" (1976, 16).

corridors of the city dweller, or the invisible shoals and reefs of the ship's pilot."[32]

The *Oxford English Dictionary* also emphasizes representation. The first meaning or sense in its definition of *map* is: "A drawing or other representation of the earth's surface . . . with each point in the representation corresponding to an actual geographical position according to a fixed scale or projection; a similar representation of the positions of stars in the sky, the surface of a planet, or the like."[33] Scholarly as well as standard dictionaries of other European languages, including French, German, Spanish, and Danish, feature related definitions emphasizing representation, surface, and Earth.[34]

There are additional difficulties with seeking necessary and sufficient conditions for the term *map*. Any particular graphic representation can be taken to be a map — even two thick, meandering intertwining lines on a sheet of paper or the carpet in the city of Eudoxia.[35] That is, a graphic we do not typically interpret as a map can be applied cartographically. Conversely, any literal map can be used in a noncartographic way — maybe as a piece of art hanging on a wall, or as wrapping paper. To determine whether a given object is a map, we therefore need to complement the representational definition with knowledge about how the cartographic object is applied and interpreted. Use matters.

A map is not a single kind or type of thing. The word *map* is not, as philosophers say, a *natural kind* term referring to things of the same

---

32. Robinson and Petchenik 1976, 17.

33. This definition concludes: "Also: a plan of the form or layout of something, as a route, a building, etc." See chapter 5.

34. See the entries "carte" in *Dictionnaire de l'Académie française*, https://academie.atilf.fr/consulter/CARTE?page=1; "landkarte" in *Das Digitale Wörterbuch der deutschen Sprache*: https://www.dwds.de/wb/Landkarte; "mapa" in *El Diccionario de la lengua española de la Real Academia Española*: http://dle.rae.es/?id=OJpDHMj; and "kort" in *Den Danske Ordbog*: https://ordnet.dk/ddo/ordbog?query=kort (all accessed December 20, 2017). Of course, other languages should be consulted, including non-European and non-Indo-European ones. The point here is that representationalism pervades definitions of our concept of "map" across languages of modern Western mapping traditions.

35. Due to its time-limited public presence and limited capacity to travel geographically, I shall set aside what Woodward and Lewis call "performance cartography," which involves maps embodied in and communicated through "gesture, ritual, song, poem, dance, and speech." They take such maps — construing the term quite broadly indeed — to be especially relevant to cartographic representations in "non-western spatial thought and expression" (D. Woodward and Lewis 1998, 3, table 1.1).

type—for example, electrons or chairs.³⁶ Many forms of representation can be construed as a map. Even when we investigate world maps, such as Waldseemüller's, Guaman Poma's, or Tom Van Sant's (as we shall below), the complexity and instability of the map concept becomes apparent.

## CHARACTERIZATIONS BASED ON PROCESS AND FUNCTION

Maps do not exist just within the realm of modest representationalism. They also function within our behaviors, our institutions, and our conscious and unconscious understanding of phenomena. Maps are not solely static, general, and abstract. An emphasis on process or function, or both, is required to complement our representational definition above.

Maps are contingent and unstable. Even once a map is completed, mapmakers often add new data, editing and improving it. Maps emerge processually from multiple social and cognitive practices involving aesthetic design; data processing; abstraction, idealization, and generalization; and political, social, and other human dimensions. According to the British geographers Rob Kitchin and Martin Dodge, maps should be understood in terms of the social practices scaffolding them. Kitchin and Dodge call for "a shift from ontology (how things are) to ontogenesis (how things become)—from (secure) representation to (unfolding) practice."³⁷ They yearn to reboot cartography as a discipline focused on dynamic *process* rather than static representation.³⁸ Perhaps a map is simply the graphical outcome of a set of cartographic practices.

Harley has provided a complementary cartographic perspective. The first section of his influential article "Deconstructing the Map" suggests two sets of "cartographic rules": those associated with "scientific epistemology" and "those governing the cultural production of the map."³⁹ The latter set of rules fascinate Harley:

> To discover these rules . . . governing the cultural production of the map . . . we have to read between the lines of technical procedures or of

---

36. See Putnam 1973; Boyd 1999; Hacking 2007b.
37. Kitchin and Dodge 2007, 335.
38. Kitchin and Dodge 2007, 342, cf. 343. Their article also presents two vignettes: a map of population changes in Ireland between two censuses (1996 and 2002; ibid., 336, fig. 1), and a thought experiment using a Manchester street map.
39. Harley 1989, 4–5. This article has been cited 1,915 times (Google Scholar, April 21, 2017).

the map's topographic content. They are related to values, such as those of ethnicity, politics, religion, or social class, and they are also embedded in the map-producing society at large. Cartographic discourse operates a double silence toward this aspect of the possibilities for map knowledge. In the map itself, social structures are often disguised beneath an abstract, instrumental space, or incarcerated in the coordinates of computer mapping. And in the technical literature of cartography they are also ignored, notwithstanding the fact that they may be as important as surveying, compilation, or design in producing the statements that cartography makes about the world and its landscapes.[40]

Geographic and scientific representations are made and used within a value-laden, political, and cultural universe in which certain aims and values are assumed and imposed. More than focusing on mapmaking processes, Harley here emphasizes the ethical and political *functions* of a map, all of which drive and constrain its content.

Do we then need a definition of a map highlighting process or function, or both? Cultural and even artistic or deeply personal modes of map thinking resist definitions based on simple, necessary, and sufficient philosophical principles.

For Harley, maps contain social structures hidden beneath an abstract space. For the cultural critic Rebecca Solnit, maps are "invitations" or "tickets" for understanding the world in a certain way, whether we are mapping, within the San Francisco Bay Area, its corporate-military complexes, or the original first nations, AmerInd territories now poured over with concrete and steel. For the philosopher and cartographer Denis Wood, maps are "weapons," serving the political and social aims of the mapmakers, who are often in positions of state, military, or corporate power.[41] I sometimes think of maps as a "theater" in which many voices, activities, and interests interact, rarely harmoniously.

Each of these diverse characterizations is reasonable in its own right and such a broad family of perspectives commits us to including ethics, aesthetics, and communication in map thinking. *Map* might be possible to define narrowly, but any definition is more helpful when it respects the complexity of any abstract representation. For me, mapping is a *representational strategy for imagining and controlling different kinds of space.*

40. Harley 1989, 5–6.
41. Wood 1992a, 66, 67, 73; 2012a, 286, 296. Solnit writes: "Maps are always invitations in ways that texts and pictures are not; you can enter a map, alter it, add to it, plan with it. A map is a ticket to actual territory, while a novel is only a ticket to emotion and imagination" (2010, 8).

Let us now turn from our conceptual and definitional analyses of maps and mappings to consider three specific maps and what they can teach us.

## Three Maps

Our many maps *perspectivize* in that they understand the world from a particular, comprehensive angle or viewpoint. They also slice, dice, chop, and grind. That is, they *partition*. Sea charts, city plans, and topographic maps divide a spatial system or whole into different kinds of objects; properties or features; and processes or dynamics. In perspectivizing and partitioning, our cartographic representations are the ultimate record of empires and of our evolving comprehension of our spatialized worlds. Maps seduce and persuade. They build and destroy worlds.

But how exactly do maps do all of this? Specifying the cartographic object only theoretically—as I have done above—is insufficient. We need concrete examples. Consider the world map as paradigmatic of maps. Three world maps challenge our representational intuitions regarding truth and empirical accuracy: Waldseemüller's, Guaman Poma's, and Van Sant's. Each of these illustrates the depth and influence of history, ethics, and culture on maps and mapping.

### WALDSEEMÜLLER'S MAP

In the early sixteenth century, a group of scholars, mapmakers, and classicists in Saint-Dié (Lorraine, France) were actively translating Greek and Latin texts to German. This group, known as the Gymnasium Vosagense, prepared a new Latin edition of Claudius Ptolemy's *Geography*.[42] Of particular interest to us are the friends Martin Waldseemüller and Matthias Ringmann.

Waldseemüller's 1507 map (fig. 1.1) was intended to accompany a new geography text, Waldseemüller and Ringmann's *Cosmographiae Introductio*.[43] This map emanated a significant number of firsts, including using

---

42. Harris 1985; S. I. Schwartz 2007; Hessler 2008; Brotton 2012, 146–85.

43. See Hessler 2008, 38–66. We recently learned that an earlier map from circa 1491 by Henricus Martellus, a German living in Florence, seems to have been an important influence on Waldseemüller. Waldseemüller used the same projection as the 1491 map (Van Duzer 2012, 2019; G. Miller 2018). As was common at the time, Waldseemüller gathered geographic locations and text in legends from multiple sources, including Martellus's map. As one example, in one

the name "America"; showing America as a separate continent; covering all 360 degrees of longitude; and rendering the Pacific Ocean as a distinct body of water.[44] *Universalis cosmographia* is known as the "birth certificate" of America.[45] It is a paradigm example of what the philosopher Saul Kripke would call a *baptism event*: an initial act of naming that originates a causal chain of use, even if our view of this history is always incomplete.[46] The map also arguably represents—and is an influential catalyst of—the very birth of modern Western mapmaking and use.[47]

*Universalis cosmographia* embodies precision and accuracy. The projection is a modification of Ptolemy's second projection to include high latitudes and the new lands. The two small inset maps at the map's head—of the old and new worlds, respectively—are rendered in terms of Ptolemy's original, narrower version of the second projection.[48] Part of

---

legend Waldseemüller used the exact same (Latin) text as Martellus, translated thus: "which we carefully include in this image in the interest of revealing the true knowledge of places" (Van Duzer 2019, 54). However, this map depicted only Africa, Asia, and Europe.

44. Crane 2002, 56; for the complex naming history of the Pacific Ocean, involving Ferdinand Magellan, among others, see Maroto Camino 2005, 72–82.

45. Schwartz 2007, 51; cf. Brotton 2012, 147. The only extant Waldseemüller map currently resides in the US Library of Congress, after having been purchased by the US government for $10 million in 2003 from the count of Wolfegg Castle, where the map had been rediscovered in 1901. At its time, the map was printed in many copies (possibly up to a thousand; Brotton 2012, 154), and influenced the work of other cartographers and cosmographers, including Nicolaus Copernicus (Hessler 2008, 54); Peter Apian and his contemporary Gemma Frisius, who expanded and annotated Apian's *Cosmographia*; as well as Frisius's student, Gerardus Mercator (Shirley 1983, 29; Crane 2002, 57–67; Hébert 2011, 32; see also chapter 4).

46. Kripke 1972, especially 96–97.

47. Shirley 1983 covers the period 1472 to 1700. Waldseemüller's map is no. 26 of 639 maps surveyed. Earlier maps shown are either medieval "T-O" maps (see Winther 2014d) or standard Ptolemaic maps. The single exception is the 1506 Contarini-Rosselli map (no. 24), which was discovered in 1922. This first known map to show parts of the New World hails Columbus. Vespucci is ignored. North America is portrayed as an extension of Eastern Asia, and Cuba and closely related islands are exhibited somewhere in the middle of the Atlantic over an enormous, and then unknown, hypothetical Austral land mass. This beautiful map was not influential.

48. See J. P. Snyder 1993, 12–14 (Ptolemy's original second projection), 33–34 (later adaptations of Ptolemy's second projection, including Waldseemüller's adaptation); Hessler 2008, 47.

the lower-left inset reads, "A general delineation of the various lands and islands, including some of which the ancients make no mention, discovered lately between 1497 and 1504. . . . All this we have carefully drawn on the map, to furnish true and precise geographical knowledge."[49]

The outline of South America is statistically more accurate than should be expected in 1507, given known historical voyages. Waldseemüller also correctly places the Florida coast near Cuba, although Florida was likely unknown to the Europeans before Ponce de León's travels in 1513.[50] Further evidence of Waldseemüller and Ringmann's careful empiricism is that they conducted the "first known systematic field survey" of the Rhine Valley around Strasbourg.[51] The representational drive of this map is strong, even if it appears exotic.

Power, politics, and imagination also pervade *Universalis cosmographia*. Although significant controversy remains over the source of the name "America," including possible Mayan or other AmerInd roots,[52] Waldseemüller's map honors the Florentine explorer and colonialist Amerigo Vespucci.[53] *Universalis cosmographia* embodies the power of baptizing a continent into the Western body of knowledge.[54] It also exudes a stark, universalizing, and even vicious European colonialism,

---

49. Hessler 2008, 17.

50. On statistical tests for the map's accuracy rendering the South American coastline, see Hessler 2006; on Florida, see Shirley 1983, 29.

51. Hessler 2008, 43.

52. Cohen 1988.

53. "Because it is well known that Europe and Asia were named after women, I can see no reason why anyone would have good reason to object to calling this fourth part [of the world] Amerige, the land of Amerigo, or America, after the man of great ability who discovered it. The location of this part and the customs of its people can be clearly understood from the four voyages of Amerigo Vespucci that we have placed after this introduction" (translated in *Cosmographiae Introductio*, Hessler 2008, 101). The Gymnasium Vosagense translated and transmitted Vespucci's narratives. Columbus, who thought these lands parts of Asia, is barely mentioned, except in a few cases: in the upper-left inset over Zephir, god of the western winds; the small inset in the middle of sheet 2 of 12 (middle sheet of left-most column), which mentions both Columbus and Vespucci as discoverers of the new land; and in a small inset over northeastern South America, as it were.

54. In commenting on how Waldseemüller's map "recognized the independence of the new lands with respect to the *orbis terrarum*," the Mexican philosopher and historian Edmundo O'Gorman emphasizes as "decisive" and "novel" the "attributing [to America] a specific being and a proper name that individualizes it" ([1958] 2006, 172; my translation).

branding the northeastern part of the South American continent thus: TOTA ISTA PROVINCIA INVENTA EST PER MANDATUM REGIS CASTELLE (All of this province was discovered by mandate of the king of Castile). A further description reveals a fundamental motivator for colonialism: "Here a greater amount of gold has been found than of any other metal."[55]

The map strives to represent, but it will never reveal all its representational secrets. Waldseemüller's abstraction practices are lost in the sands of time: the text accompanying the map is too brief; the woodblocks long rotted; the location of the printing press unclear (was it in Saint-Dié? In Strasbourg?). More importantly, this map is multimodal. It contains diverse insets. It reproduces twelve winds blowing, myriad place names, and explanations of which empire lords over which parts of the world. As the Library of Congress cartographer John Hessler argues, *Universalis cosmographia* negotiates the "intellectual space born of the science of the early sixteenth century" with "symbolic power [and] . . . mythic and semantic meanings."[56] Waldseemüller's 1507 birth certificate of America contains much more of the geopolitical and geocultural than a preliminary representation of Earth's surface. While rooted in classical civilization, this map is also an attempt—at once scientifically sober and socially violent—to break into a globalized European modernity.

Ironically, to call *Universalis cosmographia* a mere representation would be to preclude even a sound representational understanding of the map, which includes investigating the reasons for detailing certain geographic features and for choosing Ptolemy's projection over others. But a narrow emphasis on representation misses the focus of an analysis premised on process or function—the momentous political, historical, and cultural reverberations set in motion by *Universalis cosmographia*. If a map is an invitation, a weapon, a theater, or a representational strategy, we must understand in what ways this is so.

## GUAMAN POMA'S COUNTERMAP

Our conception of world maps is further stretched by Felipe Guaman Poma de Ayala's world map (ca. 1540s–1616; fig. 1.2). Guaman Poma included the map in *El primer nueva corónica y buen gobierno* (The first new chronicle and good government), a remarkable illustrated account from circa 1615 written in a mix of Quechua and Spanish. The chronicle makes deep and pained observations about sixteenth-century colonial Peru in the context

---

55. Inspect the upper-left corner of the map. Translated in Hessler 2008, 13.
56. Hessler 2008, 63.

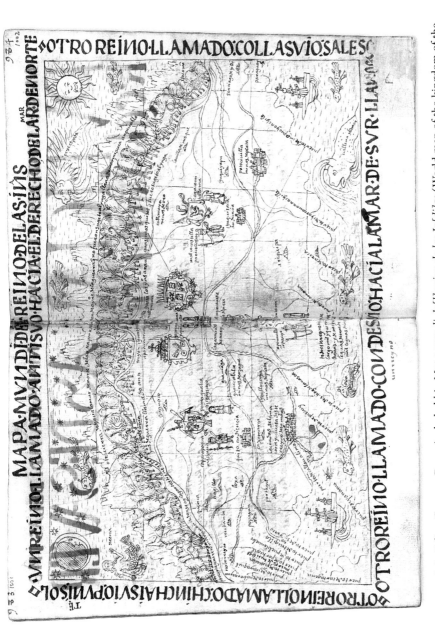

FIGURE 1.2. Felipe Guaman Poma included his *Mapa mundi de[l] reino de las In[di]as* (World map of the kingdom of the Indies) in his *El primer nueva corónica y buen gobierno* (The first new chronicle and good government), 1615. (Royal Danish Library, GKS 2232 4°.)

of "indigenous Andean resistance and consciousness."[57] The volume includes a prophetic cosmological vision of the past and future of "las Indias" and the globe.[58] Guaman Poma's chronicle remains the longest and most important indigenous critique of European imperialism in written form.[59]

The key map in this chronicle resists the orthodox colonialist and imperialist power structure, set of assumptions, and historical narrative. A crisp representational definition of what a map is can render such tensions invisible.

"The great city" of Cuzco can be found at the map's geometric center as the "head ["cauesa" (rather: "cabeza")] of the kingdom of Peru."[60] Moreover, the two diagonals divide the Incan "reino de las In[d]ias" into its four *suyus*, thus:

|  | Antisuyu |  |
| --- | --- | --- |
| Chinchasuyu | (Cuzco) | Qullasuyu |
|  | Kuntisuyu |  |

Following the conventional Andean opposition of "upper" (superior, male) and "lower" (inferior, female),[61] Antisuyu and Chinchasuyu are the

---

57. Adorno 2011, 77. Only one known copy of the chronicle exists, in the Royal Library in Copenhagen (see http://www.kb.dk/permalink/2006/poma/info/es/frontpage.htm; accessed March 9, 2014).

58. Wachtel 1973; Adorno 2000.

59. A few other indigenous AmerInd maps are known, and it is clear that rich mapmaking and map use traditions existed on the pre-Columbian continents; for instance: "Oztoticpac Lands Map," ca. 1540 (Mundy 2011); "Aztec Map of Tenochtitlan," 1542 (Brotton 2014, 104–5). Regarding the map she was charged to explicate, the art historian Barbara Mundy writes: "[The Oztoticpac lands map] shows how surviving members of [don Carlos Ometochtzin Chichimecatecatl's] noble family used maps to defend their hereditary rights and reaffirm long-standing traditions of landholding in the face of Spanish threats" (Mundy 2011, 57). These types of maps were not brought by Europeans but "were closely related to maps ... once ... made for ... the Aztec state" (Mundy 2011, 56). In his review of *Mapping Latin America* (see Dym and Offen 2011), Denis Wood tries to conceptually rein in the extension of the concept "map." That is, in labeling too many "graphics" as maps, we may be "throwing away the map to save it, ignoring its peculiar power to demonstrate its pervasiveness" (Wood 2012b, 137).

60. Overtones of world navels can be found here: "The roads not only depart from Cuzco, but also arrive at the capital city from the entire empire, somewhat like the saying that all roads lead to Rome" (Gasparini and Margolies 1980, 60). On the concept of "world navels," see Winther 2014d, 2019a.

61. Gelles 1995; Adorno 2000, 91.

*hanan*, or "upper," regions, and Qullasuyu and Kuntisuyu are the *hurin*, or "lower," divisions.⁶² In accordance with Renaissance cartographic tradition, Guaman Poma appears to orient the map with north at the top.⁶³ An alternative hypothesis proposes that *east* is at the top, as was standard on Medieval European *mappae mundi*.⁶⁴ With east at the top, Guaman Poma's centering the map on Cuzco rotates space a quarter turn.⁶⁵ Perhaps "the one thing we can say with certainty is that Guaman Poma attempts, in his mappa mundi, to combine two conceptualizations regarding the measurement and division of space, one western, one Andean, one perhaps more literal than symbolic, the other perhaps more symbolic than literal."⁶⁶ While the map's orientation is indeterminate, a clear representational desire emanates from it: Cuzco, Lima, Guayaquil, Santiago de Chile, and the Pacific Ocean are depicted in their appropriate relative positions. Moreover, parallel latitude and meridian longitude lines are added as a grid network, called a *graticule*.

Guaman Poma's map was part of his plea for grand-scale fiscal and colonial reform. He had hoped to convince the Spanish king Philip III to (re)turn the current "world upside-down" ("el mundo al rreués") to its natural order, an order brutally disrupted by *la conquista*—the

---

62. See Wachtel 1973 for a discussion of the relative placement of the four divisions.

63. Adorno 2011, 76. Indeed, along the borders Guaman Poma annotates the four kingdoms, or parts of the Incan empire. Consistent with Adorno's analysis, the right border says, "COLLA SVIO sale sol" (Qullasuyu the sun comes out). But on the top border he writes, "ANTI SVIO [to] the right of the sea to the north"; and in Quechua *anti* clearly means "east." The directionality of Antisuyu could thus be north or east. According to the "east upward" hypothesis, and as can be visualized by referring to a standard smaller-scale map, the top (Antisuyu) of Guaman Poma's map is in the direction of what we now consider to be Brazil; and the bottom (Kuntisuyu) in the direction of the Pacific Ocean.

64. Furthermore, in Quechua *kunti* means "west," and *chincha* "north."

65. Alternatively, it may only be a *half*-quarter turn. After all, in the heart of Cuzco, the cardinal directions bisect the four highways toward the *suyus*, with Antisuyu facing east (Gasparini and Margolies 1980, 60, fig. 47). Finally, that the western coast of South America actually bends around Cuzco, thereby also bending the ancient Incan *suyus*, does not make this discussion any easier. See the image at https://upload.wikimedia.org/wikipedia/commons/4/44/Inca_Empire _South_America.png (accessed April 23, 2015).

66. Adorno, pers. comm., July 2015. To put the ambiguity in perspective, consider that the name for the Andes Mountains likely derives from *anti*, or "east." Are the Andes in fact to the north or to the east of Cuzco?

conquest—and generalized imperialism and colonialism.[67] Various seaports, silver mines, monsters and mermaids, and Spanish galleons are indicated on the map. The galleons perhaps represented Guaman Poma's panic regarding AmerInd wealth nurturing Spanish greed. Nevertheless, Lima is designated as a capital, the "head of his majesty's empire." This may have been an attempt to placate the Spanish king.

The map projects Guaman Poma's prophetic *cosmovisión*: a Christian world, with the Indies occupying the first part (namely, Chinchasuyu[68]), or perhaps the third (Qullasuyu[69]). Each world region would persist separately—no mixing, or *mestizaje*, allowed. The respective capitals, or *world navels*, would be Cuzco, Rome, Guinea, and Turkey.[70] In this comprehension of space, time, power, and culture, the "synthesis" of Christian tropes and indigenous reality is "constantly threatened with an explosion." Guaman Poma emitted a "pathos-laden messianism" wherein his hopes for a redemptive, strong future for his people were combined with a dire expectation of continued tragedy.[71]

Interestingly, an attempt to imagine or reimagine, to map and remap, the world as given by power structures—what I call *cartopower*—was part and parcel of both Waldseemüller's and Guaman Poma's mappings. While Waldseemüller's *Universalis cosmographia* invites expansion in culture and knowledge production outward from Europe and toward one version of modernity, Guaman Poma's *mappa mundi* countermap encapsulates a future indigenous reorientation of modernity: a divided globe with a return to traditional indigenous culture in the Americas. In both cases, America was—and continues to be—a stage for contested representation and power. Indeed, countermaps are cartographic objects effectively resisting the standard maps: orthodox maps—those orthodox maps behind and under which lie dominant military, governmental, and economic cartopower.

---

67. Wachtel 1973, 221; Adorno 2000, 106. Although Guaman Poma had hoped that the king would inspect and publish his work, we do not know whether Philip ever saw it. However, it almost certainly reached the Spanish royal court. The manuscript eventually found its way to the Copenhagen Royal Library via a Danish diplomat who presented it to his king, Frederik III (Adorno 2011, 78). Although worthy of further discussion, I shall here set aside reasonable critiques that *both* the Spanish and Incan empires were imperialist and colonialist.

68. Wachtel 1973, 215, fig. 6.

69. Wachtel 1973, 218, fig. 8.

70. Guaman Poma 1615, 949 [963]; see Wachtel 1973 for discussion of the relative placement of the four divisions.

71. Wachtel 1973, 227; my translation.

## VAN SANT'S ULTIMATE MAP?

Is a map the more perfect and the more useful the more detail it includes? Borges's example of a one-to-one map of an empire, presented above, suggests not. One implicit idea might be that there is a golden mean of scale, and of level of detail—not too much and not too little. If one assumes that a map should be a perfect and pristine mirror of reality, however, Tom Van Sant's GeoSphere map (plate 2) could be seen to fit the bill as the ultimate world map.[72] Van Sant and his team collected thousands of satellite images taken over a period of three years and then compiled them, creating this epic visual map.

Objectivity and precision appear to suffuse this composite visualization and forces the question that if cartography is "the quest for ever more 'realistic' portrayals of the earth,"[73] then does this beautiful rendition of Earth signify the end of cartography itself? How objective is Van Sant's map, and what does objectivity—in cartography and across the sciences—consist of? Could an ultimate or total 1:1 map exist, and what would this mean for the completion and perfection of knowledge and representation?

Let us tread lightly and not overstate the objectivity of this map. Its objectivity is contextual. This is a result of a process. Intricate technologies mediate the seemingly transparent data.[74] NASA initially employed television-style cameras called return-beam vidicons rather than using the US military's more advanced imaging technology. However, these cameras were discontinued after failing spectacularly in the 1970s. The next technology of choice was the multispectral scanner, which scanned the landscape with a "banging mirror" "slam[ming] back and forth thirteen times a second."[75] This device diced and minced reality, "direct[ing] the light reflected from the earth through a prism onto a small array of photoelectric or other detectors."[76]

Creating and using images from this technology required choices and conventions. Coloring, error control, image reduction, and other forms of after-the-data processing were necessary throughout image capture and mapmaking. The GeoSphere project's imaging strategies eliminated

---

72. See the video Van Sant 2013 at 2 minutes, 17 seconds.

73. Wood 1992b, 62.

74. Hall 1992, chap. 2, "Ground Truth: Landsat Maps and the Remote-Sensing Revolution"; Wood 1992b, chap. 3, "Every Map Shows This . . . but Not That."

75. Hall 1992, 61.

76. Wood 1992b, 51.

night and seasons. They erased clouds. Ironically, this is not Earth as it actually appears. The banging mirror is not a mirror of nature.

What do we expect a map to be? Although a map is not merely an image, picture, or diagram, certain visuals can function better as maps than others. Is the GeoSphere project map—a composite of abstracted images—not a "map," as Van Sant, *National Geographic*, and others insist on calling plate 2? Given that the aggregate image can be used as a map, perhaps collating images can embody cartographic information, when intended to do so.

Van Sant's map inhabits a moment in the flow of time. It may be many things, but it is not the ultimate map. It is instead a complex achievement, eliminating important features of Earth. Uncovering the assumptions undergirding maps and scientific theories enables us to glimpse the multitude of representational paths *not* taken and to understand why certain choices were made at the expense of others.

Although there are obsessions with objectivity and accurate representation in the three main maps explored in this chapter, Waldseemüller's and Guaman Poma's maps also ooze ideology and futuristic visions. These two maps operate in dialectical tension vis-à-vis their commitment, or resistance, to colonialism. And Van Sant's map is interpretative and not the ultimate map—it hardly represents all vital features of Earth. These canonical world maps help us experience firsthand why any definition of *map must* be multidimensional.

## Conclusion

Standard representational characterizations of maps may be subsumed under analyses focusing on process and function—maps as invitations, weapons or theaters. The target of map analogizing always changes. As map thinkers, we cannot—and should not—separate representation from process or function, even when considering the map object itself.

World maps continue to evolve. A readily downloadable virtual globe such as Google Earth bursts with geographic information. Fully linked to Google Maps, Google Earth displays layered topographic and traffic information for your chosen territory. You can zoom in on a satellite view of your house or your favorite city square or beach. You can even upload your own pictures and construct map mash-ups. Like Van Sant's map, Google Earth and Google Maps rely on complex processing of satellite imagery. As in the case of Waldseemüller's or Guaman Poma's maps, geopolitics permeate Google Earth and Google Maps. Specifically, government agencies with military interests (e.g., the National Geospatial-Intelligence Agency or the CIA), private corporations (e.g., Google or Keyhole, Inc.), and small-scale programmers or hackers (e.g., Lars and Jens Eilstrup Rasmussen) intertwined and collaborated to unleash Google Maps as a powerful—simultaneously liberating and oppressing—functional public map platform.[77] Google Earth and Google Maps gave birth to *computational maps*, which encode too much information to represent at one pass. The underlying programs compress data and store it elsewhere, in the proverbial cloud. The map is more like an extended network than something you hang on a wall. Narrow representationalism hardly captures the dynamics of such multiscalar and hyperlocal mapping-as-you-go.

Such geographic and cartographic technologies place us on the cusp of yet another supposedly perfect map. But sampling limitations, errors of various kinds, and the diversity of sensory modalities militate against the possibility of any ultimate map of a complex world. Even so, next-generation computational maps may continue surprising us. Certainly, they obligate a retooling of the cartographic object concept to simultaneously include representation, process, and function.

Perhaps the real lesson from Borges's "On Exactitude in Science" is not the impossibility or futility of a one-to-one or 1:1 map, but the inevitable ebbs and flows of maps and mapping strategies as conventions, assumptions, and practices shift.[78] A new cartographer's guild will undoubtedly reconstruct another more useful map of the empire. It will probably be a computational and digitally made map—and with this new map, the world will be reconstituted and remade.

77. Zook and Graham 2007; Dalton 2013; Rankin 2016.
78. I draw a light distinction between *1:1* and *one-to-one* maps. The former relates to literal spatial scaling—say, of an immense paper map to the world it represents (see chapter 3). The latter use is much more metaphorical, as in speaking of a putatively perfect and complete cognitive mapping, by an individual person, of that person's world (see chapter 2). Borges had both in mind; when writing about Van Sant's map and Google Earth, I have more *1:1* in mind.

# · I ·
# Philosophy

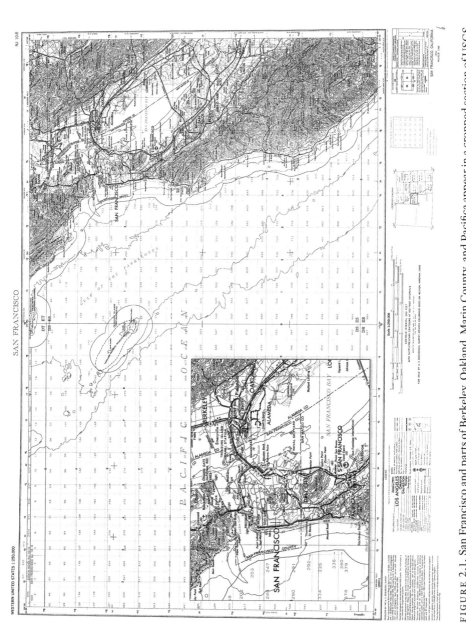

FIGURE 2.1. San Francisco and parts of Berkeley, Oakland, Marin County, and Pacifica appear in a cropped section of USGS topographic map NJ 10-8, 1956; revised 1980. See also plate 4. (US Geological Survey, Department of the Interior.)

# 2

# *Theory* Is to *World* as *Map* Is to *Territory*

The topographic map of the San Francisco Bay Area shown in plate 4 and figure 2.1 includes contour lines, heights, highways, hospitals, bridges, and beaches. As objective as it seems to be, it is an abstract and simplified map. It is not the real thing.

Whereas topographic maps represent abstracted natural and human features, thematic maps display the geographic distribution of key, richly textured properties such as human population density and temperature gradients. You could represent the Bay Area with thematic maps, including public transit, highway, or bicycle maps.[1] Or, if you were planning to buy, rent, or sell a house in San Francisco, you could consult cadastral maps showing legal title and division of land lots. Yet all such maps are still abstractions, in that almost all features of the actual landscape are omitted.[2]

What can maps tell us about human practices of acquiring knowledge, and experiencing the world? How does spatialized cognition and communication—"deep mapping," as we saw in chapter 1—influence scientific representation?

Applying map thinking to science and philosophy generates the basic version of what I call the *map analogy*:

ANALOGY 1. *A scientific theory is a map of the world.*

Construing thinking, representing, and storytelling as mapmaking processes helps us understand the way we philosophically understand reality, space and time, and ourselves.

---

1. Jack and Gay Reineck provided constructive feedback.
2. For further information on the distinction between topographic and thematic maps, see Swiss Society of Cartography 1977, 10–11; Robinson, Morrison, et al. 1995, 12–15; Slocum et al. 2005, 13; Winther 2015b.

TABLE 2.1 Three basic scientific analogies

| Analogy | Target domain | Source domain | Accentuated features |
| --- | --- | --- | --- |
| Electromagnetic lines of force : Mechanical forces | Electromagnetism | Mechanics and gravity | Geometry of force fields |
| Benzene ring : Snake biting its tail | Chemistry | Zoology or observation or nature | Spatial organization of whole structure |
| Natural selection : Artificial selection | Evolution | Animal and plant breeding | Mechanisms of selection for breeding |

## Analogy

The map analogy pervades philosophy and other fields, but what is analogy? Philosophers, psychologists, and computer scientists try to fathom analogical reasoning, intuition, and inference. Roughly, analogy involves comparing an item or situation from one domain, field, or case to another domain, field, or case—mapping or matching up features one to one. Because some features are readily comparable, reasoning by analogy aims to infer that one or more additional properties from the *source domain* also tell us something further about the *target domain*.

Consider a few analogies scientists have used to describe the world:

- Some natural laws of electricity and magnetism are like natural laws of mechanics and gravity (James Clerk Maxwell).[3]
- The benzene ring in chemistry is like a snake biting its own tail (August Kekulé).[4]
- Evolutionary "natural selection" is like artificial (human) selection (Charles Darwin).[5]

Table 2.1 compares the objects, properties, and processes of each of these analogies across target and source domains.

Analogies trade on similarity, but not on identity. They are between things that are somewhat different. Target and source domains should not be conflated. For instance, Darwin's theory can be described *as if* nature selects intentionally (as humans do when artificially selecting cows

---

3. Bartha 2010, 208; Maxwell 1890, 157.
4. Rocke 2010.
5. Darwin (1859) 1964, chap. 4, "Natural Selection," 80–130.

with the highest milk yield, or dogs with the best ability to herd sheep), and yet it would be a misunderstanding of natural selection to take this picture literally.

Throughout *When Maps Become the World*, I use the words *metaphor* and *metaphorical* very broadly. This chapter aims to give a precise and closed analysis of analogy. Metaphor is almost a figurative comparison; analogy a literal one. Now, perhaps analogical structure lies beneath many metaphors, while metaphors help us extend analogical structure into new realms of understanding and imagination.[6]

### THREE TYPES OF ANALOGY

Analogies can be divided into three types: positive, negative, and neutral.[7] The philosopher of science Mary Hesse characterizes them as follows:[8]

- Positive analogy: An object, property, or process carries over from the source domain to the target domain.
- Negative analogy: A particular object, property, or process present in one domain is absent in the other.
- Neutral analogy: We do not yet know whether a given target feature corresponds to some source feature.

For an intuitive example distinguishing among these three kinds of analogies, consider the saying "if all you have is a hammer, everything looks like a nail."[9] The source domain here is carpentry or tool use. The target domain is intended to be, roughly, human problem-solving. Positive analogies include the appropriateness of an object or idea for that which it is supposed or designed to fix: hammers force in nails, addition

---

6. What is the exact difference between analogy and metaphor? Some commentators on scientific inquiry do use the word *metaphor*, but Hoffman 1980 and Brown 2003 effectively analyzed analogy. Oppenheimer 1956 starkly distinguishes metaphor and analogy. It seems apt to cite Givón (1986): "The *metaphor* term comes from the literary analysis tradition, the *analogy* term comes from the philosophic tradition, most recently via Kant and Peirce" (100).

7. Keynes (1921) was the first to provide extensive discussion of positive and negative analogies, presenting a formal apparatus for analogical inference.

8. Hesse 1966, 1974. Bartha 2010 usefully reviews analogy.

9. This is philosopher Abraham Kaplan's "law of the instrument": "Give a small boy a hammer, and he will find that everything he encounters needs pounding" (1964, 28–29).

allows you to calculate the final price of the items in your grocery cart. Another positive analogy is that the "fit" is highly contextual. Just as you cannot use a hammer to efficiently turn a screw (but: a screwdriver!), so you cannot use mathematics to solve a complex interpersonal issue with your partner or parent (but: psychology and empathy!). Negative analogies include that tools are physical objects and ideas are abstract. A relevant neutral analogy can be captured with an open question for which I would like an answer upon further inquiry: Does the fact that a carpenter needs many tools for the job correspond to some important fact about human cognition needing to employ many paradigms and models to solve a suite of problems?[10]

The framework of positive, negative, and neutral analogies reminds us that any source and target lie somewhere on a spectrum of potential correspondence with respect to their different features. In common parlance, we tend to use the term *analogy* only to refer to a similarity between objects (such as when they both share or lack a given property). I am calling this *positive analogy*. In contrast, I use the term negative analogy the same as one might use the term *disanalogy*—that is, to show how two things are different. These terms are chosen to align with extant literature with which I intend to dialogue, including work by Mary Hesse, Paul Bartha, and Douglas Hofstadter.

Analogical arguments are especially useful for the case of a relevant neutral analogy that would extend existing knowledge once—and if—we learn that it actually *is* a positive analogy. Predicting and exploring new, fallible information is precisely the function of analogical inference in science.

As one sustained example of an influential and powerful analogy, consider Darwin's Tree of Life. The divergence from shared ancestry of all organisms can be analogized to a large branching tree, such as an oak tree.

---

10. Or consider "It is the east, and Juliet is the sun," from *Romeo and Juliet* (Act 2, Scene 2). Shakespeare is comparing a celestial body (source domain) and a person (target domain). Positive analogies here include the radiance of warmth and light, and the feelings of pleasure evoked by the object. Negative analogies include the way the objects of the source and target domain grow and remain stable—atomic fusion in the former case, food and metabolism in the case of Juliet. Finally, neutral analogies pervade scientific inquiry due to a lack of complete knowledge about the source and target, but they are harder to find in a more familiar case—perhaps the extent to which Juliet's radiance is affected by "mood clouds" could be a neutral analogy worth assessing.

Thinking of genealogy (the target domain) in terms of a living, growing tree (the source domain) makes explicit various positive analogies:

- branching of both trees and genealogies is dichotomous, in general;
- tips—whether leaves or extant species—are the living, dynamic, changing part of the structure;
- growth is constrained and limited via some kind of competitive mechanism—in the source domain, it is competition for light; in the target domain, it is competition among individuals of different varieties for distinct niches in the economy of nature.

These positive analogies stabilize and strengthen the overarching analogy.

There are also negative analogies, such as that the tree cambium is alive, as a layer inside the trunk, all the way to the roots of the tree, whereas old, extinct species are not alive. This negative analogy is assumed not to overwhelm the preponderance of positive analogies justifying comparisons across the two domains.

The surprising, neutral analogy is that, just like branching on a tree is recursive and hierarchical, and a big branch contains all the smaller branches on it, *so a genealogy is recursive and hierarchical. All species of a single branch are more closely related to each other than to a species on another branch of the tree.* This is a special type of neutral analogy—namely, an analogy of interest—because of its potential to increase scientific knowledge (it can be tested through genetics, as we will see in chapter 8). Darwin writes: "All true classification is genealogical."[11] To view phylogenetic relations as a Tree of Life helps shed light on how all species are related.

More generally, if one domain, field, or case is similar enough to another domain, field, or case in the right ways (i.e., has sufficient positive analogies), while not too dissimilar to it, then it may be justifiable to infer that it is like it in yet another way (neutral analogy). Thus, analogical inference can be represented like this:

P1. X has properties $a$, $b$, and $c$.
P2. Y has properties $a$ and $b$.
[Therefore, by analogy]
I. Y has property $c$ as well.

---

11. Darwin (1859) 1964, 420.

The conclusion, the inference about c, provides us a hypothesis—an interesting, relevant neutral analogy—that can then be tested. In this analogical inference, the two premises (P1) and (P2) provide some justification for this hypothesis I.[12] The result of empirical testing, if confirmed, will then close the analogy, making it either a positive one (both X and Y have property c) that helps guide future research, or a negative one (X has property c, but Y does not) we move away from.

### CRITICAL CAUTIONS

The history of science is replete with spectacular, innovative, and productive analogical inferences, including Isaac Newton's analogy from fast projectiles to planets in orbit and Alfred Wegener's analogy from icebergs on water to continents floating on Earth's hot inner fluid.[13] Because analogies are between distinct domains, however, they will always break down somewhere, and to some extent.

A well-known serious failure of analogy is the case of sound and light.[14] Both sound and light act like waves: they behave similarly when meeting surfaces or refracting around corners or through slits. Thus, echo and reflection, loudness and brightness, pitch and color are positive analogies. As physics has shown, however, the analogy fails when trying to infer the behavior of light traveling through a double slit (wave-particle duality) or through a vacuum (there is no "ether"). In fact, negative analogies preponderate in important ways over positive ones in this case.

Analogy is uncertain because it is an ampliative inference rather than a deductive one. As logicians and philosophers explain, ampliative inference adds knowledge in moving from the premises of an argument to the conclusion. In contrast, deduction unpacks what is already implicit—it is logically certain so long as the premises are true.[15] Ampliative inference also includes induction (in which a conclusion follows probabilistically from the evidence contained in the premises) and abduction (inference to the best explanation). Both of these are also uncertain—the conclusion could be false even if the premises are true. However, because analog-

---

12. Bartha 2010 analyzes the structure and formalization of analogical inference in detail.

13. See the list of sixteen scientific analogies in Holyoak and Thagard 1995, 186–89; on molecular analogical models, see Brown 2003; on Wegener's iceberg analogy, see Wegener 1966, 37, 39, 43, 45; Solomon 2001, 58, 62, 87–88.

14. Hesse 1966, 60–61; Bartha 2010, 14.

15. Peirce 1992b.

ical arguments have a well-understood comparative structure between source and target domains, and do not emphasize aggregating many unique cases as being of the same kind, analogy is not induction; and because the goal of analogy is not to infer the conditions necessary to give rise to known outcomes, analogy is not abduction, either.[16]

There are indeed good empiricist reasons to be wary of ampliative reasoning,[17] and particularly of analogy and abduction, which rely on information and structure going far beyond empirical data. I cautiously defend abduction and analogy by accepting that there may be assumptions we can make without evidence that are legitimate—for example, rigorously tested statistical presuppositions[18] or cognitive a priori principles. We constantly make analogies, as psychology and cognitive science teach us.[19] Indeed, if analogizing maps to theories (or models) can illuminate representational practices, then the map analogy itself provides justification for being friendlier to analogical inference as such, even for those of us with skeptical, empiricist tendencies.[20]

## The Map Analogy

The overarching analogy of interest in the present study is the map analogy, whose basic form is: *A scientific theory is a map of the world* (analogy 1). Because this analogy contains a whole set of analogies, it would be useful

---

16. Peirce 1992b usefully explains differences between induction, deduction and abduction; so does Psillos 2000. For those interested in further details, it might help to say that David Hume, Charles Peirce, and Mary Hesse are key philosophers discussing the grounds, strengths, and problems with, respectively, induction, abduction, and analogy.

17. Van Fraassen 1980, 1988; Cartwright 1983, 1989.

18. See chapter 7.

19. Analogy is of long-standing interest in the sciences of the mind. The cognitive scientist and philosopher Douglas Hofstadter has defined analogy making as "the perception of common essence between two things"; analogy, for him, is "the core of cognition" (2006; quote at 19:30). Hofstadter and Sander (2013) further develop these thoughts. Dedre Gentner and her associates endorse a "structure-mapping" theory of analogy (Gentner 2003, 106; cf. Gentner and Jeziorski 1993). See also Tversky 1977; Weisberg 2013; Nersessian 2002; Fauconnier 1997; Fauconnier and Turner 2002.

20. John Norton favors what he calls a "material" approach to induction and, by his extension, analogy. He believes there are good analogies in science, but he critiques attempts to find a "universal schema" for analogy (Norton 2016).

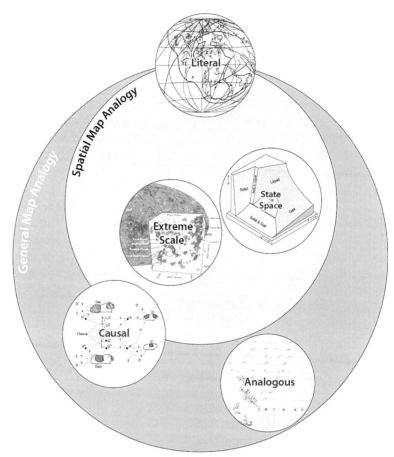

FIGURE 2.2. Euler diagram of the two map analogies and five kinds of maps. Paradigmatic examples are shown within the circle of each map type and discussed in chapters 6 (extreme-scale, state-space, literal, and analogous), 7 (causal), and 8 (analogous). (Illustration by Mats Wedin and Rasmus Grønfeldt Winther.)

to begin by surveying the conceptual landscape with a typology of map analogies, which also involves a typology of different kinds of spaces.

## A TYPOLOGY OF MAP ANALOGIES

Consider a few types of maps to which theories (or models) can be analogized: *extreme-scale*, *state-space*, *literal*, *causal*, and representations analogous-to-maps or *analogous maps*. Figure 2.2 shows a Euler diagram of some of the logical and conceptual relations among these maps. Theories and models, and perhaps hypotheses, paradigms, and the like can be analogized to such maps.

The general map analogy (the largest circle in the Euler diagram) draws on assumptions shared across many domains of representational practices.

The second-largest circle represents the spatial map analogy. With this circle, we arrive at an important target of this book: *spatialized knowledge*. Representations rendered according to the spatial map analogy are found in domains as wide ranging as cosmology, mathematics, neuroscience, and genetics, and thus encompass a carnival of scientific examples. Three types of spatialized knowledge maps are here distinguished. They do not together constitute the full range of this category.

The *extreme-scale map* zooms in to the very small, as in linear genetic maps, or zoom out to the very large, as in maps of cosmic microwave background radiation. It operates at nongeographic scales, but still represents physical objects or causal processes in actual physical space. Here the analogy is:

ANALOGY 2. *A scientific theory or model is a spatial map, appropriately scaled up or down, and theorizing or modeling is mapping in an extreme-scaled space.*

A *state-space map* renders scientific theory as movement—according to mathematical natural laws—in a mathematical space called a *state* or *phase* space. Actual and possible trajectories are traced out as a map of the theory.[21] Formal scientific theory can thus be rendered as paths and landscapes in abstract mathematical spaces. This gives us the following analogy:

ANALOGY 3. *A scientific theory or model is an abstracted map in a state-space territory, and theorizing or modeling is mapping in this abstract space.*

A state-space map can take the form of a simple block diagram or a set of multidimensional graphs. It tends toward the more complex forms, though, because what it visualizes is complex: all the possible states, or conditions, that result from the input of mathematical variables as the

---

21. This might be one way of analyzing the semantic view of theories, in which meaningful mathematical modeling is at the fore. See Winther 2016; see also chapter 6.

variables change and interact across time, in the system of interest. Most mathematical theories in science can be rendered as state-space maps.

In part because the extreme-scale and state-space map types represent, respectively, *physical* and *mathematical* spaces and territories, our Euler diagram portrays them as not overlapping. However, both involve conceptual abstraction, and concrete data is relevant to both. The two map types navigate the concrete-abstract and the data-theory dialectics, but do so differently.[22]

A *literal map* is a map at geographic scales containing at least some geographic objects, such as the extent of ecosystems in ecological maps or tectonic plates in geological maps. Our Euler diagram in figure 2.2 depicts such maps as sometimes guided by general and spatial map analogies, and sometimes outside the analogical domain altogether. This is because a literal map is a *scientific* literal map *only* insofar as it has been appropriately analogized as a scientific representation (or as part of one) — only when it assists in the scientific work of prediction, theory confirmation, data organization, and so forth. For example, a highway map is not a scientific map until we use it in a scientific project — say, an ecological one. The literal map portion falling outside the analogical universe in figure 2.2 covers protoscientific literal maps not yet analogized to become proper scientific literal maps. This is the analogy:

ANALOGY 4. *A scientific theory or model, in whole or in part, is an actual cartographic map.*

Fourth, a *causal map* shows certain causal factors interacting concretely with one another. For this map type, the analogy need not appeal to the spatial or temporal features prominent in most maps and mapping. Think of a biochemical network diagram where different kinds of molecules are causally stimulating and inhibiting other kinds of molecules.

---

22. The contrast between compositional science and formal science I first developed in Winther 2003 and published in 2006c remains pertinent here. In biology, for instance, compositional science articulates the concrete structure, mechanisms, and function — through developmental and evolutionary time — of material parts and wholes, such as the heart, liver, and ovaries of vertebrates. Formal science in this domain focuses on the relations, captured in formal laws of nature, among mathematically abstracted properties of abstract objects, such as gene frequencies or population numbers. An exploration of the similarities between compositional and extreme-scale maps, on the one hand, and formal and state-space maps, on the other hand, could be of interest both in biology and in fields outside of biology.

For instance, in the Krebs cycle, important to cellular respiration, acetyl coenzyme A—derived primarily from sugar—is broken down to produce ATP, a molecule ubiquitous to life that provides energy to countless types of biochemical reactions. The Krebs cycle has ten chemical intermediaries, or substrates, with eight enzymes facilitating each of these reactions. In a map of how one reaction causes the next, the substrates can be represented as nodes. (In addition to such network causal maps, in chapter 7 we shall see how statistical analyses can also be used to make causal maps.) A causal map follows this analogy:

ANALOGY 5. *A scientific theory or model is a map of causal connections, inferred through statistical analysis or via mechanistic experiments, and often depicted in a topographically accurate diagram.*

Such a map is generally not spatial in a basic Euclidean sense. The dimensions are not metrical, but are used to organize the causal factors. However, a causal map may be topologically accurate in the same sense as the London Underground map is—if the map shows relative positions and prominence of certain causes on particular effects. A more spatial version of a causal map would be, for instance, one showing the natural habitat area of certain species and how they interact with neighboring species, possibly by outcompeting them in an area. Or it could be a map in dance notation. The essential point of these maps is that causal interaction is the target of visual representation.

There are also maps that have even less to do with space than causal ones. For instance, much analytic, formal mathematical theory is not primitively spatial or even diagrammatic; rather, it involves definitions, analytic equations, axiomatic proofs, and so on. Even so, the scientific general map analogy is instructive for understanding mathematical theorizing and abstraction. Mathematical theory, with its laws and principles, is a map showing us where to look in nature and what to measure and manipulate. Darwin's Tree of Life also doubles as a kind of analogical map, directing our thinking of the branching origin of species, and of the common ancestry of all life (as we shall see in chapter 8). In these cases, the result is *representations analogous-to-maps*, or, more briefly, *analogous maps*. The analogy here is:

ANALOGY 6. *A scientific theory or model, even when not spatial, is a map guiding us, and theorizing or modeling is mapping.*

This typology is incomplete. There is empty space in the diagram, allowing room for map types I may have missed. Additionally, there will be gray areas here, as the boundaries of the five map types are fuzzy.[23] Mysteries and controversies regarding time are also elided here, although they bear further investigation.[24]

To be more precise, then, *mapping is a communal and personal representational effort to imagine and control the different kinds of space of distinct map types.* Successful mapping means accurately and effectively populating these managed spaces with useful representational elements (such as icons and symbols) that further our navigational goals in the broadest sense.

### USES OF THE MAP ANALOGY IN HUMANISTIC INQUIRY

The map analogy has been employed heavily in philosophy and in other areas of humanistic inquiry. To give a sense of the scope of the map analogy, Anglo-American analytic philosophy, "continental" European philosophy, pragmatic philosophy, philosophy of science, religious studies, and history are each touched upon below.

#### *Anglo-American Analytic Philosophy*

Many analytic philosophers, with their emphasis on clarity and identifying conceptual assumptions in everyday language, tend to employ map analogies with reference to any propositional attitude (i.e., beliefs, desires, and intentions). For instance, the philosopher Frank P. Ramsey

---

23. Fuzzy boundaries are generally undesirable in a Euler diagram. But that convention, if adhered to too strictly, can itself be a reification, where a formal representational style comes to dictate the view of the world by demanding clear and distinct categories.

24. Is time linear or cyclical, absolute or relative, or all of these? How physically intrinsic or phenomenologically emergent is time? Can past events be reinterpreted and reshaped? Is the future here, the past somewhere else? For our purposes, representing time could be done implicitly on single spatial maps (e.g., showing the population of a country in a table with three different decades, on a map), or explicitly across multiple spatial maps or in dynamic computational maps (such as those showing the expansion, contraction, and potential appearance and disappearance of country borders on a map of an entire continent) (cf. D. Rosenberg and Grafton 2012; Kraak 2014; Bakhtin 1981). Furthermore, time is often one formal dimension among many on a state-space map, even if its intuitive interpretation differs from that of strictly spatial dimensions.

defined "belief" as a "map of neighbouring space by which we steer."[25] The philosopher Mark Wilson views maps as models for the analysis of human reasoning, language use, and behavior. He suggests creating a pluralism of perspectives to correct for conceptual limitations: "a rich *atlas* of maps . . . each of which is dedicated to answering questions best suited to its own personality."[26] Other analytic philosophers employ the map analogy to various ends.[27] In many of these cases, diverse kinds of mental content referring to the world are taken to be usefully analogous to maps.

The philosopher Gilbert Ryle takes the map analogy beyond studies of mental content into analogizing the philosopher's practices and goals to the cartographer's.[28] In one essay, he compares the relationship between a philosopher and a competent speaker of a language to that between a mapmaker and a village inhabitant.[29] The latter member of each pair engages in concrete activity, while the former investigates that activity abstractly. For Ryle, the philosopher is a cartographer of the territory of living language. Interestingly, the same individual may alternate between being a maker and a user of an abstract map, on the one hand, and a "village inhabitant" with no knowledge of a map on the other, although this involves a marked reorientation in one's purposes.[30]

In *The Concept of Mind*, Ryle explores the "logical geography of concepts." Again, the philosopher is seen as an abstract cartographer of concepts and arguments, while "Teachers and examiners, magistrates and critics, historians and novelists . . . parents, lovers, friends and enemies" are "like people who know their way about their own parish, but cannot construct or read a map of it, much less a map of the region or continent in which their parish lies." Identifying logical connections among many different concepts about both mental states and psychological activities "has always been a big part of the task of philosophers." And while Ryle thinks some progress has been made, he takes the mind/body split en-

---

25. Ramsey 1990, 146.

26. M. Wilson 2006, 291. Wilson's argument articulated via a map analogy might be that *reasoning is mapping* (289–96).

27. See, for example, Armstrong 1968; Godfrey-Smith 1996; Ismael 1999, 2007; Millikan 1984; Sellars 1981; Stich 1990.

28. Ryle 1949, 7.

29. Ryle 1971, 440–45.

30. "Where he normally thinks of *his* home, *his* church and *his* railway station in personal terms, now he has to think of them in impersonal, neutral terms" (Ryle 1971, 441). The Ryle essay contains much else relevant to the map analogy.

gendered by the French philosopher René Descartes to have "distort[ed] the continental geography" of studies of the mind, and the conceptual framework underlying such studies. Drawing on the map analogy, Ryle states his mission in *The Concept of Mind* thus: "To determine the logical geography of concepts is to reveal the logic of the propositions in which they are wielded, that is to say, to show with what other propositions they are consistent and inconsistent, what propositions follow from them and from what propositions they follow."[31]

Ludwig Wittgenstein's philosophy is difficult to categorize, but it has had significant influence within analytic philosophy. In his terse *Tractatus Logico-Philosophicus* ([1921], 1922), Wittgenstein writes about "pictures" (*Bild*) and the world thus: "1 The world is everything that is the case"; "1.2 The world divides into facts"; "2.063 The total reality is the world"; "2.12 The picture is a model of reality"; "2.1512 It [the picture] is like a scale [*Maßstab*] applied to reality."[32] Repeatedly, Wittgenstein implies that our pictures or models *map* the world, reality, the total set of facts.

More poetically, Wittgenstein's preface to *Philosophical Investigations* describes philosophical reflection as incomplete mapping: "The best that I could write would never be more than philosophical remarks; my thoughts soon grew feeble if I tried to force them along a single track against their natural inclination. —And this was, of course, connected with the very nature of the investigation. For it compels us to travel crisscross in every direction over a wide field of thought. —The philosophical remarks in this book are, as it were, a number of sketches of landscapes which were made in the course of these long and meandering journeys."[33]

Wittgenstein's *Yellow Book* starts with the claim that "we lack a synoptic view" in philosophy, the same challenge travelers encounter in exploring "the geography of a country for which we [have] no map, or else a map of isolated bits."[34] Wittgenstein gives this metaphor a linguistic reading: "The country we are talking about is language, and the geography its grammar."[35] In asking whether "our own language is complete," in the *Philosophical Investigations*, Wittgenstein also uses a metaphor of language as an accreted city, which smacks of a map analogy: "Our language

31. Ryle 1949, 7–9. Christian Frankel provided constructive feedback.

32. These pictures can be made explicit; not all can. Wittgenstein concludes his *Tractatus* thus: "Whereof one cannot speak, thereof one must be silent." I here quote selectively from Wittgenstein 1922.

33. Wittgenstein 2009, 3ᵉ. The preface is from 1945. For fuller discussion of Wittgenstein and maps, see Wagner 2011.

34. Wittgenstein (1979) 2001, 43.

35. Wittgenstein (1979) 2001, 43.

can be regarded as an ancient city: a maze of little streets and squares, of old and new houses, of houses with extensions from various periods, and all this surrounded by a multitude of new suburbs with straight and regular streets and uniform houses."[36]

Interestingly, Wittgenstein considers "the symbolism of chemistry and the notation of the infinitesimal calculus" to be "suburbs of our language."[37] A complex city and language are both historically layered, structurally fractured, shaped by multiple modes of influences from near and far, and mappable by experts.[38]

## Continental European Philosophy

"Continental" European philosophy at large also makes extensive use of the map analogy.[39] Critical philosophers in these domains analogize language and knowledge to maps, investigating the misleading and seductive nature of representation. They play with the paradoxes of equating map (or theory or story) with the world. For Alfred Korzybski (and, later, Gregory Bateson[40]), we must not confuse abstract and concrete, theory and world, or map and territory. Recall the absurdities and hopes of Borges's story from chapter 1, or contemplate Lewis Carroll's Mein Herr telling Sylvie about the map "on the scale of a mile to the mile."[41]

Jean Baudrillard worries about the relation between simulacrum and reality in contemporary mass-market globalized society. His *Simulacra and Simulation* opens with a basic lesson drawn from "the Borges fable" (my chapter 1 opener): "Today abstraction is no longer that of the map,

---

36. Wittgenstein 2009, §18, 11ᵉ.
37. Wittgenstein 2009, §18, 11ᵉ. Note resonances with Calvino's cartographically infused *Invisible Cities* (1974).
38. Ann Lipson provided constructive feedback.
39. The study of literature cannot be explored in depth here. However, Conley theorizes and surveys "cartographic writing" in Early Modern French literature (1996, 4–7). Conley argues that, by deploying space and cartography, unusual forms of literature helped create "the new entity, the 'self' and the 'subject.'" Additionally, Moretti explores modes of mapping the location and movement of fictional characters, goods for consumption, and so forth in "literary maps" (2005, 53). See also the beginning of chapter 1.
40. Bateson explicitly builds on what he takes to be Korzybski's "original statement": "the map is not the territory," from the 1970 Korzybski Memorial Lecture, which appears as "Form, Substance, and Difference" in Bateson 1972, 454–71. Korzybski 1933, 750: "A map is *not* the territory."
41. Carroll (1893) 2010, 162–63.

the double, the mirror, or the concept. Simulation is no longer that of a territory, a referential being, or a substance. It is the generation by models of a real without origin or reality: a hyperreal."[42]

For Baudrillard, an entire social thought system is an enormous and multifaceted map that, in contrast to Korzybski, does become—or *is*— the territory. Baudrillard argues that, in contemporary mass media and urbanized culture, simulations have become simulacra completely unanchored from original reality, and now shape new kinds of reality. Today, others (our "friends" or "followers") might think of our Facebook and Instagram personalities as more real than our own, living breathing face-to-face personality. Virtual, abstract maps of us *become* us, in the eyes and hearts of others. Imagining the social map as an immense and powerful network in this way is familiar to other continental structuralists and post-structuralists, beyond Baudrillard.[43]

## *Pragmatic Philosophy*

Philosophers in the tradition of American pragmatism expanded map thinking around the turn of the twentieth century. Charles Sanders Peirce, William James, and John Dewey deployed the map analogy to illuminate representational practices and elucidate notions of reality.

In an imaginary dialogue with an interlocutor, Peirce has a land surveyor taken to task for not creating "a true representation of the land." The surveyor retorts: "It cannot . . . represent every blade of grass; but it does not represent that there is not a blade of grass where there is. To abstract from a circumstance is not to deny it."[44] As with maps, so with syllogisms. While syllogisms are sometimes critiqued for being "a purely mechanical process," they are "not intended to represent the mind, as to its life or deadness, but only as to the relation of its different judgments concerning the same thing"; syllogisms represent some aspects of "mental action," but hardly all.[45] Peirce had ample experience in surveying, mathematics, meteorology, and cartography, and even developed a creative, conformal map projection (fig. 2.3).

Peirce's close associate William James takes a more skeptical approach.

---

42. Baudrillard 1994, 1.

43. Further synoptic and erudite discussions of the map analogy in continental and critical European philosophy can be found in Lacoste 1976 and Jacob 2006.

44. Peirce 1992a, 62. See Bronner 2015 for a contemporary turn on this kind of argument.

45. Peirce 1992a, 62–63.

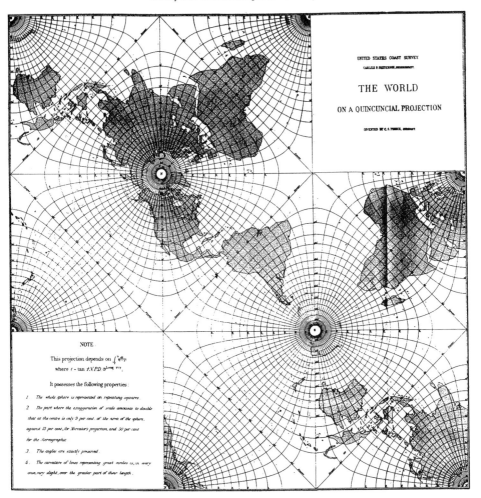

FIGURE 2.3. Charles Sanders Peirce's conformal quincuncial projection from 1879, "formed by transforming the stereographic projection, with a pole at infinity, by means of an elliptic function." (Peirce 1989, 68; map on 69.)

James recognizes that concepts are abstractions that "steer us practically every day, and provide an immense map of relations among the elements of things."[46] He thinks that single concepts may be misleading, however, where the entire "map remains superficial through the abstractness, and

46. James (1911) 1977 quoted in McDermott 1977, 243. For further analysis of concepts such as "an immense map of relations," see Winther 2014c.

[46]  CHAPTER TWO

false through the discreteness of its elements." For James, abstraction *does* imply the negation of what is left out. Indeed, for James, "conceptual knowledge is forever inadequate to the ful[l]ness of the reality to be known."[47]

Finally, Dewey highlights the complementary relationship between map and explorer: "The map is not a substitute for a personal experience. The map does not take the place of an actual journey.... But the map, a summary, an arranged and orderly view of previous experiences, serves as a guide to future experience; it gives direction; it facilitates control; it economizes effort, preventing useless wandering, and pointing out the paths which lead most quickly and most certainly to a desired result."[48] The map guides exploration, but the explorer must directly experience the journey.

## *Philosophy of Science*

Explore with me now some of the map analogies in the work of philosophers of science. We should consider this seventh map analogy, explicitly tagging models:

ANALOGY 7. *A scientific model is a map of the world.*

Throughout *When Maps Become the World*, usage generally makes clear whether analogy 1 or 7, or both, is intended.[49] For my purposes, no specific interpretation of the relation between model and theory is endorsed or required.[50]

The philosopher of science Rudolf Carnap uses the map analogy in

---

47. McDermott 1977, 245.

48. Dewey 1976, 284. Other philosophers have applied the dialectical logic of this explorer–map relationship. For instance, while Paul Feyerabend is a philosopher of science, he employs the map analogy to make a similar pragmatic point, arguing that reason is analogous to a map, and practice to exploration. Reason and practice together are mutually constitutive *"parts of a single dialectical process"* (Feyerabend [1975] 2010, 233; see also Latour 1990).

49. Models have gained increasing importance; see Frigg and Hartmann 2017; Winther 2016.

50. A survey I sent to twenty eminent scientists (of whom sixteen responded) included the question "What do you think is the difference between theory and model, if there is any?" Scientists typically distinguished usefully, but in different manners, between models and theories. They considered both important.

conveying core philosophical notions about how to ensure that science is objective and not based on subjective aspects of experience. According to him, knowledge building in science requires a step-by-step process of making *structural, definite descriptions*, relying only on a rich network of relations among scientific concepts. A map of "the Eurasian railroad network" serves as the single-longest "concrete example" in Carnap's *Aufbau*. With this example, Carnap map thinks his concept of structural definite description.[51]

In a commentary on that book, Nelson Goodman compares a philosopher's activity to that of a mapmaking metascientist, in a manner reminiscent of Ryle's analysis. According to Goodman's survey of *Aufbau*, what Carnap calls a "constructional system" is not intended to "recreate experience," "but rather to map it." Goodman explicitly appeals to map generalization practices and says that "a map is schematic, selective, conventional, condensed, and uniform."[52] Carnap approves wholeheartedly of Goodman's defense of Carnap's constructionalism—not as an effort to "copy or picture reality either as a whole, or in part, or on a diminished scale," but as a project "to represent the relations among the objects in question by an abstract schema."[53]

A subsequent case of map thinking lies at the very heart of philosophy of science: Thomas Kuhn's analogy between scientific paradigms and maps. His influential *The Structure of Scientific Revolutions* characterizes normal science as "puzzle-solving." Puzzles are difficult, yet assume that a solution already exists.[54] Researchers and scientists choose puzzles from a total class of possible problems because of this presumption and not because they are "intrinsically interesting or important."[55] Analogously, the cartographers creating the USGS map (fig. 2.1) presupposed that they could create a map that would be accurate and useful, a new map adding knowledge of territory long occupied.

51. Carnap (1967) 2003, §14–§16.
52. Goodman 1963, 552.
53. Carnap 1963, 940.
54. Kuhn says: "One of the things a scientific community acquires with a [new] paradigm is a criterion for choosing problems that, while the paradigm is taken for granted, can be assumed to have solutions. To a great extent these are the only problems that the community will admit as scientific or encourage its members to undertake" (1970, 37).
55. Central problems such as "a cure for cancer or the design of a lasting peace, are often not puzzles at all, largely because they may not have any solution" (Kuhn 1970, 36–37).

Kuhn's gift to the philosophy of science was the insight that deep innovation emerges only with new paradigms and revolutions. He used the map analogy to imagine normal science as something more refined and creative than mere puzzle solving. According to Kuhn, a paradigm

> functions by telling the scientist about the entities that nature does and does not contain and about the ways in which those entities behave. That information provides a map whose details are elucidated by mature scientific research. And since nature is too complex and varied to be explored at random, that map is as essential as observation and experiment to science's continuing development. . . . paradigms provide scientists not only with a map but also with some of the directions essential for map-making. In learning a paradigm the scientist acquires theory, methods, and standards together, usually in an inextricable mixture.[56]

Kuhn's paradigm is much more than a mental or cognitive map. *Paradigms* are socially embedded sets of assumptions and practices necessary for scientific research, and are conditions of possibility for the very existence of science.[57] Kuhn also implies that the techniques, assumptions, and materials associated with a paradigm are mapmaking resources. Thus, we can state a Kuhnian map analogy:

ANALOGY 8. *A paradigm is a map of the objects and processes taken to exist in the world, and it contains practices and assumptions for further mapping the world.*

If assurance of a solution is a foundation of the puzzle metaphor, the importance of exploration is a mainstay of Kuhn's map analogy. Like maps, paradigms are incomplete and involve the production of novelty. Indeed, the map analogy offers promising ground for rethinking oppositions between theory and practice, and between convergent (conservative) and divergent (creative) thinking.[58]

---

56. Kuhn 1970, 109; Kuhn 2012, 109. This is the only place I have discovered the map analogy in Kuhn's *The Structure of Scientific Revolutions*.

57. See Friedman 1999; Elwick 2012; Winther 2012a. Friedman 1999 is especially instructive here.

58. Kuhn 1977. Janette Dinishak provided constructive feedback. Turnbull 1993; Sismondo 1998, 2004; Sismondo and Chrisman 2001 analyze the map analogy in philosophy of science.

## Religious Studies

Map thinking also emerges in the study of religion and mythology. Consider religious studies scholar J. Z. Smith, whose inaugural lecture at the University of Chicago in May 1974 concludes thus: "For the dictum of Alfred Korzybski is inescapable: 'Map is not territory'—but maps are all we possess."[59]

Smith contrasts three maps. According to the first, there is a tendency to split the world of humans into "human beings (who are generally like-us) and non-human beings (who are generally not-like-us)"; an *us/them* distinction is thus established, analogously to "the boundaries of any ethnic map."[60] This is the standard nineteenth-century scholarly map, drawn by Westerners who "live in a post-Kantian world in which man is defined as a world-creating being and culture is understood as a symbolic process of world-construction."[61] This interpretative strategy permits only certain kinds of religions to be classified as genuine ones, especially monotheistic religions. "Primitive" religions are omitted, as they receive a "negative evaluation."[62]

A second map, defended by twentieth-century anthropologists, instead provides a "positive (even nostalgic) appreciation" of indigenous and nonmonotheistic religions as being holistic.[63] While Smith claims that this map "set[s] forth a new cartography," he fears that "it remains uncomfortably close to being a mirror image of the 'mainstream' map."[64]

Smith's third map is one he posits in contrast to these, by drawing attention to the "cargo cult" context of trade with white Westerners in which various religious stories across the Pacific developed. Smith identifies the "dimensions of incongruity"—internal logical, narrative, and moral inconsistencies and potentialities—which he takes to exist across *all* religions.[65] He interprets one Pacific religious story (from the Maluku Islands, Indonesia), of Hainuwele, the "coconut girl" born of a palm fruit, in terms of the incongruity between the indigenous economy and the white Western customs of trade. The indigenous economy is based on

---

59. J. Z. Smith 1978, 309.
60. J. Z. Smith 1978, 294.
61. J. Z. Smith 1978, 290.
62. J. Z. Smith 1978, 297.
63. J. Z. Smith 1978, 297. Smith also presents a "utopian map" (309), but this might be a fourth map, better left for another occasion.
64. J. Z. Smith 1978, 296.
65. J. Z. Smith 1978, 309.

[50] CHAPTER TWO

evening out resources among its members, but the white European traders' is based on profit. Hainuwele excretes manufactured goods, and out of jealousy, the villagers kill, dismember, and bury her. Her parts give rise to various species of edible plants, restoring the threatened tradition of equitably bartering what you yourself have grown or made.

Oceanic cargo-cult narratives like the Hainuwele origin myth are paradigmatic examples of Smith's third map. Smith concludes his lecture with exactly this type of map, which treats incongruities playfully, accepting that the ultimate meaning of life we so yearn for does not emanate from "a series of burning bushes." Human life is subtler, aritualized, and non-repetitive.[66]

In short, a plurality of maps serves as an explanatory frame for understanding religions. Each scholarly map is partial, but together they are, according to Smith, all that we possess.

## History

John Lewis Gaddis's *The Landscape of History* starts with a discussion of metaphor and the idea "that something is 'like' something else."[67] The map analogy does significant work in Gaddis's exploration of the tools and goals of the historian. As in Walter Benjamin's reflections on Paul Klee's painting *Angelus Novus*, the historian faces the past as he is propelled backward toward the future.

We can "represent" this past "as a near or distant landscape."[68] Gaddis suggests that the historian is like a cartographer. Historical narrative is like a map, he argues. Both "distill the experiences of others"; are purposive; involve "pattern recognition"; require generalization and abstraction; are not complete replications; and are premised on "selectivity," "simultaneity," and "scale."[69]

Physical and historical landscapes differ in several important ways. First, historical landscapes are "physically inaccessible." The past cannot be visited. Second, the historian, unlike the natural scientist and perhaps the cartographer, has to "make moral judgments."[70] However, both the historian and the cartographer try to "fit representations to realities."[71] Thus, "the historian too is laying down a grid, stifling particularity,

---

66. J. Z. Smith 1978, 306–9.
67. Gaddis 2002, 2.
68. Gaddis 2002, 3.
69. Gaddis 2002, 11, 15, 32; 33–34, 45–46; 33; 15, 62; 32, 34, 45, 75; 22–26.
70. Gaddis 2002, 35; 122.
71. Gaddis 2002, 114; cf. 34.

privileging legibility, all with a view to making the past accessible for the present and the future." Gaddis tells us: "The effect is both constraining and liberating: we oppress the past even as we free it."[72]

Interestingly, while reality and representation are not identical,[73] historical representations often "*become* reality" as they replace individual memories and become dominant social narratives.[74] Representation plays a persuasive role, evoking "critical judgment" and "a new view of reality" for users of maps and stories.[75] Historical narratives become the world.

## *Lessons from Uses of the Map Analogy in Humanistic Inquiry*

The map analogy suggests that our individual mental maps are shifting constellations of assumptions, beliefs, and comportments. This is so for the maps both of subjects or experiencers themselves, and of the researchers analyzing human practices. For humanists, the map is understood primarily as the systematic product of an abstract, cartographic scholarly impulse to map human practice. Experiencers being mapped often do not have access to the very map being made of them and their activities.

However, Ramsey's notion of belief as a navigational map, together with James's image of an "immense map" of concepts that "steer us practically every day," also implies that, in our personal and leisure lives, each of us operates with internal maps of concepts and sentiments. We can compare the individual elements, and the totality, of our map with the personal maps of friends, loved ones, or wise ones. We can learn to use new and alternative maps, including scholarly maps (e.g., Smith's three religious maps or Gaddis's notion of narratives as maps).

Interestingly, the cartographers Robinson and Petchenik recognize the power and ubiquity of the map analogy across the humanities: "Most cartographers are probably not aware of the basic role that students in other fields ascribe to maps as a kind of a priori analogy for a variety of basic concepts." A map "represents some other space," and "the spatial aspects of all existence are fundamental."[76] "Students" include philosophers of science such as Alfred Korzybski, Michael Polanyi, Ernst Cassirer, Thomas Kuhn, and Stephen Toulmin, the work of each of whom Robin-

---

72. Gaddis 2002, 135.
73. Gaddis 2002, 123–24.
74. Gaddis 2002, 136–37.
75. Gaddis 2002, 48.
76. Robinson and Petchenik 1976, 13–14.

son and Petchenik analyze. I share these two cartographers' worry that much of this philosophical literature is not particularly reflective about what a map actually *is* or *is good for*. We must turn to cartography, map studies, and map thinking to nuance and deepen our understanding of the map, and thereby, through the map analogy, increase our comprehension of how science, knowledge, and storytelling work.

Perhaps the most imperative point here is that map thinking invites any of us—as map inhabitant/experiencer, or as abstract researcher/mapper—to reflect critically on our plurality of personal and scholarly maps, to track ethical and political consequences, and to ask "What if . . . ?" Maps do not just make the world. They help make *other* worlds, and the worlds of *others*.

## Assumption Archaeology

The above investigation of mapping in humanistic inquiry is what I call an *assumption archaeology* of the map analogy.

*Archaeology* is here intended to evoke "notions of the repressed, the lost and the forgotten, and of the drama of discovery, which are often spatialised in terms of the relationship between depth and surface."[77] This archaeology pinpoints contingencies and logical dependencies among assumptions, illuminating the procedures and heuristics of representational practices.

Part 2 will show this concretely by applying assumption archaeology to science. A science example before then is the Hardy–Weinberg equilibrium, a canonical idealized model in evolutionary population genetics informing us that gene frequencies will stay the same across generations in the absence of evolutionary influences. We can use assumption archaeology to help uncover multiple assumptions that must hold for the Hardy–Weinberg equilibrium model to be empirically appropriate—assumptions, for instance, about infinite population sizes, no mate selection, and no migration across populations. We might end up incorrectly applying the highly abstracted model wholesale to all empirical situations, even though for the vast majority of them, Hardy–Weinberg assumptions do not actually hold. Without assumption archaeology, a model such as Hardy–Weinberg might be unjustly confused and conflated with the world, in any variety of ways.

The assumption archaeology outlined here borrows from Michel Fou-

---

77. Thomas 2004, 149.

cault, although we can find a forerunner of this view in Nietzsche and traces of the idea in Kant.[78] Like Foucault, I do not aim to "reproduce the order of time or reveal a deductive schema" in excavated representations.[79] Rather, assumption archaeology remains alert to discontinuities in time and logic, and to fractures, contradictions, and inversions in the material, practical, and theoretical scaffolding involved in making and using representations.[80] These are the sort of discontinuities and inversions that take place when Einstein's theories upend Newton's. They are the contradictions that happen when assuming a Euclidean grid geometric space with a single absolute frame of reference is no longer sufficient but must be supplemented by measures of the curvature of space, and the insistence that there is no absolute frame of reference.

Even so, unlike Foucault, I wish to also explore and excavate whole systems of representation. One way of orienting ourselves within a particular unity of representations—a family of representations—and the scaffolding surrounding such a unity is by identifying powerful assumptions: the likes of space is absolute; space is God-given; gravity is a universal force.

Identifying such assumptions enables us to understand representations and how these are rationally consistent with particular research purposes. It is important to note that, if you are aware of your presuppositions, you can share them. By identifying them and taking them seriously, you are less likely to make a mistake. You are less prone to miss that you are making a mistake, and more likely to be open to feedback from others. By excavating assumptions, in the rationality of others and in

---

78. In *Critique of Judgment*, Kant contrasts "natural history" ("the description of nature") with "archaeology of nature" ("an exposition of the earth's former *ancient* state") (Kant 1987, §82, 315). More deeply perhaps, in an essay or "jotting" with the title "On a Philosophical History of Philosophy" from Kant's unpublished works, Kant says, "A philosophical history of philosophy is itself possible, not historically or empirically, but rationally, i.e., *a priori*. For although it establishes facts of reason, it does not borrow them from historical narrative, but draws them from the nature of human reason, as philosophical archaeology [*als philosophische Archäologie*]" (Allison and Heath 2002, 417). For discussion, see Agamben 2009, 81–82; McQuillan 2010. For Nietzsche's view, see the notion of *genealogy* in Nietzsche 2008. Lucas McGranahan, Ian Hacking, and Philippe Huneman provided constructive feedback.

79. Foucault 1972, 148; cf. 26, 138.

80. Regarding scaffolding, see Estany and Martínez 2013; Caporael, Griesemer, and Wimsatt 2014.

your own research program, you will be less given to saying, "I know I am right!" All these qualities make assumption archaeology an apt tool for de-reification. There is a space of common presuppositions defining the possibilities of agreement, and by being transparent, we reveal those possibilities; mutual respect for differing perspectives may be established; and ways to productively extend our maps can be identified.

Embarking on an assumption archaeology of the map analogy, as we have done above, we can identify basic representational practices of *Homo cartograficus*—the mapping animal. In my view, three particularly powerful assumptions indicate the relevance of the map analogy to understanding knowledge building across the sciences: *spatialized knowledge, representational practice*, and *political and social context*.

As we saw in chapter 1, humans and other animals have a "deep mapping" capacity and desire. We yearn to map space and the world. Space and spatial thinking are pervasive, basic to human experience.[81] We use spatial information in everyday actions, such as following the instruction manual for a household appliance, creating a new floor plan for our furniture, or solving a Rubik's cube or playing chess.

At a higher organizational level than the individual, empires, nations, and tribes across the globe have repeatedly fought each other to gain geographic territory. A modern and more metaphorical version of this, in the United States of America, is that elected political groups repeatedly gerrymander voting districts into contorted shapes in order to concentrate voter support for building future social worlds according to their own agendas.

Many of the insights science produces concern *spatialized knowledge*. The German philosopher Ernst Cassirer even goes so far as to claim that space is crucial to most theorizing: "There is no field of philosophy or theoretical knowledge in general into which the problem of space does not in some way enter."[82] Following the map typology above, I argue that extreme-scale maps, state-space maps, and literal maps involve spatialized knowledge. Just as USGS mapmakers represent geographic space, so scientists render various kinds of spaces for the objects and

---

81. See Ingold 2000; Massey 2005.

82. Cassirer 1957, 143. Robinson and Petchenik review Cassirer's analyses of space; the development of individual spatial experience; the evolution of spatial conceptualizations in scientific theories; and the relation of space to symbolism, in light of the map analogy, or "map metaphor" (Robinson and Petchenik 1976, 7–10). Indeed, even "language may be said to gain its first foothold in the sphere of space" (Cassirer 1957, 450).

processes they aim to represent and manipulate. For instance, geneticists zoom in from the common human scale to examine DNA and RNA, and cosmologists zoom out to study the universe. You could imagine traveling through such spaces in appropriate attire, were you suitably scaled.

Any representation is an abstraction, and creating an abstraction requires *representational practices*. As will be detailed in chapter 3, these include managing empirical data for representational purposes by focusing on the essentials. *Selecting* and *simplifying* are two processes of focusing. The depth of these core representational practices thus explains why the map analogy is so useful, and such a powerful tool for understanding representational and inferential practices across the sciences.

Third, analyzing maps increases our sensitivity to *political and social context*, as well as power, especially relative to the world-making function of maps. One similarity between maps and scientific theories is that power and convention suffuse the abstraction and ontologizing, the making and using, of both scientific and cartographic representations. Some representations—like Mercator's projection, selfish gene theory, and rational choice theory—dominate others, and many promising unorthodox representations languish for lack of attention. Influence begets influence through institutional prestige, journal editorial positions, prizes, mentorship lineages, uneven interdisciplinary training, disparate economic incentives, and the unequal values ascribed to research programs and areas of inquiry. We rarely think we intend to squelch alternative, valid perspectives, but when we profit from the status quo, it can be easier to shunt aside theories and practices disfavored by those in power. Pernicious reification emerges as a result.

Such concerns continue the theme of the fallacy of pernicious reification (which chapter 4 will take up formally and in detail). Just as the map is not the territory, so scientific abstractions are not the world. Pernicious reification is often emergent. Individual researchers seldom set out to reify, but social dynamics can eliminate reasonable and promising counterrepresentations. Space and time should thus be made for those who champion representations that are systemically ignored (e.g., allowing for map projection alternatives to Mercator's, or for nongenetic explanations of the phenotype). Recognizing the social entrenchment of mainstream viewpoints and the suppression of marginalized ones must be part of any analysis of reification and of any development of de-reifying tools. Pernicious reification is a socially conditioned phenomenon caused by overemphasizing a small, privileged set of representations.

Scientific modeling and simulation often involve thinking about worlds or situations different from existing ones, and science involves indirect representational processes.[83] Scientists use modeling and simulation to learn, and do so by varying the assumptions in abstract theories and letting possibilities branch across time according to laws and algorithms. On the surface, literal maps do not seem to have these characteristics. But as I shall show in chapters 4 and 5 and throughout part 2, map users—like scientists—also imagine and consider multiple idealized possible worlds. Both cartographers and scientists describe the world by making assumptions of how it might be—and was, and will be—different.

The proof is in the pudding. The significance, ubiquity, and utility of the map analogy will be known by its fruits.

## Conclusion

This chapter has peeled away the complex content of the map analogy, identifying eight forms:

1. Basic map analogy
2. Extreme-scale map analogy
3. State-space map analogy
4. Literal map analogy
5. Causal map analogy
6. General map analogy
7. Model map analogy
8. Paradigm map analogy

I have also characterized five types of maps (and spaces): extreme-scale maps, state-space maps, literal maps, causal maps, and analogous maps. We have also seen how far the map analogy travels across the humanities and social sciences, from analytic, "continental," and pragmatic philosophies and philosophy of science, to religious studies and history, among other fields. Theory has so widely been analogized to maps because, although neither map nor theory is the world, the world is the territory that both represent at their various levels of abstraction. Map and theory share a sense of space, whether literal or abstract.

Mapping and theorizing are key exemplars of representational practices, and the map analogy provides a context for understanding the perils of pernicious reification, which occurs when the map user forgets—or

---

83. Hughes 1997; Winther 2006c, 2014a; Pincock 2007; Weisberg 2007; Bueno and Colyvan 2011.

is never told—the map's or theory's political and social context and the assumptions that ruled its making. Map analogies are powerful and illuminating, particularly when their limits are made explicit through assumption archaeology to revive the richness of multiple valid perspectives.

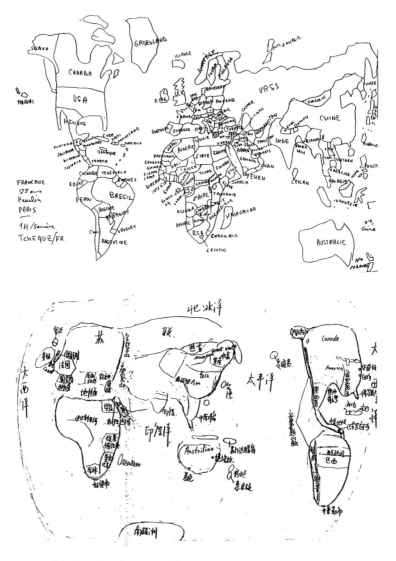

FIGURE 3.1. Tom Saarinen asked first-year university geography students from across the globe to draw, from memory, their map of the world. Saarinen believed that "the map most readily recalled would probably be the one most commonly seen and used in the map sketcher's society" (1999, 137). *Top*: Map drawn by a twenty-two-year-old male Czech/French student from Paris (142, map 1). *Bottom*: Sinocentric map (149, map 7). (Texas State University, Grosvenor Center for Geographic Education, Research in Geographic Education.)

# 3
# From Abstraction to Ontologizing

I am sitting in an airplane. I stare out the window and can see the runway. The plane slowly starts accelerating. I feel the rushing in my limbs and gut. Buildings, cars, and hills whiz by, faster and faster. As the plane climbs, it becomes increasingly difficult to identify trees, buildings, and cars below me.

The level of magnification changes. Trees and cars disappear. Buildings smear. Whole city blocks and farms transform into abstract squares; rivers and highways become generalized curves; forests become patches of green. A quilt of browns, blues, and greens emerges. It dawns on me that nothing like this happens when I ride a bike, take the train, or drive a car. I yearn to touch that solid ground on which I live and dream.

The trajectory is merciless. Soon I am above it all, looking down with sweeping vision. No distinct trees, no individual buildings, and certainly no people. Simplicity and neutrality reign. There are only the patches of color, abstracted shapes, and generalized features of a map.

Map abstraction has in it this flip from everyday human-scale perception to a detached and abstracted scale, typically immense or microscopic. Map abstraction involves representational challenges and promises. How do mapmaking and our takeoff experience resemble practices of creating scientific representations? How is choosing a map grid and projection, sprinkling pertinent symbols onto it, and selecting a color scheme like surveying the landscape from the clouds, and like producing an appropriately abstracted scientific theory or model of rich reality?

This chapter draws on cartographic practices to explore how representations are produced. My overarching goal is to examine the internal workings of abstraction and ontologizing practices.

CHAPTER THREE

## The Abstraction-Ontologizing Account

The map analogy illuminates scientific representational practices. To begin, abstraction and ontologizing link map and world. Abstraction moves from the world, via measurement and conceptualization, to a representation.[1] *Ontologizing* means deploying a representation to do work in the world, whether to change it, or to imagine the world to be a certain way and plan and act accordingly. Although feedback effects occur between them, abstraction and ontologizing can be separated analytically as two distinct processes. Both often involve theory with spatialized knowledges.

I interpret the world neither as a pregiven primitive nor as a fully conceptualized construction. In other words, both full-blown realism and full-blown social constructivism are false. If I am correct, distinguishing between map and world will be a recalcitrant problem. How do our imperfect maps generate predictive hypotheses and explanatory models? How can our limited maps build actual worlds? I suggest that the abstraction-ontologizing account requires only a minimal and intuitive distinction between map (or theory), on the one hand, and territory (or world), on the other.

My abstraction-ontologizing account is partially inspired by the *communication paradigm* of classic cartography, which was especially influential during the 1960s and 1970s. For instance, according to Robinson and Petchenik, "just as most writing assumes a reader, most mapmaking assumes a map percipient, a viewer to whom the map will convey information."[2] These two cartographers considered mapmaking and map use to be aspects of an overarching communication system.[3] As MacEachren

---

1. Both abstraction and ontologizing involve theory—abstraction is a set of practices producing capacious and generative theory; ontologizing uses theory for practices in the concrete world. Thus my dichotomy crosscuts the typical theory/practice distinction. Other map thinkers starkly separate theory from practice. Toulmin ([1953] 1960, 119–24) compares the distinction between "theoretical" and "applied" sciences to the distinction between maps and itineraries. As with Ryle's competent language speaker and village inhabitant, related in chapter 2, Toulmin suggests that practical knowledge antedates theoretical. Ingold (2007, 15) distinguishes (theoretical) "pre-planned navigation" from (practical) "trail-following" in embodied and memory-laden landscapes.

2. Robinson and Petchenik 1976, 23.

3. Similarly, see Koláčný 1969; Muehrcke 1969, 1974a, 1974b; Ratajski 1972. The inspiration of information theory was not necessarily considered an enlightened one (see Shannon 1956; Green and Courtis 1966).

teaches: "Communication came to be viewed as the primary function of cartography and the map was considered the vehicle for that communication."[4] Perhaps the main moral of the communication paradigm of cartography is that mapmaking must include critical reflection on how maps are used.

For me, mapmaking is an example of abstraction. Map use is a case of ontologizing. Although my analysis does not actually require the communication paradigm, many distinctions made by that perspective will prove useful throughout the book, including mapmaking and map use; map and world; cartographer and map user; and process and product.

## Abstraction

There are many philosophical accounts of abstraction,[5] but the map analogy permits us to see things about cartographic generalization and scientific theorizing we might not otherwise notice.

We can divide the abstraction practices of mapmaking into three stages: (I) calibration of units and coordinates, (II) data collection and management, and (III) generalization. The first two stages require material instruments and empirical protocols. They are thus analogous to scientific experimentation. In contrast, cartographic generalization is analogous to scientific theorizing. Imagine generalization as your plane's ascent—the movement from the concrete world on the ground to the increasingly abstract view from the sky. Turn now to each of these three stages of abstraction.

### ABSTRACTION STAGE I: CALIBRATION OF UNITS AND COORDINATES

Any cartographic or scientific representation requires measurement. For example, distance between two cities on a map can be measured in kilometers or miles. An experiment on a new mixture of gases will almost certainly require measuring the pressure in pascal in SI units (*Système*

---

4. MacEachren 2004, 5. MacEachren's overview of the communication paradigm, 3–6, is a pithy one. MacEachren and others build on this communication perspective, developing distinct semiotic cartographic analyses.

5. Reviewing these is beyond the scope of this book. See, for example, Cartwright 1983, 1989; O. O'Neill 1987; Nussbaum 1995; Ohlsson and Lehtinen 1997; Radder 2006; Martínez and Huang 2011; Winther 2011b, 2014c.

*international d'unités*), or in pounds per square inch (psi) in US customary units. Equally important are coordinate measurement frameworks within which to place complex data; for instance, astronomical data must be presented in spatial coordinates.

To calibrate units and coordinates, the mapmaker surveys space and time. Space surveying includes (flat) plane and (curved) geodesic surveying,[6] while time surveying includes postulating time zones.

## Plane Surveying

Plane surveying assumes a locally flat Earth and aims to determine longitude, latitude, and elevation of specific locations.[7] Surveyors use instruments such as beacons (for reference points), the surveyor's wheel (for measuring distance), the theodolite (for measuring horizontal and vertical angles), and various modern devices incorporating lasers and GPS.[8]

You may know the theodolite as the tripod-held apparatus used by land surveyors. It contains a sighting mechanism for obtaining data precisely locating salient geographic features.

Two strategies for calibrating that data involve tracing imaginary networks of adjacent triangles across the landscape. For both, the surveyor places beacons or posts to serve as anchor points that are the vertices of a triangle. Standing the theodolite at one of those points, the surveyor focuses on one of the other two vertices in the triangle to take a measurement.

In *triangulation*, the surveyor measures the angles at which the triangle's other two sides meet the two ends of the triangle's baseline, whose length has already been measured, and across which the surveyor travels. Subtracting those two angles from 180 degrees—the total of any triangle in flat, Euclidean space—reveals the size of the angle at the third point. Using trigonometry, the surveyor can then calculate the third point's position as well, because the two side legs will each have to be a precise length to form the proper angle where they meet at that third vertex. Conversely, in *trilateration* the surveyor measures the length of the triangle's three legs and then calculates the angles.

Workers who surveyed France for César-François Cassini de Thury's

---

6. Johnson 2004.

7. *Geometry* means measuring and surveying land or Earth.

8. GPS is not a panacea for challenges in surveying—errors of various kinds occur; for an example, see GPS.gov, n.d.

1744 map used both triangulation and trilateration.[9] The resulting map encapsulates a decade of careful effort during which workers "completed... 800 principal triangles and nineteen base lines."[10] Their method was both innovative and influential. It set the stage for an intensive effort to triangulate European colonies in the nineteenth and early twentieth centuries.[11]

## *Geodesic Surveying*

We know Earth is not flat. Geodesic surveying therefore assumes a curved globe, calibrating curved space through two kinds of abstractions of Earth: an ellipsoid and a geoid. Both are curved, Earth-like surfaces. Inevitable distortions associated with map projections pertain to geodesic surveying and to mapping Earth at a global scale, rather than to highly localized, flat-plane surveying.

**Ellipsoid**

Due to centripetal forces caused by Earth's rotation, our planet bulges at the equator. An appropriate mathematical representation of Earth's surface is thus a nonspherical *ellipsoid* centered on Earth's predicted center and slightly flattened at the poles. This geometric ellipsoid idealization smooths out the entire planet's surface.

After all, the twenty-kilometer vertical range between the highest Central Asian mountains on Earth (ca. 8.8 km) and the deepest Pacific Ocean trenches (ca. 11 km) is a mere 0.3 percent of the Earth's radius.

---

9. The story of the Cassini family's cartographic adventures spans four generations (Brotton 2012, chap. 9). Apparently, in 1682 Louis XIV complained upon studying Cassini's map: "I paid my academicians well, and they have diminished my kingdom" (S. S. Hall 1992, 384).

10. Brotton 2012, 314. The resulting map can be seen at https://sites.google.com/site/geohistoricaldata; search for "Cassini 1744" and zoom in on the image to see the triangles (accessed July 20, 2018).

11. See the programmatic piece Holdich 1901, in which the knighted president of the Royal Geographical Society complained that the various "colonial surveys" of South Africa were not coordinated with the "magnificent system of geodetic triangulation" carried out simultaneously by various British authorities, resulting in "internal discrepancies which absolutely invalidated the map for military purposes" (595). Surveying had to change; triangulation had to be adopted, Holdich argued. And it was. On related British mapping efforts on the Indian subcontinent, see Keay 2000.

From the perspective of the entire globe, our planet's surface is basically smooth. The current international standard ellipsoid—the World Geodetic System 1984 (WGS 84)—provides the reference latitude and longitude grid, called the *graticule*.[12]

### Geoid

The ellipsoid is smooth, but a more accurate physical model of the Earth is the lumpy, bumpy *geoid*.[13] The geoid is an abstracted, mean sea-level surface (plate 3).

We might normally think of sea level as a single constant magnitude (number) everywhere, but in fact, the oceans' surface has a topography of its own. Tides, currents, varying salinity, and the mixing of warm and cool waters are among the phenomena that push sea level up and down.

If we could do away with those influences, the ocean surface would still be uneven—it would undulate. That rolling surface is a consequence of variations in the pull of gravity due to irregularities in Earth's composition and height, and from Earth's rotation. This slightly rolling sea surface is the basis of the geoid.

To complete the geoid, surveyors have extended that undulating zero-elevation line across land by making additional measurements of gravity across the entire Earth (even from and over the top of the Himalayas), and doing complex mathematical calculations. The unevenness in the geoid's surface is less than 200 meters, despite actual height differences in rock of almost 20 kilometers from the depth of the Mariana Trench to the top of Mount Everest.

As a *Scientific American* blog post has put it, the geoid has "a surface such that if you placed a marble anywhere on it, it would stay there rather than rolling in any direction. Another way of saying it is, imagine you were an engineer traveling around the world with a level. Then wherever you go, the level would be exactly parallel to the geoid"—and exactly

---

12. To be precise, latitude and longitude are determined by using a transverse Mercator projection from the ellipsoid onto a series of flat maps. This is the Universal Transverse Mercator coordinate system, or UTM (Robinson and Sale 1969, 28–30). For this projection, the cylindrical developable surface is oriented not around the equator, but along a meridian. In cartographic argot, this projection has a *transverse* rather than an *equatorial* aspect (orientation). Because the ellipsoid is divided into sixty bands, each of 6 degrees of longitude, the amount of areal distortion caused by this projection is relatively small. A Tissot indicatrix would still be appropriate; see J. P. Snyder 1993; and chapter 4.

13. See National Ocean Service, n.d.; Castelvecchi 2011a; Fraczek 2003.

perpendicular to the gravitational vector.[14] The geoid is an *equipotential* surface. Of course, because sea level is increasing during these times of global warming and climate change, mean sea-level calibration has to be updated often.

Having done away with tides, currents, and so on to arrive at the geoid, we can now add them back to get the oceans' full topographic features. Thus, there are three types of "figures of Earth": the idealized ellipsoid, the potato-shaped geoid arrived at by countless gravitational measurements, and the full, rich topography of sea and land surfaces as compared to the geoid (plate 3).

The internationally accepted geoid data model is the Earth Gravitational Model 2008, or EGM2008. Together the EGM2008 and the WGS 84 provide the coordinate system conventionally used to assign the latitudinal and longitudinal (EGM2008) and elevation (WGS 84) locations for any kind of geographic object or feature.[15]

Even among what should be the most consistent of calibrated representations, there are important differences. Plane and geodesic surveying coordinates need to be integrated with care, through appropriate transformation equations (see the information on map projections in "Abstraction Stage III," later in this chapter) and critical reflection on technologies (e.g., the role of satellite data).

## *Calibrating Time*

For most of human history, trade, travel, and communication were primarily local and at the temporal scale of walking humans or galloping horses. Synchronizing time across two very distant locations was largely irrelevant. Time at every locale—for example, town or city-state—was simply set by the sun's predictable movement across the sky. There was no real need to know exact solar time at a location far away. Time and space were local.

Because time difference corresponds 1:1 to longitudinal difference around our earthly globe, time and space are geometrically related. Relative time differs across successive east-west coordinates at the rate of one temporal minute per ¼ longitudinal degree, or one hour per 15 degrees.

14. Castelvecchi 2011a.
15. GPS locations are calibrated with WGS 84 and, by extension, EGM2008 (i.e., satellites must themselves calibrate their location with respect to a coordinate system). See Winther 2015b. Both WGS 84 and EGM2008 are products of the National Geospatial-Intelligence Agency of the US Defense Department.

When global navigation, time-sensitive trade, and communication across large distances became significant, calibrating time became essential.

Early seafaring captains of European colonialism typically needed to know what time it was at home (or in Greenwich or Paris) to know where they were longitudinally, in the world and on their maps.[16] They kept clocks on board showing the time at their faraway homes, and compared that time with, say, local noon.[17]

Industrialization and urbanization took hold across the world in the eighteenth and nineteenth centuries. To establish clear transit schedules and prevent train delays and crashes, railroad engineers and government authorities needed to establish local mean times by averaging the variations in a day's solar length, and then choosing the time at one location (often a major city) to be the standard throughout the local region.[18]

And as telegraphic communication became commonplace during the late nineteenth century, governments—often lending their ears (and money) to capitalist barons—negotiated and agreed on a prime meridian, which was to become Greenwich. They also came to a consensus on a twenty-four-hour day with twenty-four standardized time zones, imagined as irregular, longitudinal 15-degree slivers of Earth's surface.[19] Three temporal calibrations of modernity thus emerged sequentially: longitude inference, mean local times, and time zones.

## Calibration from Mapmaking to Science

Unavoidably, both mapping and science represent by way of spatial coordinates and units. In addition to space (meters) and time (seconds), assignments of standard scientific units include temperature (kelvin), charge (coulomb), luminosity (candela), and amount of substance (mole).

For instance, defining and stabilizing the concept of "temperature" was

---

16. Sobel 1995 and Howse 1980 describe this in detail.

17. Such clocks were notoriously difficult to build, but most other solutions to the "longitude problem" were failures (Sobel 1995).

18. "A delay of thirty seconds in leaving a terminal calls for explanation, five minutes' delay means investigation, and a half hour gives apoplexy to every official from the superintendent to the lowest foreman" (Cottrell 1939, 190–91).

19. Schedule coordination and pressures from railroad companies caused Britain's official adoption of Greenwich Mean Time as sole "standard time" in 1880, and the institution of four time zones in the USA followed a meeting of the railroad power elite in Chicago in October 1883. These time zone adoptions percolated globally (Howse 1980, 135; Galison 2003, 84–155).

challenging. Hasok Chang documents how temperature was "invented." Reliably calibrating two anchoring points along a mercury or alcohol thermometer and gradating the distances between them was hardly trivial. There is no absolute thermometer scientists could calibrate against. Even if, by convention, the freezing and boiling points of pure water are taken to be the two end points, marked along a glass tube, water can boil at different temperatures at different elevations. It can also superheat (be heated beyond the normal boiling point at a given pressure), and so forth.[20]

However, conventional choices of material used (water), calibration markings (gradated degrees), and end points measured (freezing and boiling, under very specific conditions) were embedded in social discourse and consensus building. Scientists came to agree on these "sufficiently stable and uniform temperatures, which allowed the calibration of thermometers with which scientists could go on to study the more exotic instances."[21]

## ABSTRACTION STAGE II: DATA COLLECTION AND MANAGEMENT

Once units and coordinates for the representation have been selected and deployed, data has to be collected and managed in order for relevant objects, features, and processes to be abstracted out from the evidence.

Various US government agencies, including the Bureau of Land Management and the Defense Mapping Agency, have expended enormous effort to collect and compile data on which the USGS map of San Francisco at the beginning of chapter 2 is based. Just as the results of robust experiments feed into scientific theory, cartographers abstract from a variety of data, including, in this case, "aerial photographs" and "hydrographic data."[22] Because of the relatively small scale, and possibly because of government censorship,[23] data quality is unreliable and uneven.

---

20. Chang 2004, chap. 1, "Keeping the Fixed Points Fixed," 8–56.

21. Chang 2004, 23. For more about basic measurement in the natural sciences, see Houle et al. 2011; Suppes et al. 1989.

22. USGS Sheet Number NJ 10-8 map legend. Lloyd ([1988] 1994, chap. 8) distinguishes quantity, quality, and variety of data, appropriately characterized, as three distinct criteria for theory confirmation; cf. Lloyd 1987. See also Lloyd 2010, 2012.

23. On government maps being "born classified," see Doel, Levin, and Marker 2006, 605.

With this data, the USGS cartographers deemed certain bridges, lighthouses, and hospitals, in addition to the San Francisco and Oakland International Airports, worthy of being represented on the map. To make such choices, cartographers use conventions premised on epistemic— knowledge-related—criteria of empirical accuracy, predictive potential, and explanatory power, as well as on intelligibility and appropriateness to the map's purpose. In contemporary GIS mapmaking, some of this process is automated. Individual GIS layers can be eliminated at will, and software can remove certain objects from a given layer, leaving others intact—depending, say, on scale.

Unit and coordinate calibration, and data collection and management are analogous to scientific experimentation. Data collection in mapmaking and scientific experimentation requires measuring the content that maps or theories intend to represent. Map generalization principles and decisions ("Abstraction Stage III," below) feed back on earlier calibration and data abstraction stages.[24] Analogously, the expectations and biases of theory inform experiment, and experiment constrains scientific theory.[25]

Statistical analysis of variance (ANOVA) illustrates the feedback effect between theory and experiment. Scientists throughout the natural and social sciences use it. To grapple with ANOVA is to face the complexities, possibilities, and limitations of causal analysis.[26]

ANOVA helps a researcher compare the influence of multiple, interacting empirical factors on an item of interest, whether that item is a set of objects or processes, or a range of political beliefs. For example, with ANOVA, a sleep-quality study can compare much more than the effect of someone's bedtime. It can factor in other potentially explanatory variables, such as the time of the test subjects' evening meal and the amount they exercised that day. Moreover, ANOVA can check how strongly different combinations of these factors may affect sleep quality: Did test subjects sleep better on a day they exercised for ten minutes and went to bed at midnight? Or did they sleep better if they had exercised an hour and turned in at 10 p.m.? ANOVA can also measure these combinations across a different comparison, such as female versus male. This comparison of multiple factors reveals both *sources of variation* and the *interaction effects* among the factors or variables. When multiple factors have an

---

24. See Galison 1997.

25. For the multiple dimensions of experiment, see Galison 1997; Chang 2004; and of scientific practice more generally, see Rouse 2002, 2003; Chang 2011.

26. For ANOVA and causation, see Levins and Lewontin 1985; Turkheimer and Waldron 2000; Northcott 2008; Longino 2013; Winther 2014a; Pearl 2018; see also chapter 7.

influence, ANOVA helps avoid the danger of attributing too much of an effect to any single factor.

ANOVA was developed by R. A. Fisher, an important originator of modern statistics and experimental design. Fisher's key books, *Statistical Methods for Research Workers* (1925) and *The Design of Experiments* (1935), advance the theory of probabilities, theory of errors, distribution theory for tests of significance, definition of variance, and other elements in light of experimental design that is randomized and controlled.[27] For Fisher, theory without experiment is empty, and experiment without theory is blind.

ANOVA exemplifies how data and theory are intricately intertwined. As with map abstraction, background theory tells us where to look and what to look for—what factors to test in our experiments. Relatedly, what we find affects what we represent, and how. We start with curiosity about a phenomenon, explore relevant theories, generate hypotheses, and evaluate and design how best to collect data with bearing on our selected hypothesis. In working out the empirical details, we may go back and change our theoretical approach.

### ABSTRACTION STAGE III: GENERALIZATION

"In reducing the size of a map one is interested in discarding details," Urs Ramer writes pithily.[28] Numerous such map generalization protocols exist. These procedures are analogous to scientific abstraction practices. Two classic cartography textbooks provide useful starting points for understanding cartographic generalization: Robinson and Sale's *Elements of Cartography* (1969) and Muehrcke and Muehrcke's *Map Use: Reading, Analysis and Interpretation* (1998). Textbooks can be informative sources for identifying practices of a discipline. Riffing on Wittgenstein's metaphor, textbooks are ladders we climb during our training so that we may kick them away.[29]

### *Five Protocols*

Robinson and Sale interpret cartographic generalization to be "essentially a creative act."[30] For example, cartographers made the USGS map

---

27. Fisher 1925, (1935) 1971; Hacking 1988; Wade 1992; Edwards 2005; Winther 2018a.
28. Ramer 1972, 252. See also Winther 2015b.
29. Wittgenstein 1922, 6.54.
30. Robinson and Sale 1969, 53.

of the San Francisco Bay Area using a process of "cartographic generalization" consisting of the following "elements": simplification, symbolization, classification, and induction.[31] In turn, these are controlled by the objective, the scale, the graphic limits, and the quality of data.[32]

Muehrcke and Muehrcke remove induction from that list of elements. Perhaps that is as it should be, since simplification, symbolization, and classification use induction and other reasoning and argumentation processes—including deduction, abduction, and analogy—to achieve their aims. And Muehrcke and Muehrcke add two important abstraction practices to Robinson and Sale's list: selection and exaggeration.[33] This gives us the following five protocols, which I take to be central to analog cartographic generalization (before computational algorithms were used for the task).[34]

### Selection

*Selection* is the judicious reduction of content, particularly as a consequence of choosing map scale and map projection. In an important sense, all abstraction practices involve selection. For instance, simplification (below) is about selecting which information types to represent out of all that are available. However, this is a truism: all thinking and representing and emoting involve selecting some features or content rather than others.

What is particularly important in selection for cartographic (and scientific) purposes is scale selection and map projection selection. Scale is

---

31. Robinson and Sale 1969, 52. Robinson and Sale correctly take induction to be ampliative—that is, a process of "performing operations" through which "the cartographer ends up with more than the information with which he started" (58).

32. Robinson and Sale 1969, 52. Data quality is critical. In data analysis, mapmakers must, for instance, interpolate predicted feature values at a location from known values at nearby locations. See Slocum et al. 2005, 271–91; Haining 2003.

33. Muehrcke and Muehrcke 1998, 56. Erwin Raisz (1962, 38) posited three features of generalization: *combine*, *omit*, and *simplify*.

34. Robinson and colleagues distinguish between earlier *analog* cartography and contemporary *digital* cartography (Robinson, Morrison, et al. 1995, 4). For digital cartographic generalization, questions about the relation between humans and computers come to the fore (see Weibel 1991; Harrie and Weibel 2007; McMaster and Shea 1992; Longley et al. 2011; Winther 2015b; see also Ekbia 2008). My list describes canonical cartographic generalization prior to the algorithmic reduction of digital generalization starting in the late 1970s, though important similarities between analog and computational abstraction remain, as shown in Winther 2019b.

the scope or sweep, and the granularity or degree of detail, of whatever objects or processes we wish to represent.[35] Map projection is somewhat less important, as it primarily concerns cartographic representations and not representation in general.

Once we select scale and projection, many of the other representational features are highly constrained.

*Scale.* Scale is the proportion or ratio between features of the representation and properties of the world depicted. Scale might be rendered in terms of time passage; intensity or density of features; distance and size; or other parameters. Scale is a crucial abstraction selection practice impacting all other abstraction practices.[36]

In cartography, when the map scale is presented to the user, it tells us that "this distance on the map *represents* this distance on the earth's surface."[37] Map scale can be shown visually (e.g., with a gradated line representing, in total, 1 kilometer), quantitatively (e.g., 1:10,000,000), or in words.

Scale should be selected based on how much area one desires to cover, and at what level of detail, while taking presentation and communication constraints into account (e.g., a book, a large poster, or a screen with zooming capacities).

Scale constrains mapmaking generalization practices: the larger the scale, the more fine-grained, detailed, and concrete the cartographic map can be.[38] Cartographic generalization is succinctly described as "the reduction of the amount of information which can be shown on a map in

---

35. Specifically, scale is the scope and granularity of space studied and represented—and also the granularity of time. Scale is so central to the structure of the world, and to how we think about the world, that it also strongly constrains different areas of science. Whether we are talking about particle physics, biochemistry, neuroscience, anthropology, or cosmology, the boundaries of each of these are set (if somewhat permeably) by the minimum and maximum spatial scales of the objects and processes of its domain, from absurdly tiny to dizzyingly enormous. Temporal scales also vary consistently and reliably across the sciences. For instance, quantum mechanics and quantum chemistry deal in extremely short time scales; developmental biology in days, weeks, and months; geology in millennia and millions of years; cosmology in billions of years.

36. For a rigorous mathematical treatment of map scales, see Bugayevskiy and Snyder 1995, 17–20.

37. Kimerling et al. 2009, 24; emphasis added.

38. See Robinson and Sale 1969, 52; Robinson, Morrison, et al. 1995, 458; Swiss Society of Cartography 1977, 14; Harrie and Weibel 2007; Winther 2015b.

relation to reduction in scale."[39] World maps are small-scale—that is, a world map fitting on two paper leaves of an atlas could have a scale of one to sixty million (1:60,000,000).[40] The USGS map of the San Francisco Bay Area has a much larger scale, 1:250,000. In contrast, city maps are even larger in scale, typically varying between 1:25,000 and 1:10,000.[41] The smaller the scale, the more abstract the geographic map tends to be.

Although unworkable for geographic maps, 1:1 maps are possible, as are even 2:1 or 10,000:1. Nothing stops us from drawing a map grid on a huge piece of paper, twice as large as the territory (such as your kitchen). In this case, the map *amplifies*, rather than reduces, the space of the world it represents (fig. 3.2). Nevertheless, even in that "extra" room we cannot fully represent the world's complexity—all its objects and interactions, and its sounds, smells, textures, and so on. Our map might even have large empty and otherwise noninformative regions. This is because spatial unit calibration and map symbolization are two distinct stages of the overarching process of abstraction. Although spatial calibration has no logical 1:1 upper boundary, questions and challenges of symbolization remain for scales larger than 1:1.

Maps of cells, genes, molecules, or smaller objects—all the way down to the Planck length—are precisely such amplifying maps. Here the map is much larger than the territory. For example, an adenosine triphosphate molecule (fig. 3.2) has a radius on the order of 1 nanometer in length (approximately 0.7 nanometers), or one-billionth of a meter. The scale of our approximately 5-centimeter (2-inch) paper map of the spatial arrangement of the different atoms of adenosine triphosphate would thus be one-tenth:one-billionth (1:0.000000001).

Perhaps the largest scale difference in our physical territory, according to the current state of science, is from the smallest length (Planck length) to the largest length (diameter of the observable universe), a difference roughly represented by a 7 followed by 122 zeroes. Where the window of human perception lies along this scale constrains our actual mapping efforts.

39. Töpfer and Pillewizer 1966, 10. Their paper introduced the "radical law" of map selection, which calculates the number of objects or features to be represented as the square root (radical) of the ratio of scales between the source map and the derived map.

40. Esri provides a list of common map scales here: http://support.esri.com/en/knowledgebase/techarticles/detail/36360 (accessed March 20, 2015).

41. Some authors classify maps according to scale; see, for example, Greenhood 1964, 48–49; Muehrcke and Muehrcke 1998, 13, 537–46; Kimerling et al. 2009, 22–33; Krygier and Wood 2011, 94–95.

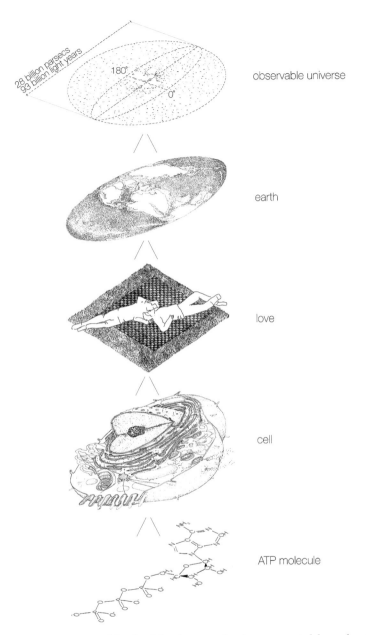

FIGURE 3.2. *Powers of Ten*. Spatial and temporal scales represented depend on the level of organization studied. The distance from Earth to the "edge" of the observable universe is $10^{26}$ meters, and time can be clocked in eons. Our planet is on the order of $10^7$ meters, and processes can be measured in geological time. Easily observable life can be measured in tens and ones and fractions of meters. Human cells, $10^{-5}$ meters on average, run on very short, molecular time scales. Molecules such as adenosine triphosphate operate according to quantum chemistry. (Redrawn from Eames and Eames 1977 by Heidi Svenningsen Kajita.)

*Map Projection.* Recall that geodesic surveying relies on curved surface models, but that trying to represent enormous areas of the curved Earth in two dimensions on a flat surface results in formidable representational challenges. A map projection transforms every coordinate on the surface of Earth's sphere—more precisely, the WGS 84 ellipsoid—onto a Cartesian, two-dimensional plane. Using *functions*, which relate each element in a set of variables to no more than a single element in another set, two highly general differential equations capture all map projection transformations:

$$x = f(u, v),$$

$$y = g(u, v).$$

Here $\{x, y\}$ are the coordinates on the Cartesian plane; $\{u, v\}$ are curvilinear coordinates of latitude and longitude on the surface of the sphere or ellipsoid; and $\{f, g\}$ are the generalized transformation functions—each map projection has a unique and possibly complex pair of such functions. These functions transform the curved surface onto a Cartesian plane.[42] We might say that the visual geometric projection, as shown by its coordinate system (fig. 3.3), is *equivalent to* its unique pair of symbolic transformation functions. In a closely related manner, we might wish to say that geometry and differential equations are here *translations across* two different mathematical languages of the same formal, mathematical content.[43]

Visually, many map projections can be thought of as if a single point of light within, on, or above an illuminated Earth were projected onto one of three *developable surfaces* wrapped around the sphere, intersecting it at one or at most two circles (arcs) (fig. 3.3). These surfaces are "developable," in two cases because they can be cut open and laid flat, and in the third because the surface is already flat:

A *cylinder* (resulting in a rectangular projection)
A *cone* (resulting in a projection similar in shape to the surface cleared by an automobile's windshield wiper)

---

42. These functions uniquely determine each map projection (see Bugayevskiy and Snyder 1995). Every pair of direct transformations also has a pair of inverse transformations, from the Cartesian plane back onto the ellipsoid surface. Moreover, one map projection can be translated onto another, often without mathematical loss (see Maling 1992, 416–29; Bugayevskiy and Snyder 1995, 251–63).

43. Frederik Stjernfelt, channeling Charles Sanders Peirce, provided constructive feedback.

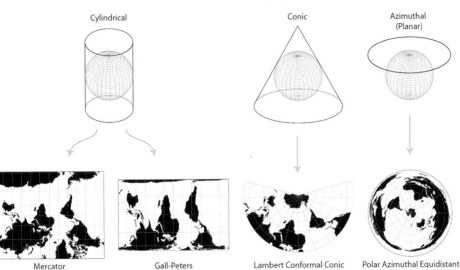

FIGURE 3.3. *Top row*: The three standard developable surfaces in their "normal aspect" (equatorial or polar). *Bottom row*: Common map projections unfolded from those surfaces. See also table 3.1. (Illustration by Mats Wedin.)

A *plane* (resulting in a circular projection, and quite often centered on the North Pole). Planar projections are also called azimuthal because an azimuth expresses direction based on the horizontal angle at which an observer at the central map point must turn from seeing one point on the horizon to seeing another.

Each map projection involves some distortion.[44]

A flat coordinate system of latitude and longitude *of some type* is always required for cartographic representation, and is contained in the projection's transformation functions. There is always a loss of angle, area, or distance, or any combination of these. The following projection types are therefore mutually exclusive:

*Conformal*, which preserves angles and shapes, but not areas or distances

*Equidistant*, which maintains scale and distance from one or at most two points

44. A *Vox* video on YouTube explains map projection and distortion vividly; see Vox 2016. The "dots" used in the video to demonstrate distortion are Tissot indicatrices. I explore projections in more detail in chapter 4.

TABLE 3.1 Basic projection types

|  |  | Developable surface used | | |
| --- | --- | --- | --- | --- |
|  |  | Cylindrical | Conic | Azimuthal (planar) |
| Map type (and metric property preserved), with example maps | Conformal (shapes and angles, within any local area) | Mercator Web Mercator[a] | [Lambert Conformal Conic, 1772] | [Stereographic, ca. 150 BCE, Hipparchus] |
|  | Equidistant (distances) | [Equi-rectangular, ca. 100 CE] | Ptolemy's 1st,[b] ca. 150 CE | Polar azimuthal equidistant, ca. 11th century, Al-Biruni (also preserves direction from center) |
|  | Equal-area (sizes) | Gall–Peters | Ptolemy's 2nd,[b] ca. 150 CE | [Lambert azimuthal equal-area, 1772] |

*Note*: Example maps whose names appear in brackets in the table are not addressed in the text; example maps without brackets are discussed in chapters 1, 3, or 4. (Several of these, and nearly 225 other projections, can be found online at Jung, n.d.).

While it is cylindrical and close to the Mercator, the Miller projection (see chapter 4), as well as the Robinson projection (see note 45 in this chapter) are *compromise* projections—custom made for a specific purpose and not derived geometrically. So they do not appear in this table.

[a]Because the cylindrical Web Mercator uses the Mercator transformation equations from the ellipsoid (rather than from an idealized sphere, as Mercator did), it is slightly nonconformal and is sometimes called *pseudo-Mercator*.

[b]Strictly speaking, both of Ptolemy's projections here have arbitrary meridians, making them "pseudoconic," and the accurate, or "saved," property not quite exact (J. P. Snyder 1993, 10–14).

*Equal-area*, in which the area of any map object is in true proportion to its area on the globe, but angles and shapes are lost.[45]

Table 3.1 tabulates a handful of map projections by the developmental surface used and the metric property each preserves.

As a dynamic depiction of the projection-making process, consider figure 3.4. The World War II architect, designer, and artistic cartographer Richard Edes Harrison, employed by *Fortune* in New York City,

45. Universe maps sometimes use equal-area projections, such as the Mollweide projection from 1805. On this classification of projections, see Robinson and Sale 1969, 205–8; Richardus and Adler 1972, 8–10; Fenna 2007, 65. *Compromise* projections trade off among the three otherwise mutually exclusive projection types. For instance, the Robinson projection compromises conformal and equal-area. Moreover, *composite* projections break apart the globe into sections (e.g., Buckminster Fuller's icosahedron Dymaxion projection, *Life* 1943).

# THE WORLD CENTRIFUGED
## NORTH-POLAR AZIMUTHAL EQUIDISTANT PROJECTION

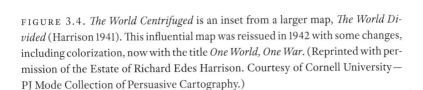

The principle of this projection may be illustrated by a dancer with a skirt in the shape of a globe upon which is inscribed the map of the world. When she whirls the skirt rises to a horizontal plane and the map on it will then resemble the map on this page. The projection has two important advantages: it shows little distortion of the Northern Hemisphere, and it nowhere breaks the continuity of the lands or seas involved in the present far-flung struggle.

FIGURE 3.4. *The World Centrifuged* is an inset from a larger map, *The World Divided* (Harrison 1941). This influential map was reissued in 1942 with some changes, including colorization, now with the title *One World, One War*. (Reprinted with permission of the Estate of Richard Edes Harrison. Courtesy of Cornell University—PJ Mode Collection of Persuasive Cartography.)

provided an accurate kinesthetic metaphor of how a polar azimuthal equidistant projection map was transformed from the globe—*The World Centrifuged*. The dancer stands as the rotational axis—a kind of world navel—of Earth.[46]

### Simplification

*Simplification* is the omission and streamlining of information, such that general features of a pattern or process are represented on the map, but unnecessary detail is abstracted away. Houses and roads can be removed, a meandering river straightened out (which can be done by algorithm),[47]

---

46. For more on the "world navels" concept, see Winther 2014d; 2019a.

47. Algorithmic reduction outputs a caricatured, zigzag line from a complex real-world line, while preserving the source line's basic properties. An early protocol of automated line simplification is the Ramer–Douglas–Peucker algorithm (see Ramer 1972; Douglas and Peucker 1973; Winther 2019b).

or many trees can be aggregated into a small homogeneous patch of green.[48] Not everything can or has to be represented on a map (recall C. S. Peirce's land surveyor dialogue from chapter 2). The more simplified a map is, the more abstract it is.[49]

Classification

*Classification* is the differentiation of objects, properties, and processes into kinds, types or categories. The fineness or coarseness of a classification system depends on design criteria, which are based on user needs, data availability, and preferred methods (and, of course, on scale and projection used). In an extremely coarse-grained system, villages, towns, cities, and megalopolises might each appear on the map as the same kind of dot. (The logically coarsest classification system would consist of a single symbol representing every type of thing.) On the other hand, the most fine-grained and most concrete system would differentiate *every* object within a certain category. Even a difference of one person in two different towns would require two distinct map signs. Neither extreme is useful.

As either maker or user of a map, maintaining an awareness of the purpose of the representation could mediate between these extremes. Philosophical frameworks for creating appropriate classification schemes are available to help with this process.[50] For instance, John Dewey and Ian Hacking critique and reimagine natural kinds. Natural kinds are classifications or categories of things that humans assume exist in the world independently of us, and even of one another. Dewey suggests rethinking the philosophical foundations of natural kinds classifications, whereas Hacking rejects the very tenability of natural kinds classifications.[51] Both, however, share the view that classifications should be thought of as purpose-driven, fallible, and local solutions carving up the universe into types of objects, properties, and processes.

---

48. An interesting *material* simplification strategy is described in *Hammond's Compact Peters World Atlas*: "Cartographers have struggled with the best way to create hillshading for hundreds of years. In this atlas the 3-D relief comes from photographing specially made plaster relief models and blending these photos with hand-rendered coloring" (quoted in Hardaker 2002, 7).

49. Selection can also be interpreted as involving other abstraction practices, including simplification. However, I view selection as more fundamental and general. Secondly, scale is a deep feature of all representations.

50. These frameworks might deploy Goodman's (1978) *relevant kinds*, Khalidi's (1998) *crosscutting taxonomies*, and Dupré's (1993) *promiscuous realism*.

51. Dewey 1982; Hacking 2007a, 2007b. For a fuller discussion, see Winther 2015b.

In the end, the map's purpose constrains the classifications. The more appropriately abstract a map is, the more adequately balanced it is between extremes: everything is not the same, yet everything is not different. As with creating a well-seasoned dish, the level of abstraction is a practical concern and a matter of taste.

### Symbolization
*Symbolization* is populating a map with signs. Following Charles Sanders Peirce, signs can be *symbols* (such as red points for cities) or *icons* (which resemble what they represent in a way that general society is familiar with—for example, a church represented by a cross).[52] The most abstract map would have few or no signs, like Van Sant's GeoSphere map (plate 2; see chapter 1), while a less abstract map would have many distinct types of symbols and icons, like the USGS map of the San Francisco Bay Area (plate 4; see chapter 2).

### Exaggeration
*Exaggeration* is the disproportionate and technically inaccurate adjustment of size and placement of map elements. Examples include increasing the width of actually scaled razor-thin rivers or freeways so that they are visible on a map. More dramatically, Harry Beck's classic London Tube map sacrifices geographically accurate location of stations by exaggerating their relative location, fixing their placements into user-friendly straight lines.[53] The purpose of exaggeration is to increase legibility, comprehensibility, and communicative power.

### The Five Protocols Analogized to the Social Sciences
My list of five map generalization protocols is neither collectively exhaustive nor unique.[54] The abstraction procedures I have just described are

---

52. Arguably, symbols are more abstract than icons. Robinson, Morrison, et al. (1995, 479) call icons "mimetic [imitative] symbols," and Muehrcke and Muehrcke (1998, 81) use the term "pictographic symbols." See also MacEachren 2004, chap. 5, "A Primer on Semiotics for Understanding Map Representation," 217–43. Monmonier (1996, 18–24) provides a symbol grammar. See also chapter 5 below.

53. See Transport for London, n.d., for the "Harry Beck's Tube map."

54. The first chapter of Goodman 1978 presents five "processes that go into worldmaking" (7); namely, composition and decomposition, weighting, ordering, deletion and supplementation, and deformation. Although there are some clear and simple mappings to cartographic categories (e.g., deformation and exaggeration), this worldmaking list can be thought of as another classification of abstraction protocols, crosscutting mine.

fairly independent, although alterations in some protocols, such as the selection of scale, will affect others.

Let us now analogize these five protocols from cartography to the social sciences. Consider three broad schools of theorizing in the field of international relations: realism, liberalism, and constructivism. Realists focus on power and security; they hold that nation-states compete on an anarchic supranational stage, where each nation-state's means-end rationality is built on fundamental and immutable laws of selfish human nature.[55] Liberalism and neoliberalism emphasize economic relations. Economically motivated cooperation and negotiation among distributed nonstate agents, such as multinational corporations and nongovernmental organizations, are here believed to spread democracy and just law, and to impact state decisions.[56] Finally, constructivism holds that "people act toward objects, including other actors, on the basis of the meanings that the objects have for them";[57] thus, even political leaders and state bureaucracies operate according to dynamic and local concepts, interests, and values.[58] Constructivists regard even basic economic concepts such as "money" and "profit" as having cultural baggage.

Cartographic generalization protocols provide insight into the theoretical abstraction of global dynamics by theorists of international relations. Realism, neoliberalism, and constructivism each *select* only certain processes from a complex system: power and security, economic relations, and cultural meanings, respectively. Consequently, a specific theory from one of these schools *simplifies* and provides a *classification*, perhaps implicit, of how agents and processes are involved at the international scale. Insofar as these theories' models are mathematical, they may be *symbolized* differently. Any theory or model of global political, economic, social, and cultural dynamics *exaggerates* certain processes at the expense of others.

### *Two Broader Approaches to Abstraction*

In addition, two other forms of abstraction—perhaps broader, more synoptic, and deeper than the explicitly cartographic five—are central to *When Maps Become the World*. In some sense, they must also be temporally, psychologically, and logically prior to the five cartographic protocols.

---

55. Donnelly 2000, 9; Buzan and Wæver 2003, 6–7.
56. Shiraev and Zubok 2016, 80.
57. Wendt 1992, 396–97. Wendt argues that governments and other power structures are "epistemic communities" (397).
58. Shiraev and Zubok 2016, 116–17.

*Perspectivizing* is seeing the world in a certain holistic, consistent manner. The San Francisco Bay Area can be perspectivized in terms of its roads or highways (e.g., a road or transit map), its bare topography (e.g., maps emphasizing contour lines), or its natural beauty (e.g., a map of natural areas, preserves, and coastlines). Mapmakers draw and imbue the map *with* a perspective; map users plumb this perspective *from* the map. The perspective lives as a set of assumptions *inside* the map.

*Partitioning* is breaking up a system into certain kinds of parts (whether they be objects, features, or processes), potentially at different levels and with different degrees of interaction, according to certain norms and practices. A *partitioning frame* dictates how to carve out the scale-dependent parts of systems that have already been perspectivized, and for which we have ample data.[59] To create the Bay Area map, USGS cartographers decomposed the area's topography and represented it as a set of contour lines. In addition, they chose to represent only some roads and highways of a complex traffic network. For any domain and problem, even after perspectivizing, there are many ways to partition. (For instance, we must use inferences to partition reasoning into premise-and-conclusion propositions.)[60] The person doing the partitioning, a *Homo cartograficus*, selects only one or a small subset of these ways.

In short, the process of abstraction consists of three overarching phases: calibration of units and coordinates, data collection and management, and generalization. Learning about cartographic abstraction assists comprehension of scientific abstraction.

## Ontologizing

Ontologizing is the act of applying or using abstract representations. It is making or imagining a world according to a map, theory, model, or other representation, whether this is done consciously or unconsciously.[61]

---

59. I develop this notion in detail in Winther 2003, 2006c, 2011b.

60. But there is a plurality of ways of representing logical arguments. Such differences might matter to how we understand the nature and power of inferences (Malmgren 2018).

61. For philosophical discussions of ontologizing, many of which are scientifically informed, see Levins 1966; Beckwith 1970; Levins and Lewontin 1980, 1985; Keller 1983, 178–79; Cartwright, Shomar, and Suárez 1995; Pollack 2005; Winther 2006a, 2006b, 2011a, 2011b, 2014b, 2014c; Mermin 2009; Derman 2011, 193–200. Philosophical action theory, having to do with volition, motive, and decision, could be useful in inquiring further into the ontologizing relation between representation and action. Bratman 1999; G. Wilson and Shpall 2016.

We use representations, and sometimes we abuse them. We abuse them when we conflate them inappropriately with the world, interpreting them as the entire world itself—whether consciously or unconsciously—for our political and other biased purposes. I understand the hazardous conflation of representation and world to be the main risk of the act of ontologizing (I call this *pernicious reification*, and explore it fully in chapter 4). This hazard could, though, also be tied to earlier processes of abstraction.[62]

Once you have a map representation or a scientific theory or model, you can launch any of a variety of ontologizing practices: representation testing, changing the world, understanding the world, and pedagogical uses.

## ONTOLOGIZING 0: REPRESENTATION TESTING

Scientific and cartographic representations must be tested and confirmed. Their predictions need to be compared with the data.[63] When we develop a map or theory, the representation should accommodate new data, more data, and more kinds of new data. This can occur as a consequence of better or different measuring devices; because of improved statistical or qualitative analyses; or due to improved ways of making theory more empirically explicit and "matchable" with the data.

Representation testing is the most basic kind of ontologizing. It is a *reversal* of the path of abstraction. It tests the representation produced by generalization against new (or differently statistically structured, etc.) empirical data.

Imagine two kinds of representational structures that are being adjusted to each other. The first is theoretical (as in science) or visual (as in cartography). It is a product that emerges at the end of the full three-stage abstraction process outlined above. The other structure is closer to observation and measurements; it is what computer or data scientists call

---

62. There is a philosophy of science tradition of seeing reification as an intrinsic risk of abstraction, as in William James's "vicious abstractionism"; van Fraassen's diagnosis of the rupture between René Descartes and Blaise Pascal; or Cartwright's "The Truth Doesn't Explain Much." On James, see Winther 2014c; see also van Fraassen 1988; Cartwright 1983, 44–53. While respecting this tradition, for philosophical simplification purposes, I here take the position that reification occurs during ontologizing, and not during abstraction.

63. See Lloyd (1988) 1994; Winther 2009c and references therein.

a *data structure*. It emerges at the end of the first two stages of abstraction described above. These structures have been called *theoretical* models, on the one hand, and *data* or *empirical models*, on the other.[64]

In cartography, a map may not represent accurate information. The legend of the USGS map of the San Francisco Bay Area is transparent about this. It whispers: "Revised information not field checked" and "Selected hydrographic data compiled from NOS [National Ocean Survey] Charts. This information is not intended for navigational purposes." Such data *could* be acquired. Furthermore, the accuracy of a highway map could be tested by extensive map-informed driving.

As a simple example from science, consider the temperature of a gas subjected to different pressures in an enclosed but flexible vessel. A computer simulation built on basic thermodynamic mathematical theory could graph a curve based on Boyle's law, which describes the relation between gas pressure and the size of the gas's container.[65] This curve would be a theoretical structure of expected states, given particular conditions of temperature and gas type.

The empirical structure here would involve data from actual gas pressure measurements. The data needs to be statistically organized, perhaps "smoothed" to minimize "noise," sacrificing some detail so that important empirical patterns can emerge. Alternatively, missing data points might be interpolated between existing ones. Testing the computer simulation representation of Boyle's law would amount to seeing whether its theoretical structure matched the empirical structure measured, held for new kinds of gases, or for tiny as opposed to immense volumes not previously tested.

In both cartography and science, representations must be retested when new and improved data become available, and also in the context of shifts in social and scientific values and aims, or changes or revolutions in research programs.

---

64. For instance, the American and Dutch philosopher of science Bas van Fraassen distinguishes between theoretical models, which "depict the 'underlying reality'" (2008, 289), and "empirical substructures" of the models (1980, 64; 1989, 227). Empirical substructures are "surface models" or "data models," and van Fraassen differentiates between these (2008, 168). On the data-phenomenon distinction, see also Bogen and Woodward 1988, and on the "observable phenomena" that the empirical substructures depict, see van Fraassen 1989, 227; 2008, 168. For further details, see Winther 2016 and references therein.

65. Boyle's law follows from the more complex ideal gas law (see chapter 6).

## ONTOLOGIZING I: CHANGING THE WORLD

Abstractions do not merely represent the world—they can also materially construct it. Representations thus perpetuate themselves by rebuilding a world that they continue tracking. The philosopher Philip Kitcher appeals to the map analogy:

> Current ventures in map-making often carry the traces of past endeavors.... Consider a straightforward example in which map-making has contributed to the alteration of the physical environment and the development of various pieces of technology. Backpacking Californians use topographical maps to explore the wilderness of the High Sierras. Older maps (together with guidebooks, lightweight camping equipment ...) made it possible for more people to experience the beauty and solitude of the mountains. In consequence, certain lakeshores became degraded from over-camping.... new kinds of backpacking technology have been introduced, such as ... "bear boxes."... The maps of today show more detail for the more remote elevations than did their ancestors, as well as recording the changes caused by human activity.[66]

As increased traveling along routes changes the landscape, such alterations must be recorded on the maps themselves.[67] By showing a clear case of direct material change in the world, Kitcher's backpacking scenario suggests a noncontroversial way in which both maps and scientific theories can change the territory they describe.[68] In comparison, genetic theory can impact bioscience. For example, gene expression maps can

---

66. Kitcher 2001, 60–61.

67. This map–world–map feedback could be interpreted as causal construction, which the philosopher of science Peter Godfrey-Smith defines as a "physical, causal intervention in the world, intervention which effects a change in external affairs" (1996, 145). This is consistent with Richard Boyd's thesis that "human social practices make no noncausal contribution to the causal structures of the phenomena scientists study" (1992, 173). Social practices can affect the world only causally and physically; according to Godfrey-Smith and Boyd, human practices cannot affect the laws of nature, as opposed to what some social constructivists might argue (see chapter 9).

68. Human-induced global climate change can itself alter the territory so quickly that geographic maps abruptly become outdated and are in constant need of improvement. Bangladesh's coastlines, interior land, and inland bodies of

guide genetic engineers in altering expressed stretches of DNA through the cutting-edge technology of CRISPR. The result may be new forms of synthetic or cyborg life that could alter our theories about gene-culture transmission, organismic development, or even intelligence. Or consider what some scholars call the *performativity* of economics. Trained economists have used mathematical theory to design products such as modern, high-tech investment derivatives, which are then sold on Wall Street.[69]

Once we recognize that scientific theory interprets and constructs the world, the question of normative values in science becomes urgent. We must ask not only how the world works but also how we would desire to see it changed in generative ways.

ONTOLOGIZING II: UNDERSTANDING THE WORLD

Map users and scientists employ representations in order to understand what exists and what happens in the world.

Each map projection has its own transformation equations, visual appearance, and proper applications. Knowledge can be usefully drawn out of projections, if the right type is consulted for the need. For instance, the ubiquitous Mercator projection is conformal and thus useful for navigation, but terrible at showing area appropriately at different latitudes. The Mercator could also be useful for explanations relying on representing country shape accurately—such as explanations of locations and frequencies of military invasions along a long coast—but it fails at explanations requiring accurate size depiction, such as explanations of agricultural production.

The range of information gleaned from the use of an appropriate map is nearly boundless. Real estate agents and attorneys consult cadastral maps during the sale of a home. Maps of the cosmos help astrophysicists identify parts of the universe relevant to studying black holes. Regular drivers and long-distance truckers consult real-time traffic maps to find routes that avoid traffic snarls. A researcher involved in an international effort to more accurately map the seafloor has said that "our major discoveries have been because of mapping."[70]

---

water come to mind, as do the shrinking Arctic ice or the US states of Florida and Louisiana, or a burning California or Australia.

69. Callon 1998; Derman 2011; Frankel 2015.

70. The marine geologist Martin Jakobsson, quoted in Frischkorn 2017.

[86] CHAPTER THREE

## ONTOLOGIZING III: CLASSROOM COMMUNICATION

Consider a more public form of representation ontologizing: map use in pedagogy and in the classroom.[71] Today's students become tomorrow's researchers, and even colonizers and homophobes. Paying close attention to the maps drawn by students thus provides insight into potential biases and pernicious reifications that may emerge in the teaching and training of map abstraction and map ontologizing. Interestingly, such biases may change over generations.

Biases vary, predictably enough, based on location—Europe dominated many individuals' idea of world geography at least until the late 1990s. One study asked first-year university geography students to sketch a world map from memory, and the great majority (79 percent) placed Europe at the center.[72] Students across the globe tended to positively exaggerate the relative size of Europe, and European countries were generally overrepresented in number and relatively correct placement. Exceptions to this Eurocentric pattern exist, of course: a small number of students exhibited extraordinary knowledge of most, even all, countries (fig. 3.1), and home regions were sometimes labeled with subnational details.

These maps are products of knowledge practices. The personal, parochial, and subjective are interwoven with the scientific, universal, and objective in how these students learned geography and acquired mapmaking skills prior to becoming university students. Many students undoubtedly accessed atlases, encyclopedias, and other resources displaying accurate geographic maps (accurate in the context of a given perspective, that is). A student's curiosity and ambition might even have led her to avidly consult teachers, experts, or peers to share geographic information. Real physical maps that are taught become cognitive maps.

Much like geography, scientific theory is learned, recalled, and used. And when the student becomes the researcher or teacher, the new scientist communicates previously established patterns of assumptions and practices. Teaching and pedagogy thus matter for the future of science and of the world it maps, and for attempts to predict, explain, and understand.

---

71. Newsrooms and art studios are other wonderful places for map ontologizing outside of technical science, cartography, and GIS.

72. Saarinen 1999, 141. Saarinen and his team collected student sketch maps at seventy-five sites in fifty-two countries, from 1985 to 1987, for a total of 3,568 sketch maps.

Drawing cognitive maps as these students did can be eye-opening. (Such an exercise—blank maps and instructions that can be copied and used in a group setting—is in the appendix.)

## Conclusion

Much like with a map projection, the world below me comes into focus. I sit in my seat. For many hours, I have stared out the airplane window at generalized colors, lines, and shapes. In and out of consciousness, I enjoyed some abstracted peace and quiet on this flight. Startled, I hear the captain announcing that we will land in approximately thirty minutes. I feel the slow descent. The color patches below acquire more internal contrasts. Shapes become more acute and neatly defined. The whole, healed, and silent abstract Earth gives way to the complex, detailed, immediate Earth.

I am amazed that this immense Earth is the world I walk on. Can the cartographic image I have been daydreaming about for hours in the airplane seat—or perhaps cartographic portraits of our home, whether Apollo 8's 1968 "Earthrise," Apollo 17's 1972 "The Blue Marble," Voyager 1's (and Carl Sagan's) 1990 "Pale Blue Dot," or Van Sant's map—serve as a moral map of connectedness and mutual responsibility? My imagination cannot help but stretch and project from the physical realm to the philosophical world, where my thoughts churn and search for parallels inside to match all I have seen outside. This too-quick return to the scale of human reality pushes me to think of the capacities and potentials we might explore around ontologizing a connected-Earth moral map. As I gear up to face a society I often feel helpless in, I cannot help but flirt with the idea of a principled map, connecting what I hold to be worthy, made into flesh among those many who represent an existence that seems as disproportionate to me as a Mercator map.

Map thinking is fertile. The abstraction process can be distinguished into three stages, and different types of ontologizing contain routes feeding back to the representations themselves. Changing the landscape merits changing the maps.

To map, model, or theorize is precisely this interplay between abstraction and ontologizing. If abstraction is taking off, and ontologizing is landing, then we see that we need both, just like we need to make as well as use maps and scientific theories and models. We must also be mindful of using representations in a generative rather than pernicious manner.

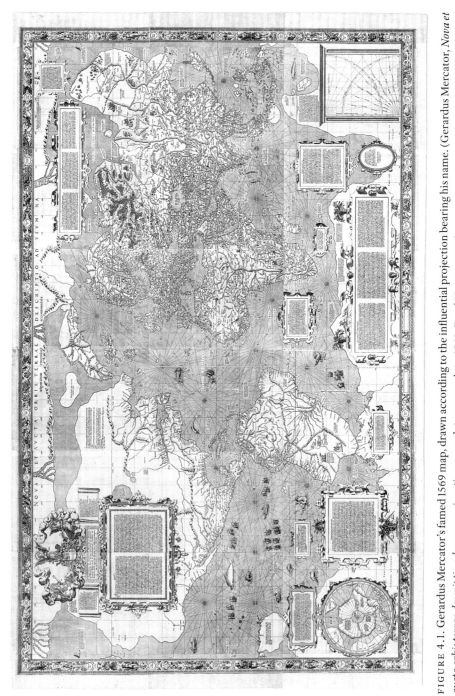

FIGURE 4.1. Gerardus Mercator's famed 1569 map, drawn according to the influential projection bearing his name. (Gerardus Mercator, *Nova et aucta orbis terrae descriptio ad usum navigantium emendate accomodata*, 1569. Basel University Library, Kartenslg AA 3-5. 10.3931/e-rara-25290.)

# 4

# Long Live Contextual Objectivity!

Imagine a world map in front of you. Does it have the standard Mercator projection (fig. 4.1) used in many atlases and classrooms, or is it perhaps drawn according to a Gall–Peters projection, stretching the continents vertically near the equator and crunching them close to the poles? Is it a chart with the polar azimuthal equidistant projection found on the United Nations flag? (See figs. 3.3 and 4.5.) Just as you cannot force an orange peel onto a table surface without tearing, stretching, or shrinking the peel, so too you cannot force the curved surface of Earth onto a two-dimensional flat map without distortion. The result is a profusion of jeopardized map projections, which are bound to conflict with one another.

Cartography teaches that adequate representation of the entire world requires a plurality of map projections.[1] Chapter 3 reviewed the basic elements of map projections; this chapter explores some philosophical implications of this representational pluralism.

I first frame two key concepts: *pernicious reification* and *contextual objectivity*. When we fall in love with our projection or theory, we treat the theory or projection as if it were a real, concrete thing that also *is* and *describes* the entire world. We overestimate our representation's capacities and promises. We show no hint of skepticism about it, or acceptance of its potential limitations. That is, our abstraction is universalized and narrowed, in addition to being utterly ontologized. We make it into a world of its own, applied beyond its appropriate context.[2] Such pernicious reification causes problems, rendering inaccessible the knowledge

---

1. See, for example, Maling 1992; Richardus and Adler 1972; J. P. Snyder 1993; Wood, Kaiser, and Abramms 2006.

2. See, for example, Muehrcke 1972, 1974a, 1974b; Muehrcke and Muehrcke 1998; Levins and Lewontin 1985; Cartwright, Shomar, and Suárez 1995; Derman 2011; Winther 2014c.

residing in other, overlooked, empirically robust, minority representations. Pernicious reification blocks the road of robust inquiry.[3]

In contrast, contextual objectivity results from the purposeful and intentional ontologization of representations within an appropriate scope. *If pernicious reification is an epistemic and practical failure, contextual objectivity is a knowledge-enhancing and concrete success. If the former is diseased ontologizing, the latter is healthy ontologizing.* My analysis is a response to the biologists Richard Levins and Richard Lewontin's worry that "the problem for science is to understand the proper domain of explanation of each abstraction rather than become its prisoner."[4]

I illustrate these two concepts through the evolution of the still cartographically dominant Mercator projection, from Gerardus Mercator's original 1569 map to the new online standard Web Mercator.

Even when we are enraptured with single theories or projections, all is not lost. I show how the tool of *integration platforms* can motivate the scientific community; philosophers, historians, and sociologists of science; and the lay public to place representations in their respective appropriate explanatory and predictive context, thereby avoiding or uprooting pernicious reification. One such integration platform I call *beyond Mercator*.

## Pernicious Reification

Pernicious reification is the ontologizing—and universalizing or narrowing, or both—of a single imperialistic abstraction. It is mistaking your abstracted map for the world in all its detailed glory. Pernicious reification is imagining Earth actually as a Mercator projection; or conceiving evolution exclusively as the process of selfish genes duking it out on the ecological stage; or insisting that *Homo sapiens* is sharply separated into racial or ethnic groups. It is believing that you can hold the entire world in your palm (rather: that the map or theory or model you hold in your palm is the entire world); and that, by closing your conceptual fist tightly around it, you fully feel the world, know it, and control it. You fail to

---

3. According to Peirce (1998a), the "sole" "rule of reason" is "in order to learn you must desire to learn and in so desiring not be satisfied with what you already incline to think." The "corollary" which followed from this, and which he feels should "be inscribed upon every wall of the city of philosophy" is: "*Do not block the way of inquiry*" (48); see also Haack 2014.

4. Levins and Lewontin 1985, 150. See also J. M. Kaplan and Winther 2013.

accept that your representation may be partial and even wrong—only you and your palm exist.

Pernicious reification takes place when, in addition to ontologizing, either, or both, of the following two practices happen:

1. *Universalizing* is holding your map, theory, model, or other representation or abstraction to encompass all, or almost all, phenomena within a given domain. You understand the abstraction, whatever its internal structure, as having extremely broad conditions of empirical application. What falls outside of this scope is considered irrelevant or uninteresting, or both, or worse: "Beyond that which we believe in, drought or insanity live."[5]
2. *Narrowing* is constraining theoretical content to a few central laws, principles, and assumptions, thereby diminishing the nuance and diversity of models and theoretical components. The internal heterogeneity of theory decreases, while explanatory power might seemingly grow.

These two components of pernicious reification are logically independent and can emerge separately. Universalizing expands the concrete, empirical scope of a theory but does not determine the degree of internal theoretical complexity. Narrowing restricts internal representational structure and may or may not also involve universalizing.[6] As the geographer Leslie Curry has stated in an often cited passage, "Geographical studies are not descriptions of the real world but rather perceptions passed through the double filter of the author's mind and his available tools of argument and representation. We cannot know reality; we can have only an abstract picture of aspects of it."[7]

We perniciously reify when we utterly ontologize it by making the single abstracted projection into the world, into Earth, into the ellipsoid, and then also universalize or narrow it, or both.[8] We universalize the Mercator projection as the only valid projection among a sea of other

---

5. Pereda 1999, 17; my translation.
6. The cognitive basis of pernicious reification requires further study. Root causes of such reification could include opinion echo chambers, filter bubbles, and cognitive and social biases eliminating the perception and evaluation of alternative—especially minority or unorthodox—representations (J. Campbell 1989; Gilovich 1991; M. D. Jones 1998; Kahneman 2002, 2011; Pariser 2011).
7. Curry 1962, 21.
8. I shall say more about ontologizing in chapter 5.

empirically valid extant or even *unconceived* projections;[9] we narrow it by ignoring projections similar to it such as the Miller projection,[10] which is a scaled Mercator projection. As a community, we are often inclined to idolize a single representation simply because it is *there*, because authorities have created and taught it, and because it is widely accepted. We then believe, for instance, that Sweden and Mexico are equal in size.

Or consider the Ming dynasty map *Da Ming Hun Yi Tu* (Amalgamated map of the Great Ming empire; fig. 4.2), which places China at its—and consequently the world's—center. This map spatially distorts (by contemporary, Western cartographic standards) the Arabian Peninsula, Africa, and Japan. Likely originating in the late fourteenth century CE, during the reign of the Hongwu emperor, founder of the Ming dynasty, the amalgamated map ontologizes, universalizes, and narrows a "world navel" vision of China.[11] Hongwu's son, his successor as emperor, ordered the construction of the Forbidden City. Nestled in the new capital of Beijing, the Forbidden City was taken to be the literal center of the universe.[12] The map ontologizes, universalizes, and narrows a vision of China as powerful and unified, and as the center of the universe.

Revival of the all-but-settled debate on why dinosaurs went extinct provides another example of pernicious reification—particularly so because the reification was so intentional. Bianca Bosker reports on it in an online article titled "The Nastiest Feud in Science."[13] Mapping the worldwide distribution of iridium in rock layers led to the physicist Luis Alvarez's widely accepted asteroid-impact theory that dinosaurs went extinct suddenly. Mapping the gradual die-off of microbes *before* that asteroid impact has reinvigorated the dissenting view, championed by the geologist Gerta Keller, that dinosaur extinction resulted from massive volcanic eruptions in India. Bosker writes that Alvarez "wielded his star power to mock, malign, and discredit opponents who dared to contradict him." (Ad hominem attacks have come from both sides of the debate, Bosker notes.) In relation to Keller's insistence on keeping open the inquiry, Bosker makes a point reminiscent of Thomas Kuhn's comments (noted in chapter 2) about research into deeper, important questions

---

9. Note that every successful scientific theory was also once an unconceived alternative. Suppressing the social or economic conditions for the emergence or development of unconceived theories is a form of universalizing (Sklar 1981; Stanford 2006).

10. O. M. Miller 1942.

11. Winther 2014d, 2019a.

12. Yu 1984; Ebrey 2010.

13. See Bosker 2018. All quotes in this paragraph are from this source.

FIGURE 4.2. Surrounding landmasses are distorted to fit around China in *Da Ming Hun Yi Tu* (Amalgamated map of the Great Ming empire). China is here a world navel.

that are more interesting than puzzles, which are known to be answerable. "Though trading insults is not the mark of dispassionate scientific research," Bosker writes, "perhaps detached investigation is not ideal, either. It is passion, after all, that drives scientists to dig deeper, defy the majority, and hunt rocks in rural India for 12 hours at a stretch while suffering acute gastrointestinal distress," as Bosker recounts Keller doing. The larger question that fuels Keller's zeal for the volcanic theory is what might be learned that could be applicable to current climate change and tragic extinction of species.[14] After all, there is a remarkably robust cor-

14. A combination of causes of dinosaur extinction—volcanism, asteroid impact, and marine regression (the exposure of seafloor as a result of geological processes)—are also possible, perhaps even likely (Archibald and Fastovsky 2004).

relation of various occasions of increased greenhouse effects and the six major mass extinctions.[15]

Pernicious reification is a failure to meet at least the first, third, and fourth principles of the "Modeler's Hippocratic Oath":

- I will remember that I didn't make the world, and it doesn't satisfy my equations.
- Though I will use models boldly to estimate value, I will not be overly impressed by mathematics.
- I will never sacrifice reality for elegance without explaining why I have done so.
- Nor will I give the people who use my model false comfort about its accuracy. Instead, I will make explicit its assumptions and oversights.
- I understand that my work may have enormous effects on society and the economy, many of them beyond my comprehension.[16]

Pernicious reification is conflating and confusing a single abstract representation with the world, thereby thinking of complex reality as *nothing but* the content of the representation.[17] The world's richness is left out, to the detriment of other justified, empirically validated points of view with which we could gather more complete knowledge and plans for action. As we will see throughout part 2, perniciously reifying our scientific representations causes problems for both the scientific community and the lay public at large.[18]

---

15. Greenhouse-effect warming of the atmosphere caused by the rise in atmospheric methane, carbon dioxide, and other gases has several major sources: human practices (today); volcanoes (in the past), and melting of methane ice (in the past and today).

16. Emanuel Derman and Paul Wilmott, "The Financial Modelers' Manifesto," reprinted in Derman 2011, 198–99.

17. See my analysis of William James on these matters in Winther 2014c.

18. Edward Tufte, a prolific data visualization expert with a background in statistics and political science, has written: "Making an evidence presentation is a moral act as well as an intellectual activity. To maintain standards of quality, relevance, and integrity for evidence, consumers of presentations should insist that presenters be held intellectually and ethically responsible for what they show and tell. Thus consuming a presentation is also an intellectual and a moral activity." From Tufte's 2006 book *Beautiful Evidence*, quoted in a review by Stephen Few (2006). (Few, who considers Tufte a mentor and calls the quote "classic Tufte," found that the book unfortunately failed to live up to Tufte's own standards of clear and pertinent presentation.)

## Contextual Objectivity

Only a map addressing a particular practical context will tell us where the boundary between San Francisco and San Mateo Counties lies, or which parts of South America and Africa were one in the deep geological past. For the first, we need a map identifying the counties around San Francisco. For the second, we have to have some inkling that land masses of various sizes on Earth's surface have moved around during geological history so that we know what kind of map to consult. Our map of San Francisco does not tell us about the continents' geology, and vice versa. Contextual objectivity is the quality resulting from good and proper application of a representation.

For either question—county line or continental drift—whether you believe you have found the answer depends on whether the map is telling the truth for what it purports to depict. We must test the map, and prove that it matches the world in particular contexts.

Using philosophical assumption archaeology, I draw on philosophy of science literature to consider two key elements or assumptions of my core concept of contextual objectivity. I shall develop these two assumptions by deploying the map analogy: *conformation* and *the essential indexical*.

### CONFORMATION

Could the USGS map of San Francisco be considered true, approximately true, or even just true enough for certain local purposes, but perhaps not true *in general*? This might appear to be an odd question. But note that different aspects and elements of the map are accurate, and fit or capture the world in distinct ways. Truth thus seems either too general or too narrow a concept for capturing such a variety of successful relations between a representation and the objects, properties, or processes it intends to represent.[19]

Framing the question in terms of truth might miss the point. One could argue that feature similarity across representation and reality, rather than the truth of propositions or claims about the world, is what matters. The philosopher of science Ronald Giere draws on the map analogy to make precisely this point: "The fit between a model and the world may be thought of like the fit between a map and the region it represents."[20]

---

19. Teller 2008; see also Lloyd (1988) 1994, chap. 8, 2012.
20. Giere 1999, 82.

A more pluralist and contextual strategy permits an abundance of terms for successful representation. Helen Longino develops a conceptual tool for the contextualism of objectivity, *conformation*:[21]

> Maps fit or conform to their objects to a certain degree and in certain respects. I am proposing to treat conformation as a general term for a family of epistemological success concepts including truth, but also isomorphism [identity of structure], homomorphism [for Longino: a more partial alikeness of structure], similarity, fit, alignment, and other such notions. Classical truth is a limiting concept.[22]

Truth in the classical philosophical sense is an ideal concept, a concept of "limit." But representations can pass various kinds of adequacy tests, such as those listed by Longino. Each abstraction—or part thereof—can be accurate in a particular way. Each can be conformational. It depends on the context of application. For instance, for a map to be contextually objective, is exact location of places of interest necessary, or is it sufficient to show their relative topology? Maps showing relative topology (for example, the London Tube map) will not conform in the same way as those intending to show exact location (for example, a London street map).

We evaluate scientific models analogously. Longino continues: "Like maps, models must be sorted into grades of adequacy in multiple categories."[23] This pluralism and contextualism about successful knowledge conditions—about objectivity—holds for parts of maps, theories, and models as much as for entire maps and models, because the scope of application of the representation delimits the parts of the world to which the representation applies.[24]

---

21. Although *conformation* and *conformal* are related in origin and meaning, Longino's sense concerns a broader range of kinds of fit than the accurate representation of angles and shape that conformal maps have.

22. Longino 2002a, 117.

23. Longino 2002a, 118.

24. Longino, pers. comm., May 2016. Stephen Toulmin ([1953] 1960) uses the map analogy to make a point complementing Longino's contextualism of conformation relations. His 1953 chapter "Theories and Maps" pioneered the map analogy in the philosophy of science, using it to argue that selection and abstraction are consistent with—even necessary for—accurate representation of the world: "The fact that [cartographers and surveyors] make a choice of some kind does not imply in any way that they falsify their results. For the alternative to a map of which the method of projection, scale and so on were chosen in this way, is not a truer map—a map undistorted by abstraction: the only alternative is no map at all." Ditto for scientific theory (Toulmin [1953] 1960, 127–29).

## THE ESSENTIAL INDEXICAL

Turning to a central distinction in discussing contextual objectivity, the map analogy demands that we explicitly acknowledge the simultaneous roles of the objective and the subjective. Diverse individuals and research communities produce public cartographic abstractions that typically respect important features of the world. And yet, mapmakers depend on their local purposes and powers to create these maps.

This logic carries over to scientific theory, as explored in Bas van Fraassen's *Scientific Representation: Paradoxes of Perspective*. On the one hand, scientific theories and models can "be written in coordinate free, context-independent form,"[25] van Fraassen avers. On the other hand, in order to apply the information contained in scientific theories, the user must be situated within the context of the theory—whether that user be a scientist designing further experiments or theorizing deeper (in my language: abstracting), or an engineer or member of the lay public applying or trying to understand a scientific theory (in my language: ontologizing).[26] To situate a user, a theory needs a pointer, an index, just as a sentence might need one in the form of a pronoun ("*He* did it" might be an accusation, "*I* did it" a confession—this use of a pronoun is *indexical*). To use a map, we must know where we are *on* it; to wield a scientific representation, we have to know where we are *in* it.

Through the concept of *the essential indexical*,[27] van Fraassen argues that maps and scientific theories are simultaneously context-independent and user-specific (i.e., personal, biased, and individual). In the moment of application, we take the map's context-independent information, make a context-bound location judgment, and use the map to satisfy our specific purposes. And since models and maps are equivalent "metaphors," we must also "locate ourselves with respect to that model."[28] Interestingly, van Fraassen develops this "inevitable indexicality of application" in light of an example from Immanuel Kant, who discusses the necessity of both having a "map of the heavens" and knowing how his hands are positioned relative to it if he wishes to predict where the sun will rise on the horizon.[29] The act of ontologizing *forces subjective indexicality onto objective*

---

25. Van Fraassen 2008, 82. See also Lloyd 1995, 1996.
26. Van Fraassen 2008, 82.
27. Van Fraassen 2008, 3, 83, 88.
28. Van Fraassen 2008, 83.
29. Kant 1992, 367; van Fraassen 2008, 80.

[98]  CHAPTER FOUR

FIGURE 4.3. *Personal Map Exercise.* Draw your profile, or have an artist sketch it. Inside your head, draw a world map, possibly in Mollweide projection; in any case, avoid the Mercator. Inside the oceans, write twenty-five words that describe you and how you perceive and build the world. (Illustration by Larisa DePalma; concept by Rasmus Grønfeldt Winther; based on a photograph by Sia Signe Sander.)

*universality* in cartography and science alike.[30] Indeed, creatively playing with the resonance between the objective and subjective, we can draw psychological and autobiographical personal maps highlighting factual aspects of our lives (fig. 4.3).

Elements for contextual objectivity include conformation and the

---

30. Daston and Galison 2007 provides another subtle analysis of objectivity; J. K. Wright 1942 discusses subjectivity in maps.

essential indexical. Importantly, each of these ingredients is consistent with the possibility of a representation being erroneous—it may fail to adequately represent its world empirically. That is, a representation would only actually be contextually objective if it were to satisfy these components; was neither universalized nor narrowed; *and* was then also empirically correct for the relevant aspects. Before testing or otherwise ontologizing the representation, the two ingredients only tell us if a given abstraction is *potentially* contextually objective, contingent on inquiry and the passage of time.

## A History of the Mercator Projection I: Gerardus Mercator

The projection Gerardus Mercator used in making his 1569 map is the Mercator projection, the canonical map projection (fig. 4.1).

The history of the Mercator projection illustrates the persistent tendency to perniciously reify, and the importance and resourcefulness of contextualizing, broadening and decentralizing, and pluralizing and integrating to achieve contextual objectivity—which is the opposite of pernicious reification.

Mercator's projection grew out of Mercator's critique of earlier projections; from the emerging need for better navigation and a clear presentation of latitude and longitude; and from a knowledgeable awareness of alternative projections.

### MERCATOR'S CRITIQUE OF EARLIER PROJECTIONS

Mercator corrected faults he identified in other projections and chart types established by the 1560s. Ptolemy's *Geography* had presented instructions for (but no illustrations of) constructing three projections: *equidistant conic*,[31] *pseudoconic,* and *orthographic azimuthal*[32] (table 3.1). Recall from chapter 3 that a map projection mathematically flattens a globe. Ptolemy's second projection, the heart-shaped pseudoconic projection, displays lines of latitude as arcs of a circle with a common center. Unlike conic projections, lines of longitude need not be straight lines, and can be curves. Ptolemy's pseudoconic projection was widely

---

31. The equidistant conic projection centered on the North Pole enables a mapmaker to minimize distortions—particularly, east-west distortions—when showing one region, such as one hemisphere.

32. See Stockton 2014.

used during the Renaissance, including by Henricus Martellus and Waldseemüller. Mercator knew this projection well, having edited a collection of Ptolemaic maps,[33] and noted its shortcomings: the meridians curve and bend toward one another, and angles are grossly distorted, making it unsuitable for navigation. Mercator was partial to none of Ptolemy's projections.

For navigation at sea, the standard in the late Medieval and early Renaissance periods was *portolan charts*.[34] Such maps frequently had compass roses and indicated sailing direction with a series of crisscrossing lines—probably a network of *rhumb lines* (also called *loxodromes*). A rhumb line is a line of constant bearing that, on a globe, spirals toward the North or South Pole, cutting every meridian (every longitude line) at the same angle.[35] Portolan charts accurately delineated the Mediterranean but represented the Atlantic coastlines of Europe and Africa much less reliably; the portolan charts stretched regions, and did not graphically represent latitude and longitude.

Gerardus Mercator wanted to improve on the Ptolemaic projections and portolan chart traditions he inherited. Ironically, the Mercator projection, which soon became ubiquitous and ironclad in its perceived authority, was a countermap in its birth. That is, it initially resisted more dominant, contemporaneous maps. It imagined another kind of world—an ocean world requiring accurate navigation.[36]

## MERCATOR'S NEW PURPOSE: NAVIGATION

The full title of Mercator's influential 1569 map in English is "New and More Complete Representation of the Terrestrial Globe Properly

---

33. Mercator edited the 1578 *Tabulæ geographicæ*, a collection of Ptolemaic maps.

34. The word *portolan* is from the Italian, meaning "harbor," and portolan navigational manuals usually included sailors' descriptions of harbors and coasts, along with the charts. The 1375 Catalan Atlas is perhaps the best-known portolan chart.

35. Whether these sea charts used a single projection or multiple projections remains unsettled. However, "majority opinion ... consider[s] ... that the portolan charts were projectionless or that any projection was accidental" (T. Campbell 1987, 385). But see Tobler 1966 and Monmonier 2004, chap. 2, which argue for a rhumb line "proto-conformal" (Monmonier 2004, 23) Mercator-like projection, while John P. Snyder (1993, 8 and 288, n19) hypothesizes an equirectangular projection.

36. See, for example, Wood 2010, 126–27; Winther 2019b.

Adapted for Use in Navigation." On this map, a straight line between two inscribed locations provides the exact compass bearing in reality. In other words, if the map indicates that you should chart a southwest course of roughly 230 degrees to get from, say, sixteenth-century Spain to Hispaniola, then you could and should follow that bearing the whole way, across the ocean. Course angles and lines on Mercator's map projection correspond to course angles and lines in the world. Nonconformal maps do *not* have this property. Were you to follow a straight line between points A and B on a polar or orthogonal azimuthal map, your route and course would be constantly distorted and you would arrive far away from your destination.

To see why, consider the rhumb lines discussed above. Mercator's magic was that he made a map as if he had peeled longitude lines off the globe and straightened them out with respect to one another so that they were all parallel. But, of course, the actually infinitely small poles were now as wide as the entire map. Specifically, Mercator's disrobed longitude lines permitted—by Mercator projection definition and via its transformation equations—the straightening out of rhumb lines, typically curved on most map projections.[37] Mercator's new projection was thus immensely more useful for navigation than any of Ptolemy's three projections, or portolan charts.

## MERCATOR'S CLEAR PRESENTATION OF LATITUDE AND LONGITUDE

The 1569 map prominently displays lines of longitude and latitude, which were rarely present prior to Mercator or even until seventeenth-century maps.[38] Mercator believed that magnetic north and true north have identical directions (that is, there is no magnetic declination) on Cape Verde,[39] so he drew the prime meridian 0° longitude through this mid-Atlantic island (fig. 4.1).

---

37. According to legend 12, Mercator was acutely aware of rhumb lines and of the importance of making them into straight lines. But we do not know Mercator's actual method of calculating the projection. Only in 1599 was the mathematics of Mercator's projection made explicit in Edward Wright's *Certaine Errors in Navigation*.

38. Wilford 2000, 91.

39. In legend 5, Mercator is also patient with those who suggested that this occurred on certain isles of the Azores.

### MERCATOR'S AWARENESS OF ALTERNATIVE PROJECTIONS

Mercator adorned the lower-left corner of his map with a fantastical inset of the septentrion, or northern, regions based on a polar azimuthal equidistant projection. This inset taught that there were four perpendicular rivers or gulfs through the Arctic ice swallowing and absorbing everything that came near, drawing all into the North Pole abyss.

Furthermore, Mercator recognized that, with his projection, "degrees of latitude would finally attain infinity" at the poles, as he says in the legend of figure 4.1. Mercator thus permitted alternatives. He accepted some of his map's limitations. Despite later protestations to the contrary—some of which I will summarize shortly—Gerardus Mercator himself did not claim a one-size-fits-all projection.

## A History of the Mercator Projection II: Post Mercator

The pernicious reification of representations in cartography and science perhaps says more about subsequent uses than about the creators' knowledge and intentions.[40] This seems to have been the case with Mercator. Later users of a representation utterly ontologize, and universalize or narrow it.[41] Representational methods become victims of their own success—their application is extended beyond reasonable and relevant empirical boundaries, and their authority grows.

Worrisomely, pernicious reification of the Mercator projection causes misperceptions about relative size, which can have dire, long-term psychological, cultural, and political consequences in the real world. For instance, Mercator's projection promises that Mexico and Sweden, or Africa and Greenland, are roughly equal in area. However, in reality, Mex-

---

40. Wood, Kaiser, and Abramms make this point bluntly: "People denigrate the Mercator projection as being out-of-date, obsolete, distorted, and an archaic holdover from our imperialist/colonialist past. We've even heard the Mercator decried as racist. The projection itself is *none* of these things. The objection *should* be directed against the projection's *misuse* as a world map when it was originally intended for the purpose of navigation. When the Mercator projection is misused it is not the projection that is at fault. Blame the mapmaker! Or the publisher! Or the user!" (2006, 20).

41. And often the world made is dominated by a central or master image (Holton 1977; Myers 1988; MacEachren 2004; Monmonier 2004; Wood, Kaiser, and Abramms 2006, 88–92; Winther 2014d, 2019a).

ico is four times bigger than Sweden, and Africa fourteen times bigger than Greenland. In 1923, the geographer John Goode characterized the Mercator projection as a pedagogical "crime" because, in it, "population density, density of existing forests . . . comparison in size of states and empires, all are untrue and inexcusable."[42] The USSR effectively received free Cold War intimidation power from Mercator.

In 1974, the German historian and cartographer Arno Peters denounced "Mercator's map." He chastised it as "the symbol of its age; of the era of the Europeanisation of the world."[43] As a political counterpoint, Peters defended a "new cartography" with a "revolutionary character."[44] Peters claimed to have apparently "combined in a single projection" ten "attainable map qualities," including "fidelity of area," "fidelity of axis," "proportionality," and "universality."[45] According to Peters, only one projection was valid. Most of the cartographic establishment, however, reacted by treating two projections as automatically and inherently invalid: Mercator's and Peters's (figs. 3.3 and 4.5).[46]

The equal-area projection Peters promoted should properly be called the *Gall–Peters projection*,[47] since it was originally designed by the clergyman James Gall. Gall had written, "Mercator's projection sacrifices Form, Polar distance, and Proportionate area, to obtain accurate orientation for the navigator; whereas, to the geographer, Form, Polar distance, and Proportion of area are more important than Orientation."[48] Critics have thus blasted the Mercator projection for geographic, psychological, sociological, pedagogical, and political reasons.[49] Finding a projection that satisfies critiques from all perspectives, however, is impossible. No single map can fully represent the world.

42. Goode 1923, x.
43. Peters 1983, 63.
44. Peters 1983, 150.
45. Peters 1983, 105–18; table on 114.
46. See, for example, Robinson 1990. See also Monmonier 2004; Wood 2010.
47. Peters 1983, 1993. The Gall–Peters map also has detractors who appeal to considerations of the aesthetic (e.g., it is "reminiscent of wet, ragged, long winter underwear hung out to dry on the Arctic Circle"; Robinson 1985, 104) and the functional (it "is simply a poor map"; Monmonier 1995, 10).
48. J. Gall 1856. On this point, see also MacEachren 2004, 313–15.
49. Consider that some studies report Mercator-like distorted understandings of country and continent sizes in undergraduate students (Saarinen 1999; fig. 3.1 above); others do not (Battersby and Montello 2009). Discrepancies in studies may be due to discrepancies in map exposure during different cultural time periods (Battersby, Finn, et al. 2014, 97).

TABLE 4.1. Comparison of the number of map projections developed during five periods of Western civilization (with percentage of total for all periods)

|  | Wikipedia | Snyder | Wood, Kaiser, and Abramms |
|---|---|---|---|
| Twentieth century | 40 (56%) | 160 (65%) | 11 (55%) |
| Nineteenth century | 13 (18%) | 55 (22%) | 5 (25%) |
| Enlightenment era (1670–1799) | 5 (7%) | 16 (6%) | 1 (5%) |
| Renaissance era | 6 (8%) | 8 (3%) | 3 (15%) |
| Classical era | 7 (10%) | 8 (3%) | 0 |

*Sources*: Wikipedia 2017; J. P. Snyder 1993, 53–54, table 1.1, 92–93, table 2.1, 150–54, table 3.2, 277–86, table 4.2 (160 projections is an estimate from Snyder's detailed table); Wood, Kaiser, and Abramms 2006, 138.

*Notes*: Up to 60–70 percent of all map projections ever developed appeared in the twentieth century. Although each of the three independent sources cited in the table displays a different total, the percentages are similar across the three sources.

Percentages are rounded to the nearest integer and thus do not add up to exactly 100 percent.

The plurality and number of map projections exploded in the twentieth century (table 4.1)—beginning well before Peters—along with an increasing diversity of other scientific theories, models, and hypotheses. New theories and models included relativity theory and quantum physics, genetic mapping, and high-temperature superconductivity, to name a few. In cartography, compromise projections became common in various atlases; John P. Snyder lists the projections in sixteen atlases from as early as 1916, and Monmonier lists the projections in twelve atlases that are more recent.[50]

Recall Richard Edes Harrison, whom we met in chapter 3. His 1944 *Fortune* wartime atlas showcased the polar azimuthal equidistant projection, which Mercator had already recognized. The United States now had "a new neighbor": the Soviet Union. While the USSR seemed far away on a Mercator projection map, it was quite close on a map using the polar azimuthal equidistant projection (fig. 3.3).[51] The war's airborne campaigns also informed the atlas, titled *Look at the World: The FORTUNE Atlas for World Strategy*. Harrison accompanied more-traditional projections with his "air age" *perspective maps* based on Earth's curvature. The atlas's introduction bemoaned standard map projections whose "viewpoint is at *infinity*—one is not over a particular point on the map, one is over

---

50. See J. P. Snyder 1993, 180–81, table 4.1; Monmonier 2004, 128, table 9.1.
51. Henrikson 1979, 174.

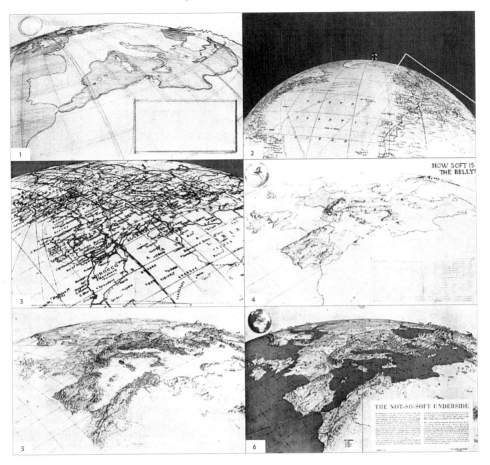

FIGURE 4.4. According to *Life* magazine, the cartographer Richard Edes Harrison's mind was a "compendium" and his office "a welter of maps, almanacs, geographies, encyclopedias, atlases in all languages, globes in all shapes and sizes" (Harrison 1944a, 56, 61). The maps are simultaneously "romantic and pictorial" and "precise instruments of geography" (61). The original captions are as follows: (1) "First step"; (2) "The globe is photographed"; (3) "Close-up of the area"; (4) "A color sketch"; (5) "The Scrub sheet"; (6) "Final product" (61). (Reprinted with permission of the Estate of Richard Edes Harrison.)

all points simultaneously"; and he worried about the fact that "maps are a collection of symbols and generalizations," which the untrained map reader can all too easily conflate with the world.[52] However, not all cartographic hope was lost—Harrison's perspective maps (fig. 4.4) respected

52. Harrison 1944b, 11–12. On R. E. Harrison in general, see Schulten 1998.

geometry by imagining the viewer floating a few tens of thousands of miles over Earth, and staring at the long horizon.[53]

Harrison considered standard rectangular maps to be useful, but he contextualized them with his perspective maps because "everything a man sees about him is in perspective. It is the normal view." He broadened and decentralized by critiquing "north up" maps and by also emphasizing globes—"too rarely is the globe allowed to play its essential complementary role." Harrison humbly and self-consciously presented his perspective maps as a supplement to—not as a substitute for—rectangular maps and standard atlases.[54] By attacking neither the Mercator nor the polar azimuthal equidistant projection, he wished to block pernicious reification of any one map or map projection, and seek contextual objectivity.

Toward the end of the twentieth century—partly in response to the controversy stirred by Arno Peters—cartographers started rallying against rectangular maps in general.[55] The American Cartographic Association (ACA) published three booklets on map projections with telling titles: "Which Map Is Best? Projections for World Maps," "Choosing a World Map: Attributes, Distortions, Classes, Aspects," and "Matching the Map Projection to the Need."[56] According to J. B. Harley, by 1990 Peters's sustained critique against cartography had the community "scrambling to close ranks" against the Gall–Peters projection in order "to defend their established ways of representing the world"; according to Denis Wood, in a "hysterical overreaction," cartographers condemned *all* rectangular maps.[57] Common to the diagnosis of both commentators is the observation that the community rejected Arno Peters's efforts.

Into the twenty-first century, as we saw in chapter 1, GIS transformed cartography. If the projection of a GIS data set is known, computers can easily translate it into other projections, although myriad visualization

---

53. Harrison 1944a, 61.

54. Harrison 1944b, 10–12.

55. See, for example, *American Cartographer* 1989.

56. The author of the first two of these is listed as Arthur H. Robinson (Robinson 1986, 1988); Robinson and John P. Snyder are listed at the editors of the third pamphlet in the series (Robinson and Snyder 1991). As mentioned in chapter 3, Robinson was the author of the classic cartography textbook *Elements of Cartography*, later coauthored (see Tyner 2005 for a history of the six editions of this textbook).

57. On scrambling, see Harley 2001b, 201. On hysteria, see Wood 2010, 127, n61. Robinson, heavily involved in the Committee on Map Projections that wrote a resolution against rectangular maps, reported on the resolution's passage in a journal essay (see Robinson 1990).

challenges and ontological choices remain.⁵⁸ Ongoing research into the plurality of map projections includes error analysis and tracing the consequences of conflating a projection with the round Earth.⁵⁹

Twentieth-century map projection frenzy notwithstanding—and despite Mercator's absence these days from many North American printed atlases and elsewhere⁶⁰—the Mercator projection persists in common wall maps. More important, GIS and online mapping services such as Google Maps, Bing Maps, and ArcGIS Online employ a *Web Mercator projection* based on Mercator's transformation equations. These equations render Earth in a cylindrical, near-conformal projection.⁶¹ The Web Mercator projection continues exaggerating the size of economically and politically dominant regions.

Web Mercator has become the online and digital cartographic representation standard. This is perhaps so because "north is always the same direction"; it simply "look[s] right"; it "allows for simpler (and therefore quicker) calculations . . . [and] continuous panning and zooming to any area, at any location, and at any scale"; and it "allows close-ups (street level) to appear more like reality."⁶² Mercator's projection is being reproduced, referenced, and reified more than ever.

## Integration Platforms

We have witnessed the dangers of perniciously reifying representations. However, maps can also be made more contextually objective. That is, we can use various criteria of empirical adequacy or accuracy (conformation) and bring to the forefront user centrality (the essential indexical).

One way to achieve contextual objectivity is through what I call *integration platforms*, which permit contextualizing and broadening our representations, among other anti–pernicious reification practices. We come to understand that every representation is limited, appropriate

---

58. Longley et al. 2011, 141.

59. Chrisman 2017.

60. For an investigation of atlases in the US, see Monmonier 2004, 128, table 9.1; for world map use in UK media, see Vujakovic 2002.

61. See, for example, Brotton 2012, chap. 12; Strebe 2012; Battersby et al. 2014; Winther 2019b.

62. The first two quotes are from Strebe 2012; the third quote from Battersby et al. 2014, 88–89; and the last quote from a forum response by Google representative Joel H., August 4, 2009, https://productforums.google.com/forum/#!topic/maps/A2ygEJ5eG-o.

only for some contexts. An integration platform can be a comparative database or spreadsheet table or group forum, or some other sort of data management device through which we—ideally communally—compare and contrast multiple representations, and the multiple assumptions of the representations used. The community, if not the individual, explores each representation's strengths and weaknesses, context of application, and potential for explanatory and predictive improvement. Integration platforms can also extend particular representations in new ways, including translating and transforming one representation into another.[63]

## A BEYOND-MERCATOR INTEGRATION PLATFORM: BLOCKING PERNICIOUS REIFICATION AND SEEKING CONTEXTUAL OBJECTIVITY

Recall from chapter 3 that a representation emerges from a sequential series of abstraction practices. We normally see the world from one perspective at a time, but we can also integrate multiple vantage points into a larger worldview. We understand the world more completely through such an integration. This pluralistic view lies at the heart of an integration platform.

*Integration platform* might be a new coinage, but it is not a wholly new concept. Respecting a plurality of maps permits us to undo—or even preemptively block—pernicious reification. Pluralism and integration platforms have been a strong presence among some map thinkers. For example, for their appendix B, Wood, Kaiser, and Abramms create a simple integration platform by listing the characteristics of the Mercator and seventeen other map projections in tabular form.[64] I shall call this a *beyond-Mercator integration platform*. This counters pernicious reification by showing the strengths, weaknesses, and contextual uses of each projection. Representations are not interpreted as absolutes vying for exclusive domination. Exposure to a plurality of representations in cartography or science encourages normative philosophical pluralism.

The appendix B table in Wood, Kaiser, and Abramms evaluates seven assumptions of each projection. They differentiate among what I consider to be ontological assumptions, metric layer assumptions, and aim and value assumptions (table 4.2).

Moreover, as I see it, Richard Edes Harrison's wartime atlas described above also contributed greatly to a beyond-Mercator integration plat-

---

63. Yang, Snyder, and Tobler 2000; Winther 2019a.

64. See Wood, Kaiser, and Abramms 2006, 127–31; see also Krygier and Wood 2011, 80–93.

TABLE 4.2 Map projection assumptions used in Wood, Kaiser, and Abramms (2006, appendix B)

| Ontological assumption | Metric layer assumption | Aim and value assumption |
| --- | --- | --- |
| • Shape of the world | • Grid lines straight?<br>• Right angles? | • Qualities<br>• Useful attributes<br>• What is shown well<br>• What is shown poorly |

form of map projections and world maps. The atlas sold well and doubtless inspired discussion among the public.

In various ways, integration platforms help us to appreciate a plurality of perspectives, and to reverse or block pernicious reification.[65]

## *Universalizing versus Contextualizing*

As opposed to the attempted universalizing of pernicious reification, an integration platform enables us to de-universalize or *contextualize*. As we saw above, contextualizing involves systematically comparing maps with other maps, theories with other theories, or models with other models in order to pinpoint the assumptions, practices, and purposes of each. The captain of a clipper ship would not consult the Gall–Peters projection map because it does not show the true shape of the shoals the captain must navigate and does not conserve the bearing angles so valuable for crossing seas and oceans—all of which the Mercator does. Airplane pilots prefer maps that preserve azimuth angles from a central point such as the North Pole when they fly over that point, rather than a Mercator projection, which explodes to infinity at the poles. Similarly, through contextualizing we come to understand the strengths and specific purposes of a given scientific model, and when alternative models might be more useful. For instance, different evolutionary models are used for different population sizes.[66]

---

65. Here I focus on the relation between different representations in an integration platform, and how we can contextualize and broaden them with respect to each other, given their differing contexts of application. But of course investigating the nature of the ontologizing process when multiple representations are simultaneously made or imagined into worlds is also key to reversing or blocking pernicious reification. I turn to *pluralistic ontologizing* as one such set of ways as part of my multiple representations account of representation in chapter 5.

66. Winther 2006a; Winther, Wade, and Dimond 2013.

## Narrowing versus Broadening and Decentralizing

As opposed to narrowing, an integration platform allows us to *broaden* and *decentralize*. We broaden maps and models by considering other centers on which to focus the map (*world navels*) or by questioning the supposition that a center is even necessary.[67] In the case of Mercator, we might consider China, India, or the Americas more central than Europe; or we might turn the projection upside down, or compare it to the Miller projection. In science, we could think of a theory not as a single model, but as a decentered yet robust family of models, where each model has unique assumptions, content, and purpose.[68]

## Assumption Archaeology

We have available to us assumption archaeology for identifying a wide variety of map projection assumptions. Imagine sprinkling small circles of equal size onto the WGS 84 ellipsoid at different latitudes and longitudes before projecting the ellipsoid onto a flat surface. A Mercator map with such Tissot indicatrix circles (more generally, Tissot ellipses) will continue showing them as circles—there is no angular distortion. However, they increase in size as we go north or south, and become immense near the poles (fig. 4.5). On a Gall–Peters projection, Tissots will have equal sizes across different latitudes, but will not stay circular.[69]

The indicatrix detects metric assumptions, including whether a map is conformal or equal area. We are even able to mentally subtract the flat-map distortion to get the correct two-dimensional surface of the sphere or ellipsoid.[70] More generally, the indicatrix identifies kinds of worlds made through the map; it registers some of the ontological assumptions of the map.[71]

The story of Mercator's projection, from the 1569 map to Web Merca-

---

67. Winther 2014d, 2019a. On centers and "world navels" in cartography, see also Wood, Kaiser, and Abramms 2006, 88–92.

68. For discussion of semantic and pragmatic views, see Winther 2016.

69. Explained tersely in Tissot 1881, 14–15. Papadopoulos 2017 provides further explication of Tissot.

70. George Blumenthal provided constructive feedback.

71. Digital formats are making possible new integrations of map projections. Tobias Jung has created an impressively comprehensive and visually straightforward integration platform online (see Jung, n.d.). By selecting any two of the 226 projections offered, the user can generate not only Tissot indicatrix ellipses for comparison, but a map showing the two projections' differences in superimposed, colored silhouettes.

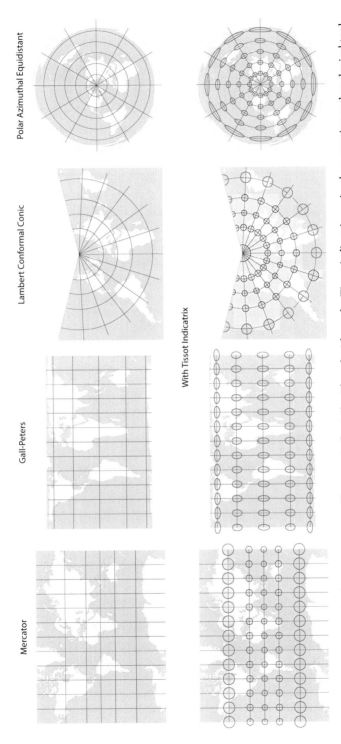

FIGURE 4.5. Four standard map projections (first seen in fig. 3.3) with and without the Tissot indicatrix, a visual assumption archaeological tool. (Illustration by Mats Wedin using Natural Earth, http://naturalearthdata.com/.)

tor, is thus a narrative of pernicious reification *and* of perennial, minority, outsider attempts at integration and contextual objectivity.[72]

Analogously to the beyond-Mercator integration platform, we can engage in careful contextualizing, decentralizing, and assumption archaeology of scientific theories and models. In mathematical models, this could be accomplished by explicitly stating underlying model assumptions and surveying the parameters within which the model is supposed to hold.[73] In visual models of organismic morphology, say, assumptions could be revealed by including an explicit scale bar to indicate relative lengths, or by declaring the level of resolution of a scanning electron microscope (SEM) image.[74]

Integration platforms demonstrate the limits of particular explanations and predictions, and lead to alternative or replacement models that can move past those specific limits. Ultimately, such platforms help rein in pernicious reifications of scientific theories (including most claims about characteristics based on everyday notions of race and ethnicity). In morally laden domains—such as military applications of physical theory, or insurance and health care applications of genetics—it is urgent to address the risks of pernicious reification.[75]

## PHILOSOPHICAL ASPECTS OF INTEGRATION PLATFORMS

In winding down the chapter, three philosophical aspects of integration platforms deserve investigation: granularity pluralism, representational pluralism, and end-of-inquiry consistency.

---

72. Note that the argument of map projection distortions is for world maps, or at least for maps of very small scale. Is there an argument for pluralism for large-scale maps? Put differently, could there be a *single* right geological map of, say, New Guinea? Scale matters, and we would still need to consider different thematic geological maps. Just as there is one right geometry and algebra for a Mercator projection, so there might be one right geological map for a given scale, a given set of properties, and a given territory. But just as the Mercator only appropriately depicts some metric properties of the globe, so a geological map shows only some aspects of the (same) territory. The same is true of a weather, ecological, or highway map: any of these would show only *some* aspects of the territory, and only at a particular scale. We thus also require pluralism of large-scale maps.

73. For instance, ferreting out assumptions about population size would be useful for identifying the pertinence of either the Fisherian or Wrightian model of evolutionary processes (see, Winther, Wade, and Dimond 2013; Winther 2006a, 2018a).

74. See, for example, Van Syoc and Winther 1999.

75. On pernicious reification in the field of genomics of race, see J. M. Kaplan and Winther 2013; Winther 2014b, 2018a.

## Granularity Pluralism

*Granularity* refers to the coarseness or level of detail of a representation. The station map that New York City subway commuters live by every day is far less granular than the one (or ones) transit workers follow. The latter locates not just stations, but railway switches, escape stairs leading to sidewalk grates, repair yards, and call boxes for contacting police.[76]

We cannot always discern the relative granularity of a group of map types. Ronald Giere, who consistently relies on the map analogy in defending the purposiveness of scientific representations, suggests we "imagine . . . four different maps of Manhattan Island: a street map, a subway map, a neighborhood map, and a geological map."[77] Giere's fellow philosopher of science John Ziman has likened scientific representations to various maps of London that people use to move around: highway maps, bus route maps, and a London Tube map.[78] Although Giere's list ranges from street to geological maps and Ziman's stays within the domain of transportation, how the granularity compares between each philosopher's maps or between the two different ranges is unclear. They all might be at comparable levels of granularity in relation to their own subject, or there might be wild variations in the granularity.

By contrast, we know that a geological map of Manhattan will generally be far more granular than a geological map of North America, and a London street map will, all else being equal, be more granular than a map locating London among European cities.

## Representational Pluralism

More abstract still is representational pluralism, which involves choices about representational modes. For instance, map projections can be represented geometrically and visually, or through analytic mathematical transformation equations. Either mode expresses the same mathematical content.[79] A rich representational pluralism also spans the sciences.

---

76. The plot of *The Taking of Pelham One Two Three* rests on the more granular maps that govern operation of the New York City subway. The 1974 film earned not one but two remakes: a 1998 made-for-TV version and a 2009 film (*The Taking of Pelham 123*).

77. Giere 1999, 81.

78. Ziman 2000, 129–30.

79. From a mathematical perspective one can translate among map projections (whether geometrically or analytically expressed). However, if we take ontological assumptions or pragmatic conventions and practices into account, we are much less free in our choice among map projections.

For example, scientific knowledge can be captured with mathematical, diagrammatic, and narrative models, and with cartographic maps.[80]

To see how representational pluralism works, consider the difference between maps and diagrams, two kinds of representation deploying space differently. A typical, literal map uses space to represent and explain an external referent or world, often in metric fashion.[81] In diagrams, topological rather than metric relations are often what matter.[82] Distances among all points or all elements in a diagrammatic space can be loosely defined. Spatial relations of a diagram typically indicate logical or geometric relations or implications, or causal relationships. As the American philosopher Elisabeth Camp aptly puts the point, "Where pictorial and cartographic syntaxes use concrete spatial structure to represent concrete spatial structure . . . diagrammatic systems often use concrete spatial structure to represent highly abstract structure."[83]

Compared to sparse scientific diagrams, literal maps are typically multimodal. Maps communicate with language and symbols, including markings, typographic characters, and color. In their multimodality, maps can be closer to pictures or images than to scientific diagrams.[84] Diagrams are typically much more structurally abstract and less multimodal, though they need not be—diagrams in the fields of architecture and design, and in experimental scientific manuals, can be multimodal.[85]

---

80. For a general overview of representation pluralism across the sciences, see Frigg and Hartmann 2017; Winther 2016. For detailed discussions pertinent to the biological sciences, see Hull 1975; Beatty 1980; Griesemer 1990; Winther 2006c, 2011b, 2012b; Weisberg 2013; Leonelli 2016.

81. Personal or fictional maps, while not often metric, typically still directly refer to objects and properties of a hypothetical world, within which our spatial biological and phenomenological bodies move.

82. For instance, a metric is not required for Euler or Venn diagrams, which primarily represent logical relations among different sets and their respective elements (Shin, Lemon, and Mumma 2014).

83. Camp 2007, 159.

84. Greg Myers (1988, 239) articulates a spectrum from photographs to drawings, maps, graphs, models, and tables. Along this and other spectra, maps are often intermediary in their level of abstractness. Moreover, maps are not a single kind of thing (MacEachren 2004).

85. For instance, compare definitions of *diagram* and *map* in Bertin (1983) 2011, 193, 285. Camp (2007) argues that thinking and reasoning need not be linguistic and propositional but can also occur via diagrams or maps. Although Silvia De Toffoli and Valeria Giardino (2014) briefly analogize maps and diagrams, their main argument is that knot diagrams in topology are dynamic and that these ground knowledge-seeking actions.

We thus use maps and diagrams for different purposes and in distinct contexts.[86] In acknowledging this fact, however, the present study faces a *dilemma*: the more we distinguish maps from diagrams (or other representational forms), the smaller is the scope and power of the map analogy; but the more we see maps as similar—that is, analogous—to diagrams, the less interesting, important, or indispensable a cartographic analysis seems to become.[87] In short, by arguing for the uniqueness of maps, we elevate cartography; but we also increasingly narrow the source of the map analogy, making the map typology from chapter 2 more rigid and less effective.

### End-of-Inquiry Consistency

What is the ultimate relation among multiple representations? Must we be able to integrate representations? If not, must they at minimum be mutually consistent? What does the *end of inquiry*—to use a concept Charles Sanders Peirce developed—look like, if we could inquire so thoroughly as to reach consensus?[88]

On the one hand, you could argue that there will be an ultimate consistency among a plurality of representations. Giere states that "different perspectives on a single universe should, in principle, be compatible," drawing on the map analogy in defending the "one world" methodological rule: "Proceed as if the world has a single structure."[89] Additionally, Kitcher claims that "the representations that conform to nature (the true statements, the accurate maps, the models that fit parts of the world in various respects to various degrees) are jointly consistent."[90]

On the other hand, you could dissent and say that there cannot—or might not—be ultimate consistency. Consider the analyses of three philosophers of science: Nancy Cartwright speaks of "a dappled world, a

---

86. See Giaquinto 2007. Interestingly, there is significantly less published on the philosophy of diagrams outside of the formal sciences and mathematics. But see Perini 2005.

87. Silvia De Toffoli provided constructive feedback.

88. Charles S. Peirce: "The opinion which is fated to be ultimately agreed to by all who investigate, is what we mean by the truth, and the object represented in this opinion is the real" (1992c, 139).

89. For "different perspectives," see Giere 2006, 80; for the "one world" hypothesis, see Giere 1999, 82–83.

90. Kitcher 2002, 570. The more realist early Kitcher had written about "ideal explanatory texts" (1993, 106), but a later Kitcher argued that there was no "ideal atlas"—consistent representations, yes, complete and objective, no (2001, 82).

world rich in different things, with different natures, behaving in different ways"; she continues: "the laws that describe this world are a patchwork, not a pyramid."[91] "The complexity of the natural world is such that a single unified picture of the world is not possible" is how Helen Longino puts it.[92] Ian Hacking presents a whole taxonomy of senses in which one may posit either unity or disunity in the sciences.[93]

Perhaps there is an ultimate unity of representation (Giere and Kitcher). There will likely be a radical disunity (Longino and Cartwright).[94] In the end, integration platforms are sufficiently robust to encompass not only a plurality of maps, but a plurality of philosophical stances on end-of-inquiry consistency.

Map thinking contributes to developing integration platforms as a way to overcome or prevent pernicious reification. In turn, integration platforms encourage further, and deeper, map thinking, serving as loci for making progress on puzzling philosophical questions about pluralism and consistency.

## Conclusion

With Mercator's 1569 map and its associated projection as a connecting thread, I have explored the epistemic and political dangers of pernicious reification, as well as the longings of contextual objectivity. Sets of map projections provide one easy-to-understand example of integration platforms. Indeed, perhaps the main lesson of an integration platform is that, as communities and as single individuals, we must *be wary of utterly reifying the single, imperialist map, theory, or model most suffused with*

---

91. Cartwright 1999, 1. The title of part 3 of Cartwright's 1999 book makes the map analogy directly: "The Boundaries of Quantum and Classical Physics and the Territories They Share."

92. Longino 2002a, 142; cf. Longino 2002b.

93. See Hacking 1996. Hacking distinguishes among three kinds of commitment to unity or disunity. One of these is the metaphysical, another the methodological. In the first, arguments revolve around whether the world and phenomena are simple and unified, or complex and various. The methodological commitment's scientific and philosophical debates revolve around what the "one scientific method applying across the board in the natural and human sciences" actually is (43).

94. Alternatively, there may be a kind of "mosaic" unity, or an "integrative" disunity—such a complex middle ground is one of my burdens in this book, and is explored in S. D. Mitchell 2003; Craver 2007; Brigandt 2010.

*power*. I have tried to show this with the beyond-Mercator integration platform. Cartography has provided content, applied methods, and stories. Philosophy has guided our study of cartography. Let us now return to the quandaries and challenges surrounding the very nature of the representation relation, and what ontologizing is.

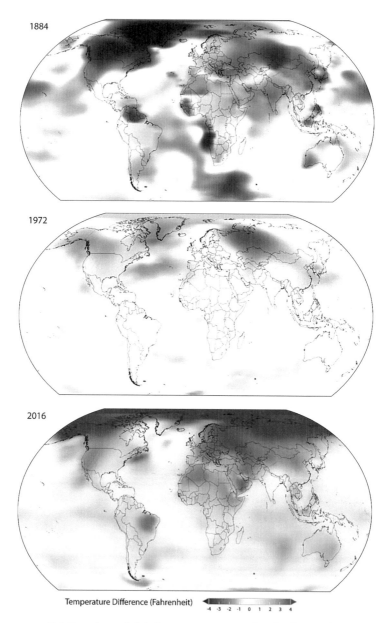

FIGURE 5.1. Snapshots of global temperature anomalies for three different years. An anomaly is a local deviation from the average (defined as the mean of all global temperatures, at every sampling point on the map, from 1951 to 1980). The Climate Time Machine on NASA's website shows that temperature patterns in the late nineteenth century were robustly consistent, despite Krakatau's eruption in 1883. Temperature changes in the most recent forty-four years massively swamp changes in the eighty-eight years prior. See plate 5 for the appropriate, contextually objective color experience. (NASA / Scientific Visualization Studio.)

# 5

# Projecting Maps into Our Worlds

Let us say you are strolling through the Louvre in Paris. For the first time in your life, your eyes fall on a delicately carved marble statue of a woman with short hair, bare chest, missing arm, and cloak-covered hips and legs. Aphrodite, the Venus de Milo, stuns you. But what is being represented? Is Aphrodite a mythological figure? Or is she a historical person? Does the sculpture exemplify Greek religion? Or are you, in your heart, representing the woman you love?

In 1992, the musician Prince released an album referred to, for convenience, as *Love Symbol*. What did the record title and artist name "⚤" represent? For the public, was it a stand-in for the artist, his music, his style? For ⚤, was the symbol primarily a fusing of the Venus and Mars symbols—signs of femininity and masculinity—thereby representing a fluid sexual identity?[1] Or does the symbol embody a plea for freedom from the recording industry?[2] "⚤" can represent the artist, his opus, his sexual identity, or the cultural history of African American oppression in the USA.

Whether reflecting on a classical sculpture or a pop star icon, it becomes clear that no simple representational relation to a pregiven world exists. So too for maps and scientific theories. Just like linguistic, artistic, and architectural representations, cartographic and scientific representations live within a complex, functional ecology of human use and purposes.

Despite the best efforts of philosophers, no single kind of representation-

---

1. Hooten 2016; Rhodes 2016.
2. The Artist Formerly Known as Prince (AFKAP) stated that he did not "own" his own music: "If you don't own your masters, the masters own you." Following the counsel of attorney Londell McMillan, who has represented many African American entertainers, the name change was a step toward independence, negotiating a way out of the contract with Warner Bros (Norment 1997, 130). AFKAP stopped being, in his own words, "a slave" (132).

world relation has been found and articulated.[3] As I show in this chapter, we are better off using many different analytical perspectives to think about representation and how it works. I explore three viewpoints in particular: *isomorphism, similarity,* and my own *multiple representations account.* Each of these philosophical accounts addresses only certain layers—or sets of assumptions—of representations: the metric, symbolic, and ontological layers, respectively. With the map analogy very much in mind, for each kind of layer I analogize from cartography to science.[4]

The previous two chapters form an argumentative arc. Chapter 3 investigated the basics of how cartographers abstract and map users ontologize—imagine and make worlds—so that you and I can map think, and thereby understand abstraction and ontologizing in science. Chapter 4 drew on the tools of integration platforms and assumption archaeology to discuss ways to attain contextually objective representations, while blocking or avoiding pernicious reification. Now we face the challenge of understanding how we comprehend the world via an abstraction, and how representing relates to ontologizing—to intervening and to experimenting.[5]

This chapter analyzes the variegated relations between representation on the one hand, and data or world on the other. As I hope to show, my multiple representations account ties together questions about world imagining and world making that have been our object in the previous two chapters, and that pertain to the nature of science and its role in society.

## Two Canonical Philosophical Accounts of Representation: Isomorphism and Similarity

In order to draw on the map analogy to illuminate representation relations, let us start with an example of cartographic representation.

---

3. Craig Callender and Jonathan Cohen (2006, 77) attempt to "dissolve" and "reframe" the philosophical discourse on representation with at least three positions: the various accounts of scientific representation on offer need not conflict; philosophical accounts of representation are actually very much about pragmatics; and we must explore fields outside of philosophy of science in order to understand representation. I concur.

4. My research interests include developing similar or analogous analyses of representation in art, music, and architecture, but doing so is beyond the scope of *When Maps Become the World.*

5. On intervening, see Hacking 1983; on experimenting, see Galison 1987, 1997.

San Francisco lures over twenty-five million tourists each year,[6] and for those with a mind turned toward science and the oceans, the wonderful tide pools that appear during the Pacific's low tide are a strong draw. Those lucky enough to have local friends have often been told to visit the ones at Pescadero State Beach, which is approximately a one-hour drive south of the city.

As with the start of many adventures, a logical first step is to consult a map in order to plot your trip. For variety and precision, let us say we check two computational maps. If you look at plate 6, you will see that the first image is Google Maps. It offers transportation routes and travel recommendations, showing you exactly how to get to Pescadero State Beach by way of Pacifica on Highway 1. The second is a dynamic, scalable, topographic map built by the California Department of Parks and Recreation (DPR) using the ArcGIS software developed by the Environmental Systems Research Institute (Esri). This map shows the terrain as well as the location and extent of natural preserves.

Consider these maps more broadly. What—and how—do map features such as contour lines represent? What about names, symbols, and colors? What might we learn when we apply insights about cartographic representation to other kinds of abstractions, such as scientific theories?

In this section, I focus on two key layers, or families of assumptions, regarding (1) metric properties and (2) symbols. These layers are amply exemplified in the two digital maps showing Pescadero. And philosophical analyses of isomorphism and similarity pertain, respectively, to these two representational layers.

## THE ISOMORPHISM ACCOUNT

The term *isomorphism* is a composite of the Greek words for "equal" and "form." In mathematics, isomorphism is a unique, one-to-one pairing of elements across two abstracted structures.[7] (For instance, we could map the set of all whole numbers from 1 to 100, labeling dots drawn on a piece of paper.) The isomorphism account of representation applies

---

6. Given a population of approximately 870,000, this amounts to roughly an additional 8 percent of the total native San Franciscan population on any given day.

7. In mathematical terms, this specific type of pairing is inversely symmetrical (*bijective*).

particularly well to the metric spatial layer of maps. This layer includes assumptions about map scale, map projection, and contour lines.[8]

As addressed in chapter 3, scale may be comprehended as a translation device. Scale transforms distances in the world to distances on a physical map, whether expansive and large, as in cosmological distances, or tight and small, as in molecular distances (fig. 3.2). Scale makes map space and world space isomorphic.[9]

As we also saw in chapters 3 and 4, map projections transform a curved 2-D surface onto a flat 2-D surface. Furthermore, in so doing, metric space *within* a single projection is distorted in a manner that can be represented with different scales. For instance, the linear scale of a Mercator projection along a given latitude becomes larger as we move away north or south from the equator. That is, we must multiply the equatorial scale by a *scale factor*, which is unique for each map projection. For the Mercator, the scale factor is a function of the secant of the latitude: $0° = 1$; $30° = 1.15$; $45° = 1.41$; $60° = 2.0$; $85° = 11.47$; $90° = \infty$.[10]

Third, contour lines on a map represent a horizontal slice of the world at a particular height. The curvier a contour line, the more meandering the terrain at that height. The closer together distinct contour lines are to one another, the steeper the terrain. And the more uniformly spaced the contour lines across many such lines, the more even the terrain, going up or down. As examples, think of landscapes of dunes or rolling hills, which can be fairly even and are represented with uniformly spaced contour lines. In contrast, mountainous regions on land and on the ocean bottom frequently consist of sudden cliffs and gorges, occasional plateaus, and overall dramatic terrain shifts, represented with dizzyingly changing spacings between contour lines.[11]

Mapmakers employ the values of the lines (sometimes notated as numbers) to depict exact elevations horizontally, and contour curviness and relative distance to visualize terrain shape vertically. Map users know that if contour lines touch on the map, they will find a cliff when they go hiking, and should bring specialized gear like climbing ropes and safety harnesses.[12]

---

8. Calibration — the first stage of abstraction, discussed in chapter 3 — is part of surveying, and is not a relation between representation and world (or data).

9. More technically, this isomorphism is achieved with the transformation equations characterizing the map projection (see chapter 3).

10. J. P. Snyder 1987, 45, table 7.

11. I explore the mapping of the ocean depths in Winther 2019b.

12. Imhof 2013, 132. Imhof's influential tour-de-force on topographic maps, *Cartographic Relief Presentation*, addresses contour lines at length, and mixes

Returning to the Pescadero maps, scale is implicit if not explicit in Google Maps, allowing you to calculate the distance from San Francisco to Pescadero, thereby permitting you to plan how long it will take you to get there. And the ArcGIS map shows contour lines, so you know where you will be walking along the flat terrain of marshes and where you will have to trek uphill.

Isomorphism in cartography thus includes 1:1 mappings of map distance to world distance; of spatial distortions in one map projection relative to distortions in another; of distinct scales in different parts of a single map projection; of generalized contour lines to imaginary, abstracted lines of equal elevation in the world.

Turning to map thinking, isomorphic relations between scientific representation and world can be analogized to isomorphism between cartographic representation and its terrain. Isomorphism is a crucial limit-case or ideal kind of scientific representation. Think of it as a gold standard guiding mathematical scientific theory toward success.[13] For instance, Newtonian theory predicts the motion of a cannonball through the air under idealized conditions. That prediction is a theoretical curve (or theoretical model). The experimental curve (or data model) falls a bit short, because conditions are not totally controlled and there is air friction. The best experimental curve is the one we get, roughly, by shooting many cannonballs at a particular angle with roughly the same amount of powder (etc.), plotting out the trajectories, and smoothing the composite data. Then we compare theoretical and experimental curves. We will see that the theoretical curve is "longer" and more arc-like than the experimental curve. They are not isomorphic. This forces us to rethink our theory—adding into our equation, for example, a factor accounting for air friction. An eventual perfect mapping between the theoretical model and the empirical "curve-fitted" model would be an isomorphism.[14]

---

design-driven heuristics and aesthetics with exquisite attention to precise topographic detail.

13. On the relationship between the isomorphism account of representation and mathematical reasoning, see van Fraassen 1989, 2008; Suárez 2003; Winther 2016, §3.3.3.

14. One energetic supporter of isomorphism as a way to understand scientific representation is Bas van Fraassen. For him, to interpret a theory is to establish isomorphic relations between theoretical models and empirical models, which themselves are isomorphic to the data, the observable phenomena. (For van Fraassen's distinctions, see chapter 3, note 64.) Strictly speaking, theoretical and empirical models are quite general categories used throughout science and need not necessarily be tied to an isomorphic account of representation.

[124]    CHAPTER FIVE

Whether in cartography or science—and perhaps in the case of anatomically plausible Venus de Milo—isomorphism is a key type of representation relation requiring explicit formal theoretical criteria and careful quantitative empirical measurements. Because of that, the isomorphic account is especially useful for metric spatial properties of maps and theories.

We now turn to other sets of assumptions of maps and theories.

## THE SIMILARITY ACCOUNT

Recall from chapter 2 that analogy works via shared properties. Maps and theories, projectiles and planets, icebergs and continents—each of these pairs shares important features. This partial identity across domains, fields, or cases authorizes us to make useful inferences from, for instance, what we know about maps to new and potentially informative properties of scientific theories.

What if this sort of analogical relationship holds not only between two concrete cases, or two abstract domains, but between world and representation? That is, what if the representation relation itself is one of analogy?

The similarity account of scientific representation argues that representation and whatever is being represented can be similar in any number of properties.[15] They are analogous. Overall similarity is the degree to which world (or data structure) and representation share features. Properties can be of many kinds: colors, icons, words, and mathematical symbols of the abstractions, and their respective meanings or referents in reality. An analysis of representation in terms of similarity is particularly instructive for grasping the *symbolic* layer of maps: assumptions about how different colors, icons, symbols, words, and so forth represent.

Starting with cartography, to understand similarity it is helpful to contrast inherent (*natural*) and stipulated (*conventional*) similarity. Cartography textbooks commonly appeal to this distinction: some symbols "*resemble* the feature attribute being symbolized," such that "we *naturally associate* [them] with feature attributes."[16] Such a symbol "'*mimes*' the

---

15. Hacking avers: "I am too brainwashed by philosophy to hold that things *in general* can be simply, or unqualifiedly alike. They must be like or unlike in this or that respect" (1983, 137).

16. Robinson, Morrison, et al. 1995, 479; emphasis added. Furthermore, Muehrcke and Muehrcke write: "The goal is to produce symbols map users will recognize through natural association" (1998, 81). Slocum and colleagues argue

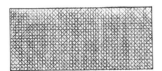
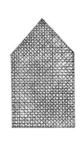
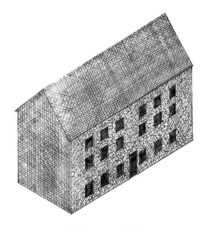

1. Geometric symbol     2. Mimetic icon     3. Pictographic icon

FIGURE 5.2. Three manners of portraying a house on a map: (1) geometric symbol; (2) mimetic icon; and (3) pictographic icon. Panel 1, especially, can be resized appropriately if the mapmaker wishes the relative area to be representative of the territory. (Illustration by Heidi Svenningsen Kajita.)

thing it represents."[17] On a map, a small picture of your home—or even an icon with house-like properties, perhaps showing various windows, floors, and so on—represents your actual home via inherent similarity (fig. 5.2, panel 3). In contrast, an abstract geometric square represents your house by stipulated similarity (fig. 5.2, panel 1).

A smaller group of cartographers collapses the distinction by proposing that all map icons are irreducibly, even fully, stipulated. Denis Wood argues that "there are no self-explanatory signs; no signs that so resemble their referents as to self-evidently refer to them."[18] In other words, symbols are suffused with convention, such that representation must be a communally shared, intended similarity.[19]

Another approach is to multiply distinctions, identifying further types of signs. Charles Sanders Peirce, whom we met in chapter 2, is perhaps best known for his seminal work in the field of semiotics, including his

---

that a symbol is "*intended to look like* the phenomenon being mapped" (Slocum et al. 2005, 63; emphasis added).

17. Kimerling et al. 2009, 133; emphasis added. The choice of such icons is not always obvious: "What, for example, is an obvious icon for a vista or an overlook? And what if there are 10 such cryptic symbols on your map?" (133).

18. Wood 1992b, 110; 2010, 80.

19. See Wood and Fels 1986; Krygier and Wood 2011.

three-way distinction among symbols, icons, and indices.[20] On paper or screen, the first two are particularly relevant.[21] Map icons are inherently similar to their target (e.g., mimetic [fig. 5.2, panel 2] or even pictographic [fig. 5.2, panel 3]), and readily evoke their meaning.[22] Map symbols—narrowly construed[23]—are arbitrary in that they are conventional and have been agreed upon, lacking an inherent, natural relationship to their targets in the world (e.g., the geometric symbol [fig. 5.2, panel 1]). However, that does not eliminate their informative value once meaning has become entrenched and can be communicated.

We establish symbol–world analogy by appealing to features—natural, stipulated, or both—that are shared between the symbolic layer and the world. We do this whether we are traditional cartographers, less orthodox mappers eliding the inherent/stipulated distinction, or map thinkers in Peirce's triadic tradition.

Returning to the Pescadero maps (plate 6), symbols in Google Maps include color codes and numbers for highways, and icons for accidents and roadwork. And the ArcGIS map uses color to indicate different kinds of California state parks, such as natural preserves or cultural preserves, or to show underwater portions of a state park.

Now, map think to try to understand scientific representation as similarity. The large, three-dimensional metal model that Francis Crick and James Watson built at University of Cambridge in 1953 was a physical model, and, I believe, a material pictographic icon of the structure of DNA. Two strands of alternating deoxyribose sugars and phosphate groups form a double-helical backbone. Inside this backbone, one of four kinds of bases is connected to each sugar. The bases pair in just two of ten possible ways: adenine with thymine, and guanine with cytosine. The double helix makes a full turn approximately every ten base pairs.[24] In a

---

20. For one place among many where Peirce addresses this three-way distinction in detail, see Peirce 1998a.

21. For a "primer on semiotics," see MacEachren 2004, chap. 5.

22. Certainly many people would recognize panels 2 and 3 of figure 5.2 as house icons. Other symbols begin their lives by being similar, and then evolve with usage. For example, on a rectangular smartphone, the icon for a telephone is a C-shaped device no one uses anymore. Children today learn these symbols without knowing where they originate. Ben Bronner and Lucas McGranahan provided constructive feedback.

23. In semiotics, *sign* tends to be the more general term and concept, covering symbols, icons, and indices. In cartography, *symbol* is taken to cover especially symbols and icons.

24. Further details can be found in chapter 8.

PLATE 1. *Earthsea Map*, by Ursula Le Guin. (Copyright © 2012 by The Inter Vivos Trust for the Le Guin Children. First appeared in *A Wizard of Earthsea*, published by Houghton Mifflin in 1968, and reprinted in 2012 by HMH. Reprinted by permission of Curtis Brown, Ltd.)

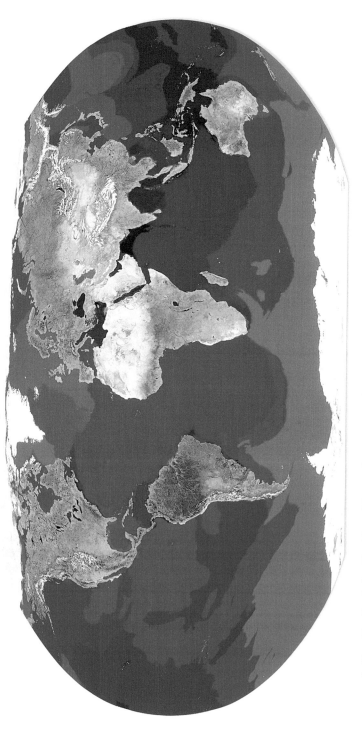

PLATE 2. Tom Van Sant's *Satellite Map of the Earth* attempts to depict the "real world"—not the every-country-a-different-color of traditional schoolroom wall maps. (Tom Van Sant / Geosphere Project, Santa Monica; courtesy of Science Source.)

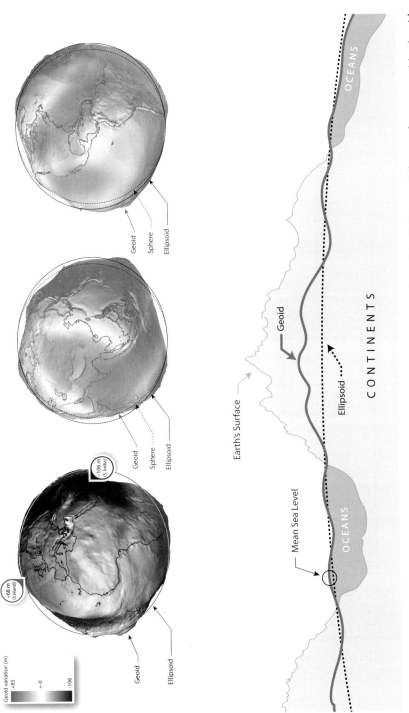

PLATE 3. The geoid, ellipsoid, and sphere, and their respective relation to mean sea level. *Top left*: The biggest differences between positive (geoid extending above the ellipsoid, in Iceland) and negative (geoid below, in southern India) are shown. *Top middle and right*: The conceptual differences among the geoid, ellipsoid, and sphere. *Bottom*: One way to define *mean sea level* is as the averaged, idealized ocean surface. (Earth Gravitational Model [EGM2008] released by the National Geospatial-Intelligence Agency [NGA]; illustration by Mats Wedin and Rasmus Grønfeldt Winther.)

WESTERN UNITED STATES 1:250,000

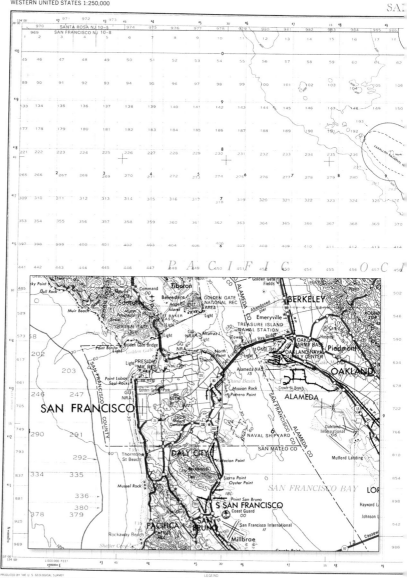

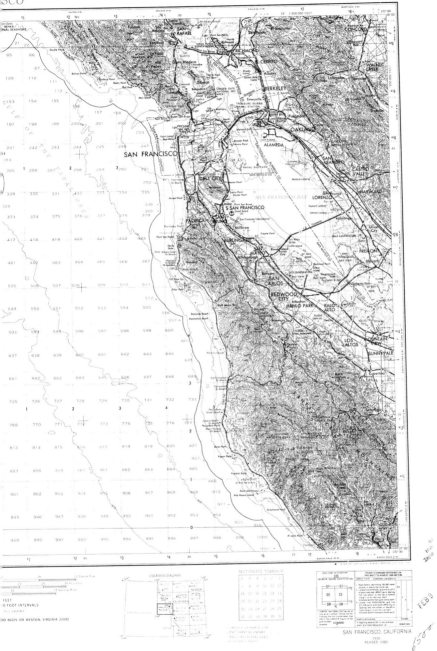

PLATE 4. San Francisco and parts of Berkeley, Oakland, Marin County, and Pacifica appear in a cropped section of USGS topographic map NJ 10-8, 1956; revised 1980. (US Geological Survey, Department of the Interior.)

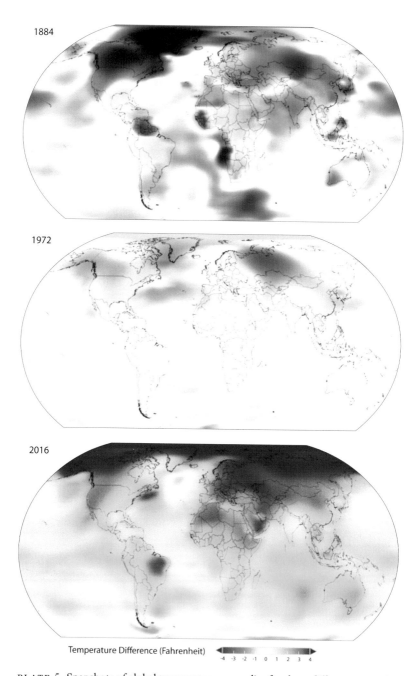

PLATE 5. Snapshots of global temperature anomalies for three different years. An anomaly is a local deviation from the average (defined as the mean of all global temperatures, at every sampling point on the map, from 1951 to 1980). The Climate Time Machine on NASA's website shows that temperature patterns in the late nineteenth century were robustly consistent, despite Krakatau's eruption in 1883. Temperature changes in the most recent forty-four years massively swamp changes in the eighty-eight years prior. (NASA / Scientific Visualization Studio.)

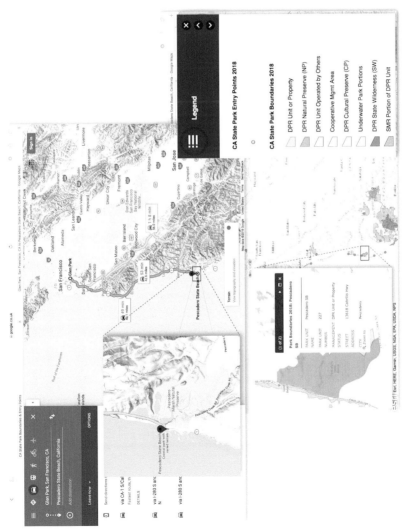

PLATE 6. Two digital, computational maps with different purposes. *Top left, with insets*: Google Maps shows how to get from Glen Park in San Francisco to Pescadero State Beach. (Map data © 2018 Google.) *Bottom right, with insets*: A California Department of Parks and Recreation map, powered by Esri's ArcGIS software platform, shows the natural terrain at the park. (Map data by Google, © Esri, courtesy of State of California, Department of Parks and Recreation.)

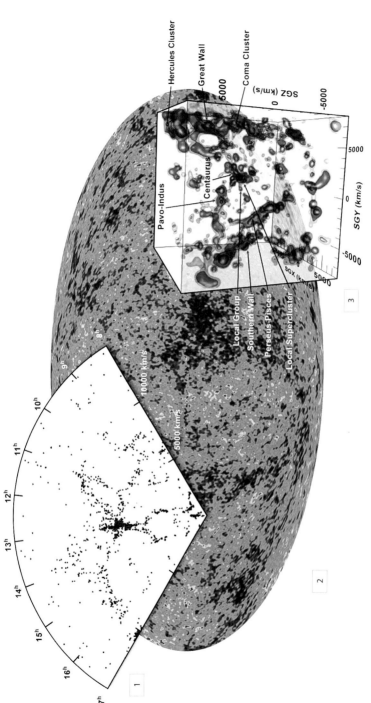

PLATE 7. Three extreme-scale maps in cosmology. (1) An intensely clustered set of galaxies resembling a stick figure was revealed in 1986. This map contains more than a thousand galaxies. (2) The Wilkinson Microwave Anisotropy Probe (WMAP) nine-year map shows the cosmic microwave background (CMB) in Mollweide projection. (3) A 3-D map of the local universe, at a scale similar to that of inset 1 (the head and upper body is the Coma cluster; the stretched arms correspond to the Great Wall; interpretation by Courtois, pers. comm., January 2019). *SGX*, *SGY*, and *SGZ* indicate supergalactic coordinates (*SG*). (Inset 1: De Lapparent, Geller, and Huchra 1986. © AAS. Reproduced with permission. Inset 2: NASA / WMAP Science Team. Inset 3: Courtois et al. 2013, 69. © AAS. Reproduced with permission.)

PLATE 8. Pablo Carlos Budassi's *Observable Universe Logarithmic Map* (sketch level). From left to right, remarkable celestial bodies are arranged according to their proximity to Earth. Objects depicted include local planets, quasars, galaxies, and the CMB. Astronomical objects appear at enlarged size to show their shape. The depiction is logarithmic along the x-axis, with distance ten times larger every 4–5 percent of the total horizontal extension of the map. (Image created by Pablo Carlos Budassi, 2018. His full universe map can also be seen at https://www.facebook.com/pablocarlosbudassi.)

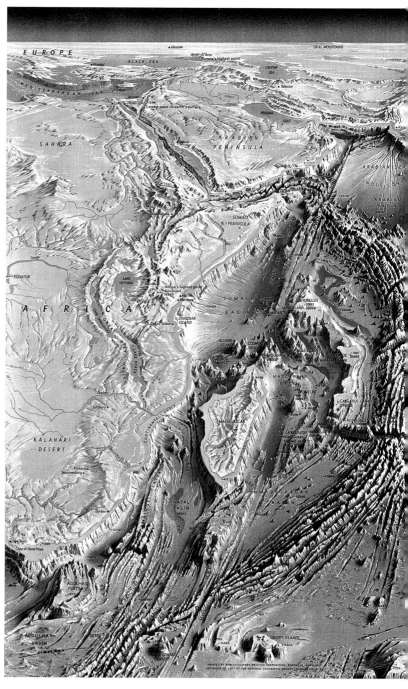

PLATE 9. The *Indian Ocean Floor* panorama by Heinrich Berann, Marie Tharp, and Bruce Heezen was a foldout map in the October 1967 issue of *National Geographic*. Subscriptions to that magazine numbered six mil-

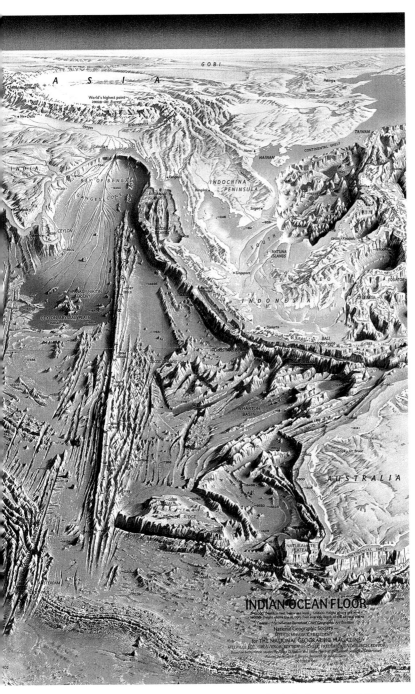

lion in the USA alone. This panorama shows both the lowest and highest points on Earth's land surface. (Heinrich Berann / National Geographic Creative / National Geographic Image Collection.)

PLATE 10. In *Model Organisms*, genetics, physiology, sex, and life intertwine. Biological researchers use knowledge of zebrafish, pigs, *E. coli* bacteria, the nematode worm, *Arabidopsis thaliana*, mice, yeast, and *Drosophila* to make inferences about the genetics, physiology, and development of related species. The Mercator map is an idealized abstraction, and a cropped photograph of the head and shoulders of the 1972 *Playboy* centerfold model Lena Söderberg is a standard "test image" for many computer science visual compression algorithms (see http://lenna.org/). Her hair is Mendel's peas. *Model Organisms* does not perniciously reify genetics: as much as possible is shown of the structures in which epigenetic and environmental interactions occur. (Illustration by Larisa DePalma; concept by Rasmus Grønfeldt Winther.)

significant sense, Crick and Watson's material representation was the first accurate icon of real, tiny DNA molecules.

Mathematical symbols carry a different kind of similarity, as they do not—indeed, cannot—straightforwardly look like what they represent. A formal theory or model of predator and prey populations (of, say, lynx and hares in the wild) represents properties such as population size, and potentially other quantifiable properties in the world.[25] It remains unclear whether the isomorphism account or the similarity account provides the best picture of how mathematical models represent in science, but isomorphism is more stringent and seems closer to scientific practice, even if similarity also has conceptual resources for accommodating formal symbolic representation.

Whether in cartography or science, and probably in the case of , similarity is a key type of representation relation establishing shared features, and it is a kind of analogy between representation (and its parts: colors, icons, and symbols, for example) and world or empirical structure. Similarity is especially useful for symbolic properties of maps and theories.

The metric and symbolic layers of maps are related to another layer thus far set aside: the *ontological layer*. While the first two are concerned with assumptions about how to interpret and present the world in the map itself, the ontological layer consists of assumptions about the nature of reality. It is thus only through the imposition of *this* set of map assumptions that we ontologize—that is, that we test representations, change the world, understand the world, and teach knowledge (see chapter 3).

## The Multiple Representations Account

If isomorphism and similarity relate primarily to how a representation matches or fits the world (or the data), my multiple representations ac-

---

25. Drawing on the work of the Nobel laureate and cognitive psychologist Amos Tversky, the American philosopher Michael Weisberg (2013, 145) proposes that a representation and target are similar when they share important features. He specifies degrees and respects with which representation "fits" the world target (Weisberg 2013, 148, equation 8.10). His account of similarity is a kind of analogy. It resonates with Ronald Giere's use-centered similarity account: "Anything is similar to anything else in countless respects, but not anything represents anything else. It is not the model that is doing the representing; it is the scientist using the model who is doing the representing" (2004, 747). For further details on scientific representation as similarity, see Winther 2016 and references therein.

count explores the process of how map *becomes* world. It relates to the ontological layer.

Mapmakers *perspectivize* reality. That is, they impose a holistic, consistent perspective to make sense of a complex and finicky world. This perspectivizing helps produce (and is guided by) a map's ontological layer, which is its set of assumptions about the world's objects, properties, and processes. The ontological layer coded into the map by the maker eventually becomes the layer of *use* and *application*. This layer describes—and provides prescriptions for—the potential world (or worlds) that the map and its users interpret and see; structure and build; and then live and love in. For example, a visitor to Pescadero State Beach who spends hours looking into tide pools and watching the tide change will have a different experience than a person who hikes the beach and examines the fauna there, in part because both think and act according to different maps.[26]

According to the multiple representations analysis, the first stage or moment is *ontologizing*, when we make the abstraction the world. The second moment is *merely-seeing-as*, when we are aware that our abstraction is not the world. Practice enriches theory in the third, synthetic moment of *pluralistic ontologizing*. Here a scientific community or the lay public uses various representations to measure, change, or understand the world, with the ability to test different representations for these purposes side by side.[27]

Pernicious reification is being stuck at the first or second stage. To ontologize without moving toward merely-seeing-as is to reify one's abstraction perniciously. One way to push past this sticking point is through the use of assumption archaeology, which may denaturalize the assumptions underlying the abstraction. And moving from merely-seeing-as to pluralistic ontologizing can be achieved using integration platforms.

The ultimate goal is the higher, communal synthesis of contextual objectivity: not only is your or my representation one among many possible ones, but ultimately your or my world is one among many mapped worlds.

---

26. My analysis in this chapter concerns primarily map users and the process of ontologizing. In chapter 3, we saw how mapmakers perspectivize and create (and simultaneously use) a map ontology via abstraction practices.

27. My multiple representations account is one possibility among many for analyzing and understanding the ontological layer of representations.

ONTOLOGIZING

Ontologizing occurs when an individual takes the internal world of a map or scientific representation to *really be* the world it intends to represent. We saw this in great detail in the second part of chapter 3, though there I set aside whether the ontologizing was ultimately generative or pernicious. Ontologizing is the expression of a map's *cartopower*, or its power to create a world, telling us how to think about the world, how we should act, and what we can hope for. We think to ourselves, "Oh, this is how I should think about and imagine Copenhagen, San Francisco, Mexico City, Caracas, Paris, the Mid-Atlantic Ridge, or Raja Ampat, Indonesia — or Earth — and act with respect to it." Cartopower is the representation's intrinsic plea, seduction, or argument, via the operation of the ontological assumptions of the ontological layer, to interpret the world one certain way, and only that way.[28]

A map's cartopower can lead to its pernicious reification, if we only engage in, and stay with, the cartopower it exudes. Let us look more deeply at how ontologizing works in cartographic and scientific contexts.

Few political, social, and geographic tensions evoke passions as strongly as the Israel-Palestine (Israel-Arab?) conflict. Indeed, words break down even in the attempt to describe this historical complex. Meanings of *territory, nation, homeland, colony, occupation, citizens, settlers, expatriate, refugees, Israeli, Arab,* and *Palestinian* are unstable, politically contextual, and historically situational.[29] Multiple violently opposed perspectives abound, many of which seem to wish to destroy the others, or at least act as if they do. External government powers, such as the USA and the European Union, or neighboring Arab states, tend to adopt alliances with warring sides. Of course, many people on all sides genuinely want peace.

Upon the crumbling of the Ottoman Empire in the aftermath of World War I, the legal status known as the British Mandate for Palestine took over Palestine and Jordan (known then as Transjordan), with France given the League of Nations Mandate control of Syria and Lebanon. Palestine was to be a "national home for the Jewish people," as stated in the Balfour Declaration, contained in a one-page letter from the British foreign secretary to British banker and Zionist Lord Rothschild:

---

28. Wood 1992b and Harley 2001a are essentially disquisitions on cartopower.
29. I shall return to this point in addressing migration maps in chapter 7.

His Majesty's government view with favour the establishment in Palestine of a national home for the Jewish people, and will use their best endeavours to facilitate the achievement of this object, it being clearly understood that nothing shall be done which may prejudice the civil and religious rights of existing non-Jewish communities in Palestine, or the rights and political status enjoyed by Jews in any other country.[30]

Note the explicit statement of no harm intended toward "non-Jewish communities in Palestine" or "Jews in any other country." Much of this text survived editing and redrafting, making it into the official League of Nations (LN) Mandate for Palestine as the second paragraph of the preamble.[31] The British officially took control of Mandatory Palestine on September 29, 1923.

Mandatory Palestine did not survive long after World War II, because in 1947 the British proclaimed their intent to give up their LN mandate, referring the situation to a committee of the newly minted United Nations, the UN Special Committee on Palestine.[32] The UN was the successor organization to the LN, with a much greater membership of nations, but with an organizational inheritance from the LN.

Panel 3 of figure 5.3 shows the situation in 1923. In all of this, there was a heated discussion within both the LN and British government about whether Transjordan was part of Palestine. Ultimately, article 25 of the Mandatory Palestine legal instrument set the framework for not deeming Transjordan part of the "Jewish national home" (article 2), as long as freedom of religion, education, movement, and taxation were respected (articles 15, 16, and 18).[33]

Maps become the world, through their cartopower. Panel 1 shows the "Palestine Plan of Partition" adopted by the UN General Assembly on November 29, 1947, as annex A to resolution 181 (II).[34] It effectively became a plan of parturition that gave birth to the State of Israel. This map recommended separate Arab and Jewish states, with Jerusalem controlled by an international coalition. Perhaps unsurprisingly, Western powers and the USSR voted overwhelmingly for resolution 181, as did Jewish authorities—Zionist and non-Zionist alike—while Arab League countries, India, Iran, and Pakistan voted against it, with the UK

---

30. Balfour 1917.
31. Council of the League of Nations 1922, article 25.
32. Newsom 2001, 77.
33. See also Bertram 2011, 265.
34. United Nations General Assembly 1947.

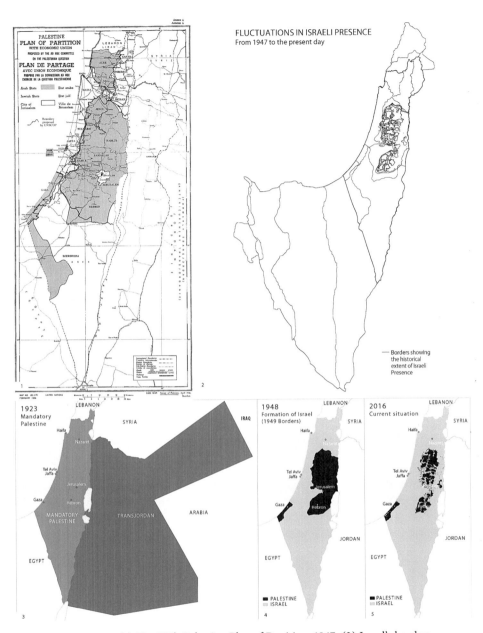

FIGURE 5.3. (1) The UN's Palestine Plan of Partition, 1947. (2) Israel's borders as they have expanded and contracted since statehood. (3) Mandatory Palestine. (4) Israel according to the 1949 Armistice Agreements (Independence declared in 1948). (5) Israel and Palestine in 2016, where the darker shade in the West Bank shows areas "A" and "B," according to the Oslo II Accord (1995). (Map 2 by Malkit Shoshan [2012, 40] with Vital Tauz; reprinted courtesy of Malkit Shoshan. Maps 3, 4, and 5 by Mats Wedin.)

[132]  CHAPTER FIVE

and China, among others, abstaining.³⁵ To quote the sociologist Baruch Kimmerling: "As the British umbrella was removed, the Arab and the Jewish communities found themselves face-to-face in a zero-sum-like situation."³⁶ On May 14, 1948—the day before the British Mandate was to end—David Ben-Gurion announced the "Declaration of the Establishment of the State of Israel."³⁷ The next day, Egyptian, Jordanian, and Syrian forces entered territory that was technically Israeli, and the ten-month-long Arab–Israeli War ensued, at the end of which Israel had established itself as a nation, and controlled significantly more land than indicated in panel 1 of figure 5.3 (compare with panel 4).³⁸ In the aftermath of this war, there were seven hundred thousand or so Palestinian refugees in Israel, and hundreds of thousands of Jewish refugees in Arab countries, many of whom fled to Israel.³⁹ While the ontological layer of this Plan of Partition for Palestine acted as a kind of birth certificate of Israel drawn in blood, it was actually a draft—suffused with cartopower, and used for ontologizing—of a territorial vision of a new, Jewish country.

As a further cartographic exemplar of ontologizing, consider Denis Wood's insistence (which I share) that every map teems with ontological assumptions and world-making capacities, which lead to pernicious reification if solely focused upon. In *Rethinking the Power of Maps*, Wood declaims: "Mapmakers are not cognitive agents parachuted into a pregiven world with a chain and a theodolite, to measure and record what they find there. Rather, they're extraordinarily selective creators of a world—not *the* world, but *a* world—whose features they bring into being with a map."⁴⁰ Wood's observation resonates with the philosopher Nelson

---

35. United Nations Committee on the Exercise of the Inalienable Rights of the Palestinian People 1979. For a map of how UN member countries voted, see UNGA 181 Map.png on Wikimedia Commons: https://commons.wikimedia.org/wiki/File:UNGA_181_Map.png.

36. Kimmerling 2004.

37. Jewish People's Council 1948.

38. For a perspective partial to Palestinians, see Palumbo 1991.

39. To this day, most Muslim or Arabic countries, except Egypt and Jordan, do not officially recognize Israeli passports or the existence of Israel as a state. For a map showing the various categories of recognition, see https://en.wikipedia.org/wiki/International_recognition_of_Israel#/media/File:CountriesRecognizing Israel2018.svg.

40. Wood 2010, 51; footnotes suppressed. Wood presents one way to analyze cartopower: the *posting*, which he considers the basic enactive unit of the map. A posting "asserts an equivalence between an instantiation of some conceptual type

Goodman's statement that "if worlds are as much made as found, so also knowing is as much remaking as reporting."[41]

Turning now to the scientific domain, consider the cartopower of two influential theories of disease: humoral theory and germ theory.

Galen's well-known second-century CE theory, with roots in Hippocrates and other Greeks, explains disease as an imbalance of the four humors (blood, phlegm, black bile, and yellow bile). In the case of epidemics, users of Galen's theory combined the humoral theory with miasma ("pollution," "stain," "that which defiles"[42]) theory, providing "a kind of seductive universal model" for disease.[43] Through the complex ontology of the theory, humors were linked to stages of human life and, famously, to temperaments such as sanguine and phlegmatic. Galen's theory also guided and justified actions such as bloodletting and, for avoiding epidemics, cleaning city waste or building cities far from swamps, lest malaria ("bad air") ensue. This was the standard theory of disease in Western civilization for approximately sixteen centuries.

The narrative changed dramatically with the development of the germ theory of disease from 1860 to 1880 by Robert Koch and Louis Pasteur. The world became populated with a plentitude of pathogenic microorganisms. The germ theory of disease influenced further research and received support, for instance, from the Parisian physician Alphonse Laveran's discovery, while working in Algeria, of the malarial parasite *Plasmodium falciparum* in 1880 and the English doctor Ronald Ross's discovery, while working in India (1881–1899), of the mosquito as malarial vector. By the close of the nineteenth century, "the discoveries of Pasteur and Koch had precipitated a search for a bacterial cause for many diseases."[44]

A new map of disease influenced science and public health. The case of malaria shows how germ theory exerted a normative representational power, analogous to a map's cartopower, critiquing and mostly replacing

---

(a *this*) and a specific location in the world (a *there*)." Wood 2010, 53; cf. Wood and Fels 2008, 28. Every map represents a selective world, populated with the kinds, laws, objects, and processes imbued with visibility and relevance by the mapmakers, and located by them on the map. A fuller examination of the logic of postings would have to turn to the field of semiotics—for example, the work of C. S. Peirce, Louis Hjelmslev, and Roland Barthes.

41. Goodman 1978, 22.

42. The Online Lidell-Scott-Jones Greek-English Lexicon (http://stephanus.tlg.uci.edu/lsj/; accessed September 12, 2018), s.v. "miasma."

43. Byrne 2004, 141.

44. Cox 2010, 2.

Galen's earlier theory. The world changed. In both cases, map or theory ontology became world ontology.⁴⁵ Even so, the terrain has not settled, and ongoing pluralism matters, as we shall see in the next sections. While Galen's theory has been rightly discarded, biomedicine continues to appeal to a combination of internal physiological dynamics and environmental forces. We should not solely ontologize the germ theory of disease. Its success notwithstanding, microorganisms should not be perniciously reified as the sole causative agents of pathologies. Environmental factors such as pesticides or lead, as well as genetic predispositions and social stresses, are also involved in disease etiology.

## MERELY-SEEING-AS

In dialectical contrast to ontologizing, a user "merely-sees-as" when he or she comprehends that the content of the map, model, or theory is one limited way of viewing, and acting, in the world. The user does not take the representation *to be* the world.

Recall the cartographic cliché "The map is not the territory" (MINT), put forth by Alfred Korzybski and Gregory Bateson.⁴⁶ MINT teaches us that, by comparing disparate representations without taking any one of them too seriously, we get a better idea of the territory. One influential cartographer, Phillip Muehrcke, has consistently emphasized MINT in critiquing the pernicious reification of maps.⁴⁷ He and Juliana Muehrcke put it clearly in their *Map Use*: "We began this book with the statement that maps mirror the world.... A mirror is a useful tool, but it shows only a piece of reality. No one would confuse its reflection with the real thing. Yet a surprising number of people treat maps' reflection of the world as if it were reality."⁴⁸

For Muehrcke, two forms of map pernicious reification—or failing to merely-see-as—are *ignoring external complexity* and *exaggerating internal simplicity*.

First, maps necessarily omit complexity: "Elusive but important features such as the touches, sounds, smells, linkages between people and the natural realms (biosphere, lithosphere, atmosphere) and, in general,

---

45. For examples of linking mathematical model ontologies and world ontologies in population genetics, ecology, and developmental biology, see Winther 2006a, 2006b, 2011b, 2014b, 2015a, 2016; J. M. Kaplan and Winther 2013; Winther and J. M. Kaplan 2013.

46. See chapter 2, note 40.

47. See Muehrcke 1972, 1974a, 1974b; Muehrcke and Muehrcke 1998.

48. Muehrcke and Muehrcke 1998, 523–24.

the vivid, meaningful experiences of life seldom find overt cartographic expression."[49] The problem is that a map exercises a kind of existential authority. We deem whatever has been eliminated from the map as less real, even nonexistent.[50] Map users are human: fallible, selective, and biased. They yearn to take and see the world to be a certain, very specific way.[51] Representational simplicity helps us understand reality, but also leaves much of the world's richness and complexity unmapped. We should not "ask a map to be the same as reality; if it were, it would lose its unique clarifying function."[52] Users should tirelessly and willfully remember external complexity, not representable on the map.

Second, the purely internal characteristics of maps are often overblown: a "stylized mountain symbol is not meant to imply that the mountain actually will look like that to the hiker."[53] By overinterpreting map symbols, we universalize the map from the inside. We may then ontologize this overinterpretation of the map, making it real by living and acting according to it.

As we saw in chapter 1, Mark Twain satirizes Huck Finn's overinterpretation of the map he was taught—Indiana as pink.[54] Muehrcke calls this the "error of map reading."[55] The biologist Richard Levins has articulated similar ideas about scientific models: "Individual models, while they are essential for understanding reality, should not be confused with that reality itself."[56] In all of this, the insistence is on sharply distinguishing map from territory, abstraction from world.

PLURALISTIC ONTOLOGIZING

In the third stage of my account, we realize that our worldview is just one of many worldviews. We *communally*, *deliberately*, and *mindfully* respect

---

49. Muehrcke 1974a, 22.
50. Recall this as William James's worry in chapter 2.
51. Muehrcke and Muehrcke 1998, 533.
52. Muehrcke and Muehrcke 1998, 534.
53. Muehrcke 1974a, 17.
54. Gieryn 1999, vii, also discusses this passage from *Tom Sawyer Abroad*.
55. Muehrcke 1974a, 17; cf. 1974b, 37. Muehrcke refers us to the pragmatist philosopher Abraham Kaplan, who diagnoses various "shortcomings of models" (A. Kaplan 1964, 275–88). Kaplan drew inspiration from the philosopher of science Ernest Nagel, who had argued that "a model may be a potential intellectual trap as well as an invaluable intellectual tool" (Nagel 1961, 115).
56. Levins 1966, 431.

multiple representations, including those that might be unfamiliar, or those that we take for granted since they have been handed down to us from authority figures. Assumption archaeology or integration platforms are particularly useful for this stage.

Return to Richard Edes Harrison, from the last two chapters. "Our hope," his introduction says, "is that the maps contained in this atlas, more especially the perspective maps, will help free many readers from the geometrical and psychological shackles of conventional maps . . . [which] contain many of the conventional symbols, generalizations, and exaggerations common to the maps to which we are accustomed."[57] Harrison's World War II atlas illustrated geopolitics and perhaps helped persuade the American public to support a greater internationalism that changed their world, and *the* world (Ontologizing I, from chapter 3). Learning about the wealth of alternatives to the Mercator projection allows us to apply Mercator mindfully, understanding the world (Ontologizing II) through Mercator's eyes and recognizing that his purpose was to improve ocean navigation by representing the world conformally, rather than to ensure European domination. And I could imagine a high school or university lesson (Ontologizing III) comparing Mercator's sea-world conformal point of view with Gall–Peters's 1970s and '80s anti-colonialist, equal-area perspective, as well as with the World War II global point of view made flesh with the UN flag. Powerful student dialogue of the role and nature of differing perspectives and representations could ensue. Such pluralistic ontologizing of cartographic representations can be analogously done on scientific representations—also via integration platforms—as we shall see below and in chapter 8.

Returning now to figure 5.3, we can see how multiple maps can serve as integration platforms for critical dialogue and political negotiations about, for example, two-state versus one-state solutions, as it were, in their various forms. Pluralistic ontologizing can take place. The integration platform of figure 5.3 can perhaps be used for discussion rather than for war. Panel 2 of that figure shows the dynamic, diachronic borders of Israel over the ensuing years, including territory gained as a consequence of the 1967 Six-Day War, as well as ongoing conflict in the West Bank (fig. 5.3, panel 5).

Panels 3, 4, and 5 illustrate the extent of Israeli boundaries at different times. Of course the historical, military, economic, political, religious, and cultural contexts are complex and contentious. Narratives—official, scholarly, and vernacular—of the history are many. But museums and school programs could be engaged to show the multiple perspectives of

57. Harrison 1944b, 12.

these maps, perhaps trying from a higher point of view to dispel hatred-filled enemy pictures of opposing sides. As I have urged elsewhere, "A growing repository of centralized world-navel maps could be a most useful exercise for critical diplomatic, pedagogical and museological reflection."[58] Each map needs to be accepted and negotiated, so as not to pay homage to the destruction of opposing macrocosms.

Suffused with empirical precision and sociocultural cartopower, single maps can become the world, but can also be part of generative integration platforms working toward a world in which we learn alternatives to war, dominion, religious sectarianism, and the profit-motives of high-level organizations such as the military and oligarchic corporations.

Finally, let us return to plate 6. The maps depict the territory in different ways. DPR's ArcGIS map views the world in terms of California state park "entry points" and "boundaries." Objects in shades of green are simultaneously a perspective on the world and the world itself. As you stare at that DPR map on your screen at your friend's house, in anticipation of the tide pools, you imagine the coast around Pescadero State Beach as an existing object with which you can, and will, interact. Later, as you drive and see the curves and feel the winding road that your companion follows on a smartphone, you live the Google map *as the world*. Thus the two maps become the world. Success here is navigating the world appropriately, according to a given map. Unlike the similarity and isomorphism accounts, only in the multiple representations analysis do representational success and failure emerge as concrete and embodied interaction between users of representations and their world.

Examining cartopower analyzes how maps become the world through use and planning. MINT warns of the error of identifying representation with world, and thus demands distinguishing map from territory. The wise, contextual use of maps amounts to seeing the world as populated by efficient highway routes (Google Maps) or nature reserves (DPR), among many other objects and processes. Importantly, this synthetic step of pluralistic ontologizing is accepting that any one of these worlds is just that: *one among many worlds*. The reality they represent contains nature reserves, highways, and so much more.

In all of this, countermapping—the introduction of alternative, minority representations, explored in chapters 1, 7, and 8 of this book—also helps pluralistically ontologize dominant maps, theories, or models. Countermapping brings to light a plethora of alternative maps, each with its own ontology of, say, Israel or Palestine, European or indigenous Americas (Waldseemüller versus Poma), or a colonialist or noncolonial-

---

58. Winther 2019a, 10.

ist Africa.⁵⁹ Even a single countermap heightens awareness, as Poma's map did. It can now be brought into an integration platform, or invite the assumption archaeology of dominant maps. Bringing more worlds into existence—even within a family of perspectives, or from dramatically outsider, minority perspectives—grants map users more alternatives in their choice of world to live in and leave to posterity. This may require deliberation and heated argument; hopefully, though, it ultimately brings negotiation and mutual acceptance.⁶⁰ When many partially accurate maps with unique perspectives, scales, or aims are compared and contrasted, the interplay of their cartopower blocks or overcomes the conflation of any single map with the world.

In short, map thinking thus invites a kind of cognitive doublethink: even as we ontologize a representation, we can also take it to be merely a view and an angle on the world—a seeing-as. *Homo cartograficus* expresses an urge to represent and then act accordingly. The doublethink of ontologizing and merely-seeing-as can be generative, when it leads to pluralistic ontologizing and contextual objectivity. Such progress can be helped along with integration platforms and assumption archaeology. Pernicious reification occurs when we *fail* to engage in the second or third stages of my multiple representations account, and think of our single represented world as the only world possible. And once we have taken the third step, the whole process can repeat.

## CLIMATE CHANGE AND MULTIPLE REPRESENTATIONS

As a scientific case study of my multiple representations account, consider climate change modeling. While the positive correlation between global greenhouse gas atmospheric content and global mean temperatures seems obvious according to basic physics, it is not a relation that has been, or should be, assumed and taken lightly.⁶¹ It needs to be measured empirically; explained by geophysics; predicted by reasonable models;

---

59. The artistic map Alekbu-Lan imagines an answer to the question "What would Africa have looked like if Europe hadn't become a colonizing power?" Countries ("Islamic states, and native kingdoms and federations") here "have at least some basis in history, linguistics or ethnography." Jacobs 2014. The philosopher Charles Mills also considers this question and its implications for racial classifications and categories, in the video-recorded lecture "Does Race Exist?" (see Mills 2010).
60. Winther 2019a.
61. Romm 2016.

and tested.[62] Although many climate change models differ wildly in their theoretical structure, they all, roughly, find that a clear positive correlation trend exists between the amount of atmospheric $CO_2$ (and methane, nitrous oxide, and other greenhouse gases) *and* temperatures of the atmosphere and oceans (see plate 5 and fig. 5.1).[63] They determine such a correlation even when adopting a variety of ontological assumptions about the climate system's basic components, including *radiation* (basic input and output of solar radiation), *dynamics* (wind and ocean movement of heat energy around the globe), and *chemistry* (consequences of atmosphere, ocean, and land chemistry).[64]

The climate change consensus is a clear case of contextual objectivity reached through a massive integration platform of many disparate models, each postulating somewhat different worlds. Climate model building was already afoot by 1896 with calculations by the Swedish chemist Svante Arrhenius, who predicted increasing temperatures with the increase in atmospheric $CO_2$ caused by burning fossil fuels, though he overestimated.[65] The consensus about the reality of climate change has had quite a long time to gel, and has ample independent data supporting it. Very few scientific conclusions—in genetics, medicine or economics, say—have received such stringent theoretical and empirical support.[66]

62. Lloyd 2010, 2012.

63. Interestingly, the analogy with a greenhouse is actually negative since, in a greenhouse, sunlight causes surfaces to heat and the closed-off air inside warms (i.e., convection). However, in the atmosphere, certain gases emit infrared radiation that they have absorbed from Earth, reflecting some radiation back down to Earth (and some to outer space), thereby warming the atmosphere (i.e., radiation). For a history of the greenhouse effect, see M. D. H. Jones and Henderson-Sellers 1990.

64. McGuffie and Henderson-Sellers 2005, 49.

65. Already in the 1820s, the French mathematician and physicist Jean-Baptiste-Joseph Fourier was the first to postulate the possibility of an atmospheric greenhouse effect—he correctly argued that Earth was hotter than expected, given our distance from the sun.

66. More than most other justifications and establishments of accepted scientific theories, this one has been particularly slow and has demanded much data (Oreskes 2004). Graph or map the trend line of average global temperatures from 1880 to the early twenty-first century against many potentially independent factors, such as solar or volcanic activity, or atmospheric levels of aerosol, ozone, or greenhouse gases. From this trend graph—the underlying statistics of which I shall turn to at the end of chapter 7—it is eminently clear that, while natural geological processes have made next to no difference, gases that humans have produced explain the upward trend, with greenhouse gases explaining by far most of it (Roston and Migliozzi 2015). Douglas Karpa provided constructive feedback.

Explore now how ontologizing and merely-seeing-as work in this case. The cultural anthropologist Myanna Lahsen, now at Wageningen University in the Netherlands, interviewed climate change modelers, providing a snapshot of the first and second steps of my multiple representations account:

> INTERVIEWER: Do modelers come to think of their models as reality?
> MODELER A: Yes! Yes. You have to constantly be careful about that.... You spend a lot of time working on something, and you are really trying to do the best job you can of simulating what happens in the real world.... you start to believe that what happens in your model must be what happens in the real world.
> 
> ...
> 
> MODELER B: ... If you step away from your model you realize "this is just my model." But ... there is a tendency to forget that just because your model says x, y, or z doesn't mean that that's going to happen in the real world.
>
> ...
>
> MODELER E: What I try to do [when presenting my model results to other modelers] ... is that I say "this is what is wrong in my model, and I think this is the same in all models, and I think it is because of the way we're resolving the equations, that we have these systematic problems." And it often gets you in trouble with the other people doing the modeling. But it rarely gets you in trouble with people who are interested in the real world.[67]

Each of these modelers recognizes their model as one among multiple perspectives, and implies that their model should be treated as such. In pluralizing and integrating representations, pluralistic ontologizing happens. Via ontologizing, merely-seeing-as, and ontologizing pluralistically, a *single* consensus ontology or worldview of a clear positive correlation between atmospheric greenhouse gases and temperature was achieved.

This consensus is neither imperialist nor perniciously reified. Rather, it is contextually objective. Many models and idealized worlds were built, respecting the bounds and variety of rich data available. Proper abstraction procedures of calibration, data collection and management, and generalization were met. Moreover, this is not a field prone to fashion or low or vague standards of evidence. We again find that a healthy, diverse scientific (and amateur) community using many kinds of models, measurements, and metaphysics provides a check on the dangers of pernicious

---

67. Lahsen 2005, 908, 909, 911; italics suppressed.

reification lurking at, particularly, the individual researcher or research program level, of someone—especially someone with power—insisting on their favorite model or preferred world.

Furthermore, nonempirical arguments about the high risk of not accepting the climate change consensus view must also be considered (such as that there is no planet B). Practical, ethical, and nonevidentiary considerations should be factors in an overarching calculation of how climate change models are contextually ontologized in decision-making and acting. Indeed, even personal context matters in the application of knowledge about the clear and present danger posed by climate change and other horsemen of the apocalypse: pollution, habitat loss, overkilling, and invasive species.

I have no children. My sister has two (one of whom drew part of fig. 9.2). Visiting her loving family of four in Denmark reminds me of what makes life worthwhile. My niece and nephew are an incarnate message—a signal from the future, when my or your heart will no longer be beating—of why we need to do all we can to think creatively and act insistently toward social and environmental preservation and improvement.

Perhaps my conceptual frame of ontologizing, merely-seeing-as, and pluralistic ontologizing as three stages or moments in how we think of the ontological layer of representations acting can provide some tools for discussing the strengths and weaknesses—as well as the hopes and promises—of models of climate change, among many other scientific theories and models. It is my humble and explicit contribution.

## Conclusion

Each of the three accounts of scientific representation—isomorphism, similarity, and multiple representations—has clear relevance to cartographic and scientific representation. Map thinking has helped us see how each account pertains to distinct sets of assumptions, or representational layers. This pluralism of philosophical analyses—which constitutes a "metarepresentational pluralism"—resists and contradicts attempts to defend a sole, universal account of scientific representation.

The limits of the three philosophical accounts deserve further investigation, beyond the bounds of the present study. As we saw in chapter 2, analogizing scientific theories and models to literal maps is ubiquitous and intuitive. But how do the spatial, symbolic, and ontological layers apply to the *other* four map types, and four map analogies? The ontological layer pertains to all map types. The spatial layer, however, applies only to maps also following the spatial map analogy (fig. 2.2): extreme-scale, state-space, and some literal and causal maps. The symbolic layer pertains even less clearly to map types such as analogous maps and causal maps. Clarifying the role of the three representational layers to all map types is an exciting prospect for future investigation.

Another limitation of my study is that there are other layers beyond the metric, symbolic, and ontological, and other accounts of scientific representation. For instance, understanding artistic and linguistic representation in the cases of Aphrodite de Milo or ☿ could draw on each of the three layers, but might also require reflecting explicitly on the role of emotions or highly personalized phenomenology, and whether these fit into the three layers, or compel us to extend a particular layer, or add more layers.

Several extant accounts of ontologizing from the social sciences resonate strongly with mine, as do some philosophical analyses of modeling and representation.[68] For instance, Ronald Giere also draws on the map

---

68. The British philosopher of economics Mary S. Morgan catalogues how economists work and think. According to Morgan, an economist's mathematical model—with its selective and miniature world—becomes an interpretative "lens." Economists merely-see-as. Subsequently, the model's small world becomes full of the very "things" economists find and see in the actual world, such as "rational economic man" (M. Morgan 2012, 406). At this point, economic theorists ontologize, which leads—according to Morgan—to something akin to my pernicious reification, when alternatives are ignored. Also relevant is sociologist Michel Callon's analysis of performativity, or language or theory that causes action or change (for example, an immigrant taking a citizenship oath). In analyzing the relation between economic theory and economic processes, Callon argues that "economics, in the broad sense of the term, performs, shapes and formats the economy, rather than observing how it functions" (Callon 1998, 2; references suppressed; see also Frankel 2015, 543–44). I might have edited "rather than observing" to "and observes." I am also sympathetic to Baudrillard's exploration of the "hyperreal."

Turning now to the philosophy of science, Tarja Knuuttila (2011, 263) approaches models as "epistemic tools"—artifacts used in gaining knowledge. Her practice-oriented approach to models, according to which models play other roles apart from being representational, complements mine. The models-as-mediators approach of Morgan and Morrison (1999) is also highly practice-based, and it is pluralistic about the nature and functions of models and other representations. Mauricio Suárez (2015) presents a subtle argument for a minimalist interpretation of representation, according to which analysts of science (e.g., philosophers of science) do not need to appeal to anything above and beyond actual representational practices in order to make sense of them. This frees us up, Suárez implies, to investigate scientific practice without philosophical shackles. Where my account might differ is that I explicitly focus on practices surrounding

analogy in defending a kind of "perspectivism." However, he insists that this analysis is uniquely correct, replacing any other.[69] His perspectivism is thus philosophically *monistic*, whereas my account is pluralistic, also at the philosophical level, which is why I have reviewed and respected isomorphism and similarity.

Perhaps no one has articulated the spirit of my account of the ontological layer better than the Dutch Golden Age painter Samuel van Hoogstraten: "How valuable a good map is, in which one views the world as from another world thanks to the art of drawing."[70] And while the multiple representations analysis helps us understand how representations imply and help us construct worlds, it remains uncertain how useful this account might be to the ontological layers of cognitive maps, personal-geography maps, map art, or other nonstandard maps.

Even so, I have tried to make explicit how maps become the world through multiple kinds of world imagining and world making, and how that can be done *generatively and correctly* (and thus in a contextually objective manner) or *dangerously and incorrectly* (and thus in a perniciously reified way), and why we might wish to focus on the ontological layer of cartographic and scientific representations in making sense of how an abstraction denotes or represents something else.

These are ultimately grand questions about the relation of theory to experiment and data; the predilections and limits of our mind in understanding the world we walk, live, and love in; and the methodological paths we should probably take if we wish to make progress toward a more ecologically and socially sustainable future. In part 2, I provide scientific grounding for the philosophy that has been my focus thus far.

---

ontologies and how worlds are taken to be. Further philosophical conversations about representation, practice, and ontology are desirable. Natalia Carrillo and Mauricio Suárez provided constructive feedback.

69. Giere 2006, 72–80. For another kind of perspectivism, see Haraway 1988.

70. Samuel van Hoogstraten 1678, 7, quoted in Alpers 1987, 74; requoted, from Alpers, in Jacob 2006, 11.

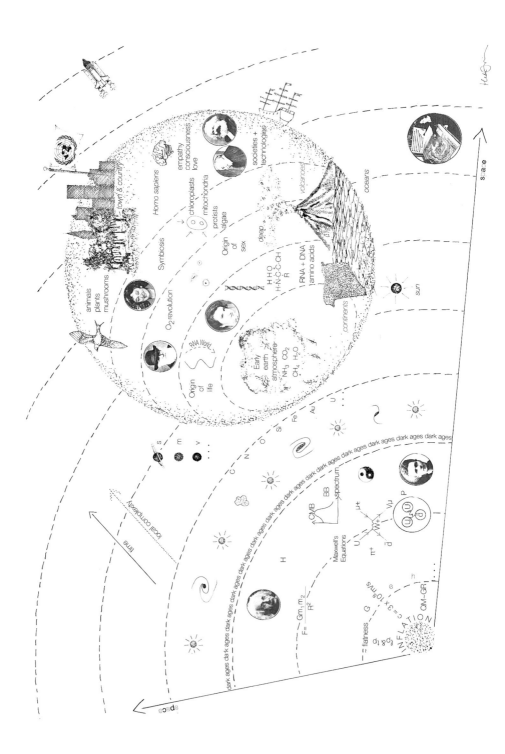

*Universe Particulars*. The arc graticule is intended as didactic temporal milestones for particularly important events or characteristics of the universe or Earth. The two dimensions of this figure should not be taken as literally space, but as rough almost "qualitative" proxies for three-dimensional space. Following the order of the arcs, our visual history can be told thus: (0) The universe experienced a massive expansion of space itself (inflation) in its first moments. (1) The Planck length ($l_p$) and Planck time ($t_p$) are the tiny scales of space and time at which effects of quantum gravity seem to matter, and the earliest universe was at these scales, hence QM ~ GR, since quantum mechanics and general relativity, and their associated processes, must be consistent and interwoven at these scales. Light, which is taken to travel at a constant velocity $c$ ($3 \times 10^8$ m/s), journeys one Planck length in one Planck time. (2) The four fundamental forces, gravity, electromagnetism, the weak force, and the strong force, are indicated in the second arc, in four languages: mathematical, linguistic, Feynman diagram, and a compositional, part-whole diagram. Albert Einstein (developer of special and general relativity theories), Marie Curie (discoverer of radioactivity), and a hint of Niels Bohr's (one key, early developer of quantum mechanics) Principle of Complementarity grace the arc with the black-body spectrum radiation of the Cosmic Microwave Background. (3) Following an unevenly scaled, chronological order, the universe's actual dark ages precede star and galaxy formation. (4) Inside stars, eventually, elements heavier than the earliest hydrogen and helium are formed, including carbon, nitrogen, oxygen, and so forth. (5) Zoom now to Earth, sharing the Sun with Saturn, Mars, and Venus, etc. Earth is roughly 4.5 billion years old, and its early atmosphere consisted of ammonia, carbon dioxide, methane, water, etc. (6) Life emerged, perhaps first as RNA. Rosalind Franklin's research was essential for elucidating the structure of DNA. (7) Sex—the union or meeting of life with life for reproductive or replication purposes—followed soon thereafter, geologically speaking. Charles Darwin had much to say about that topic. An oxygen revolution caused by single-celled microorganisms changed the fundamental composition of the atmosphere. Meanwhile, oceans had formed early on, and once Earth had cooled sufficiently, continents were shaped and changed through plate tectonics, with volcanoes appearing at places where plates meet. Marie Tharp systematically drew Earth's tectonic anatomy (Winther 2019b). (8) Like sex, symbiosis—the union of vastly different forms of life to create an organized and collaborative higher-level unit of life, as occurred with the origin and stabilization of mitochondria and chloroplasts—is fundamental to life and was an evolutionary process championed by researcher Lynn Margulis (Winther 2009c). (9) Multicellular animals evolved, at least some five hundred million years ago, and green plants at least seven hundred million years ago. Eventually, with the emergence of humans some few millions of years ago, human consciousness, politics, and economics arrived upon the scene. Karl Marx and John Maynard Keynes have some opinions on all of this, and are here seen discussing the price of gold. Urban and rural, not to say wild, habitations persist. But with the rise of capitalism and the spread of colonialism roughly a few hundred years ago, and with ongoing ecological collapse and climate change, the prospects for global governance (e.g., United Nations) and survival are debatable. Space exploration may be an option, but the future, as ever, remains uncertain. (UN emblem logo reprinted courtesy of United Nations. Picture of Lynn Margulis: Walter Oleksy/Alamy Stock Photo. Picture of Albert Einstein 1947: Photograph by Orren Jack Turner, Princeton, NJ; modified with Photoshop by PM_Poon and later by Dantadd; Wikimedia Commons. Picture of Karl Marx 1875: John Jabez Edwin Mayall; Wikimedia Commons. Picture of Marie Curie circa 1898: Wikimedia Commons. Picture of John Keynes 1946: International Monetary Fund; Wikimedia Commons. Picture of Charles Darwin 1881: Elliot & Fry; Wikimedia Commons. Picture of Rosalind Franklin: Pictorial Press Ltd/Alamy Stock Photo. Picture of Marie Tharp: Reproduced by kind permission of Lamont-Doherty Earth Observatory and the estate of Marie Tharp. Illustration adapted by Heidi Svenningsen Kajita; concept by Rasmus Grønfeldt Winther.)

# · II ·
# Science

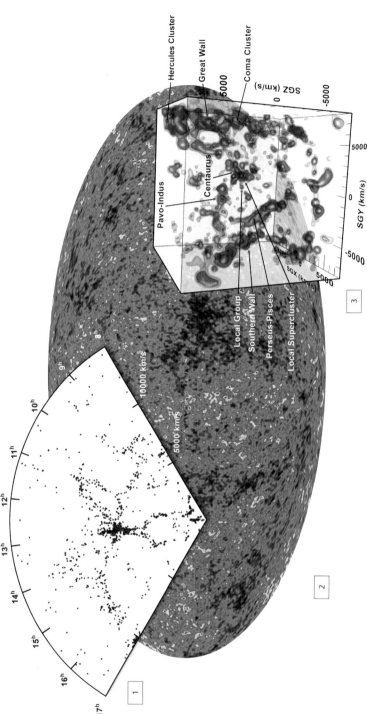

FIGURE 6.1. Three extreme-scale maps in cosmology. (1) An intensely clustered set of galaxies resembling a stick figure was revealed in 1986. This map contains more than a thousand galaxies. (2) The Wilkinson Microwave Anisotropy Probe (WMAP) nine-year map shows the cosmic microwave background (CMB) in Mollweide projection. (3) A 3-D map of the local universe, at a scale similar to that of inset 1 (the head and upper body is the Coma cluster; the stretched arms correspond to the Great Wall; interpretation by Courtois, pers. comm., January 2019). *SGX*, *SGY*, and *SGZ* indicate supergalactic coordinates (*SG*). See also plate 7. (Inset 1: De Lapparent, Geller, and Huchra 1986. © AAS. Reproduced with permission. Inset 2: NASA / WMAP Science Team. Inset 3: Courtois et al. 2013, 69. © AAS. Reproduced with permission.)

# 6

# Mapping Space

In 1992, a group of astrophysicists astonished the world by producing a map of the universe as it appeared fourteen billion years ago. It was like time traveling to the deep past and snapping a picture of an infant universe. This picture of our early universe came not from visible light, but from microwave radiation, similar to the microwave radiation you might use at home to make popcorn.

Look up into the night sky and the space between the stars looks completely black. Even through a telescope (a telescope built to detect visible light), this blackness is like a dark ocean. But with the development of radio telescopes came the detection of something else: a faint, cold aura of microwave radiation from every direction in our universe. That aura was the basis for that first cosmic microwave background (CMB) map in 1992, a map improved in 2013 (see plate 7 and fig. 6.1).

In this first chapter of part 2, we will journey through a selection of maps, starting with this cosmological map of the CMB, to show how ubiquitous map thinking is in science, even in the precise, mathematical sciences of various kinds of abstract space. We will match up a different field of the physical and earth sciences with four of the five map types explored when introducing map analogies in chapter 2. Cosmology is filled with extreme-scale maps of both the CMB and more-differentiated astrophysical structures. Geology relies on literal maps, some of which constituted the earliest evidence for the theory of continental drift. Thermodynamics and physical chemistry teem with state-space maps that provide insight into the phase changes of matter, such as from liquid to gaseous nitrogen. And mathematics provides examples of the general map analogy. (Causal maps—so important to the life and social sciences—will be addressed in the next chapter.)

CHAPTER SIX

# Extreme-Scale Maps in Cosmology

The largest thing humans can map is our universe. There are paradoxes and qualifications involved in mapping the biggest existence in space and time, with the fastest, yet finite, travel velocity that we know of: light. For instance, we cannot measure and know what lies outside the limits of the universe that we see with incoming light, including whether there are other universes. Respecting these many brain teasers, let us focus here on how astrophysicists and cosmologists have mapped what the universe was like when it was much, much smaller, an infant. And let us consider how researchers map and measure its ongoing growth.

## THE UNIVERSE'S BABY PORTRAIT

According to the Big Bang theory, the universe came into being when, from a tiny point, space appeared and expanded with unimaginable speed, increasing the universe's volume by a factor of $10^{78}$ in just $10^{-32}$ seconds. These sixty exponential doublings seem inconceivable.

After this initial expansion, called *inflation*, the universe continued to expand more slowly. It was a high-energy plasma soup made up of subatomic particles, such as electrons, protons and neutrons, all traveling *relativistically*—that is, at or near the speed of light. Astronomers say the universe was *energy dominated*, as opposed to being ruled by matter.

This fairly homogeneous plasma cooled as it expanded, and from three minutes to twenty minutes after the Big Bang, subatomic particles began coming together to form the nuclei of the lightest elements—primarily hydrogen and helium.[1]

Gradually, over the next fifty thousand years, energy dominance gave way to matter dominance. By then, only photons and neutrinos were still traveling relativistically. The universe was still too hot for stable atoms to form, which meant charged electrons and protons zipped about, scattering photons. This constant scattering kept photons from getting far, so the universe was unilluminated.

As the universe continued expanding and cooling, electrons began combining with protons to form hydrogen atoms. By about 377,000 years after the Big Bang, almost all charged subatomic particles had been bound into small atoms.

1. Helium was also formed during the Big Bang, but almost all heavier elements were formed in the heart of stars much later, after several billions of years, though evidence now suggests that some stars could have formed earlier, after just 250 million years post–Big Bang (Hashimoto et al. 2018).

Without electrons zipping around to scatter them, photons were free to travel. The dark, foggy universe cleared and became transparent. For the first time, *there was light*. Photons could travel through space for nearly infinite distances. Although clearing has occurred throughout the universe, the CMB we detect now (inset 2 of plate 7) is from this foggy *last scattering surface*—that is, from slightly (in cosmological terms) before clearing was complete—because it has traveled more than thirteen billion years to reach us.[2]

The CMB photons come to us today from every direction, causing this early stage of the universe to appear to us like the inside of a spherical surface, or shell. In fact, no matter where we might stand in the universe, the effect would be the same, because when we view the seeming "shell," we are looking into the past. The shell's radius is how far the CMB photons have traveled since they were last scattered 377,000 years after the Big Bang. But the radiation that constitutes the CMB is in fact everywhere now and was everywhere in the past.[3]

Two satellites collected the data for the two main CMB mapping projects: NASA's pioneering Wilkinson Microwave Anisotropy Probe (WMAP) from 2001 to 2010 and the European Space Agency's more data-intensive and precise Planck project, from 2009 to 2013.[4] Researchers created the CMB map by taking all the ambient microwave radiation recorded by those satellites and computationally and statistically scrubbing away any radiation from galaxies and other celestial objects, including our own Milky Way, and the slight microwave radiation from terrestrial sources. What was left was that baby picture—a snapshot of a long-gone time, the near-birth of our universe. WMAP established the universe's age at 13.77 billion years to within a half percent accuracy; proved the effectively Euclidean flatness of space; and determined the relative percentages of regular matter, dark energy, and dark matter.[5]

WMAP also showed that the CMB, and therefore the ancient universe, was stunningly smooth, uneven at no more than a factor of about 11:1,000,000. If Earth were that smooth, the top of Mount Everest would

---

2. Castelvecchi 2011b; Coles 1999.

3. For some details on CMB and the early universe, see Ryden 2016, 142–83; Weinberg 2008, 101–48, 149–85. For distribution of photon frequencies of CMB, see Goddard Space Flight Center, n.d.

4. Bennett et al. 2013; BICEP2/Keck and Planck Collaborations 2015.

5. See National Aeronautics and Space Administration, n.d.a. According to the dominant cosmological theory of inflation, space is nearly flat because the exponential inflationary expansion of space diluted any primordial curvature (cf. Steinhardt 2011).

be only 70 meters above the bottom of the Mariana Trench—a distance equivalent to the height of an ordinary twenty-one-story building you might see in a city.[6] In reality, it is roughly 12 miles or 20 kilometers from the bottom of the Mariana Trench to the top of Mount Everest. Earth is not smooth, but the universe was and is.

The CMB map in plate 7 shows the CMB's unevenness by using color differences to indicate slight differences in temperature, which is a proxy for density. Hotter areas are taken to be denser areas that are the seeds of the eventual formation of stars, galaxies, clusters, and other structures at every imaginable scale.[7]

CMB maps can teach us about the features of our early universe, but they also raise philosophical questions about representation, such as: What is represented? What is idealized away? What remains essential?

As mentioned earlier, CMB maps are created by scrubbing away data such as microwave radiation emanating from our galaxy's center. CMB maps are the result of multiple iterations of data processing and statistical analysis. Like Van Sant's composite satellite image "map" of Earth (plate 2), CMB maps are, in fact, complex achievements, not naive, simple snapshots of the universe.

Nonetheless, it is intriguing to consider that an extreme-scale CMB map is, like an ordinary photograph, a literal impression of the recent arrival of ancient photons. Is it a map or is it a photograph? The CMB photons are actually tiny remnants of the early universe that have traveled more than thirteen billion years, smashed into detectors on satellites, and then—transformed and represented by *Homo cartograficus*—appeared as the CMB map on photon-emanating screens and paper, including the CMB map of plate 7. The map creation process in this case detects, abstracts, and represents microwave radiation emanating from the universe's birth.

### THE UNIVERSE GROWING UP (AND OUTWARD)

The expansion of the universe has slowed considerably since its inception, but it is ongoing. If we draw dots on the surface of a balloon and then blow up the balloon, we can see that, from the perspective of a single dot (us), all the other dots are moving away, in two dimensions. And they are moving away not only from us, but from all the other dots.

Now imagine those dots are galaxies and the universe is expanding.

---

6. Ryden 2006, 185. The average temperature of the CMB is 2.725 Kelvin, and it varies by only about 30 microKelvin (ibid.).

7. Thomas Ryckman and Neil Peart provided constructive feedback.

All the galaxies are moving away, or receding, from us and from one another. (What is tricky is that this visual metaphor needs to be extended to three dimensions, and not just the two of the balloon surface—raisins in an expanding loaf of bread could provide such a metaphor, though a loaf of bread has a center, something the universe does not.) As a galaxy recedes, the light waves it emits become what we call *redshifted*; in this case, it is the *cosmological redshift*. Redshift occurs because the space through which the light waves are moving is itself expanding.[8] Given light's constant velocity, this lengthening moves the light through longer-wavelength positions on the electromagnetic spectrum, from blue-hot, say, to cooler red. In fact, what is now the CMB was originally visible plasma; it shifted over time to microwave, which has even longer wavelengths than visible red light.

The amount of redshift can be assessed by the relative displacement of known hydrogen absorption lines in a spectrogram of the light reaching us. By measuring the amount of the redshift, we can tell both how quickly the galaxy is moving away—its *recessional velocity*—and how distant the galaxy is.

The farther away a celestial object is, the faster it recedes. The correlation between the amount of redshifting and distance from us is approximately linear, particularly for local structures. For instance, a galaxy 10 percent farther from us than another galaxy recedes 10 percent faster that that other galaxy because of the expansion of space. This large-scale relative motion of galaxies and other celestial objects is called the *Hubble flow*. Thus, to know an object's recessional velocity is to know its distance from us. (A celestial object might reach additional speeds due to the pull of gravity—the curvature of space—around other objects, but that is a separate, local velocity and is called *peculiar velocity*.)[9]

Cosmological redshift is converted to distance so that astronomers can tell, for example, that the Hercules cluster (inset 3 of plate 7) is (ap-

---

8. This effect of the expanding universe has often been explained in terms of the Doppler effect, with the everyday example of sound waves compressing and producing higher siren tones as an ambulance approaches, then lengthening into lower tones as it drives away. But this is different from cosmological redshift. Doppler effect redshift *does* occur celestially, but it occurs when light waves are emitted by a moving source (such as a galaxy moving in its peculiar velocity, as opposed to the general expansion of the universe). In contrast, cosmological redshift occurs as space expands *after* the wave has been emitted, as it travels through expanding space (Rothstein 2015; Swinburne University of Technology, n.d.; Odenwald and Fienberg 1993).

9. This is the Doppler effect redshift; see note 8 above.

proximately) five hundred million light-years from Earth. We can use this distance data to map structures such as galactic filaments, clusters, and *cosmic flows* in our local universe. These structures are indices of gravitational fields causing the local-scale peculiar velocity cosmic flows (measured by Doppler effect redshifts).[10] They teach us about the mass distribution and gravitational processes of the universe in general.

Until the mid-1980s, astronomers largely assumed galaxies were more or less randomly distributed throughout the universe. Notwithstanding a cluster here or there, they assumed a certain homogeneity. After all, the cosmic background radiation was smooth as silk. It made sense that any clustering that had taken place within that early smoothness would have evened out.

But, in 1977, astronomers began a series of systematic surveys in which they measured the redshifts of everything in the sky that is visible from Earth. The astronomer Margaret Geller and her colleagues Valérie de Lapparent and John Huchra mapped the positions of one thousand galaxies and discovered the eye-catching shape of a stick figure, seen in inset 1 of plate 7.

In a transformative moment, Geller recognized the stick figure's significance.[11] Galaxies, she realized, were distributed not randomly, but in patterns. Geller described the distribution of galaxies in the stick figure survey as resembling a cross section through suds in a kitchen sink. The celestial structures are spread as though across the common walls of adjacent bubbles, surrounding the voids inside the bubbles. This foamlike pattern is now called the *cosmic web*.

More recently, Hélène Courtois, R. Brent Tully, and their team used calculations of cosmological redshift, distance, and peculiar velocity to create three-dimensional maps of galactic motion, tracing graceful sweeps resembling the flow of water. Doing so, they identified the Milky Way's location as being on the very tip of one arm of an enormous but delicate-looking supercluster of over one hundred thousand galaxies

---

10. Cosmic flows and other smaller-scale peculiar velocities are distinct from the Hubble flow. Peculiar velocity can be in any direction, including (some vector) toward us. Hence once an astronomer subtracts the Hubble flow (perhaps through use of a comoving frame) from a moving galaxy, galactic cluster, or other celestial object, it can have *blueshifted* peculiar velocities. For instance, our nearest galactic neighbor, the Andromeda galaxy, approaches our Milky Way, as indicated by its blueshifted radiation. Our neighbor will collide with us in some 4.5 billion years.

11. Taubes 1997; cf. Geller and Huchra 1989.

that they named Laniakea.[12] (In inset 3 of plate 7, Laniakea includes the Pavo-Indus, Centaurus, and our local superclusters—all moving toward a central "great attractor.")[13]

Courtois, Tully, and collaborators have emphasized the importance of *cosmography*—the mapping of the cosmos. They write: "We argue that maps, with names for features, promote a familiarity and specificity that contributes to physical understanding."[14] A dynamic representation of specific structural features of the universe creates a visual image that makes those structures more real and intuitive, especially to nonexperts, and inspires new questions, insights, and curiosity. The Argentinian artist Pablo Carlos Budassi has created a particularly striking cosmographic map (plate 8).

## COSMIC-SCALE MAPS AND THE ABSTRACTION-ONTOLOGIZING ACCOUNT

In considering cosmological maps, we can apply the three-stage analysis of abstraction laid out in chapter 3 as the abstraction-ontologizing account:

- calibration of units and coordinates
- data collection and management
- generalization

### Calibration of Units and Coordinates

The CMB map shows that the universe, when viewed at a large enough scale, has remained surprisingly *isotropic* and *homogeneous*. This idea—that the universe, at sufficient scale, is roughly the same in all directions,

---

12. "From the Hawaiian; *lani*, heaven, and *akea*, spacious, immeasurable" (Tully, Courtois, et al. 2014, 73).

13. See Tully, Courtois, et al. 2014; Nature Video 2014.

14. Courtois et al. 2013, 1. Waldseemüller's map was a cosmography, in name and intention; Weinberg 1972 has an entire chapter titled "Cosmography" (407–68), which includes sections on the Friedmann–Lemaître–Robertson–Walker (FLRW) metric, redshifts, and measures of distance. In response to a query about what effects maps might have had on her, Courtois wrote that "since childhood I was taught how to read all details and derive information from maps" and that today she "frequently spend[s] time gazing into various maps . . . : marine maps, road maps, star maps, globes, mural paintings, etc." (pers. comm., December 2017).

[156]    CHAPTER SIX

and that mass and energy are continuously distributed—is the core of the *cosmological principle*.

The CMB map also shows that the universe is Euclidean (with only a 0.4 percent margin of error).[15] Without such Euclidean flatness, either a collapse or a relatively quick expansion would have occurred. That is, a "closed" universe has a 3-D geometry equivalent to that of a sphere, with its positive curvature, and ends with a "Big Crunch"; an "open" universe has a 3-D geometry equivalent to that of a saddle, with its negative curvature, and ends with a "Big Chill" or an accelerating "Big Rip." Had the universe been significantly open or closed, we would never have been here—at least, not in *this* universe.[16]

In 1922, the Russian physicist Alexander Friedmann developed equations from the perspective of Einstein's theory of general relativity that describe the expansion of space since the universe's birth. Today, practically all cosmological models assume the Friedmann–Lemaître–Robertson–Walker (FLRW) metric, which incorporates Friedmann's equations.[17] Analogous to a map graticule (chapter 3), this calibration metric assumes the geometry of our isotropic, homogeneous, and flat space, enabling scientists to map the space and time of the entire, expanding universe. It is therefore relevant both to the CMB and to our local universe, depicted in plate 7. The assumptions of isotropy, homogeneity, and flatness allow the FLRW metric to be used to determine *comoving distance*, which is generalized into a *comoving frame*. A comoving distance preserves relative distance among objects and events in the universe, even as the vacuum of space expands.[18] It factors out the Hubble flow.

---

15. See National Aeronautics and Space Administration, n.d.b. Note that space is not flat around massive objects that significantly warp space-time, such as black holes or even our sun.

16. This is one "fine-tuning" problem in cosmology, analogous to throwing a rock in the air neither too slowly (such that it falls back to Earth) nor too quickly (such that it escapes Earth's gravitational field). Ours seems to be a highly unlikely universe. See Lightman 1991, 59–62. The phrase *Big Bang* was coined in 1949 on BBC radio by British astronomer Fred Hoyle, as an arresting term for a concept he in fact rejected. Regardless of its origins, the term has stuck: after organizing an international competition in 1994 to ennoble the universe's opening act with a better name, Carl Sagan and media collaborators declared they had found no better alternative. Big Bang it is (Strauss 2016, 218).

17. For more details on the FLRW metric as an exact solution to Einstein's field equations of gravitation, see Ryckman 2017, 22, 196, 235–36, 375–79; Weinberg 2008, 1–9, 34–45.

18. Interestingly, the frame is defined for a given rate of the expansion of space (equivalent to a given Hubble parameter). As this rate changes, the frame

*Mapping Space* [157]

We can understand comoving distance by once again using the balloon metaphor. In a comoving frame, distance between dots—the *proper distance* at a given point in cosmological time—is calibrated to balloon expansion so that distance remains constant between different groups of dots, even as the balloon expands. (Displacement among dots caused by one of them moving independently of the balloon's expansion at a given moment in time is then an indication of peculiar velocity—that is, a deviation from the commoving frame.)[19] These and more-complex metric calibrations establish the distance coordinates (and various kinds of horizons) for geometrically and physically proper maps of the universe in both space and time.[20]

To move from mapping the entire universe to mapping everything within roughly five hundred million light-years of Earth—or even to the approximately fifty galaxies in a neighborhood of some ten million light-years, called the *local group*—is to go from the most extreme scale to a smaller, but still gigantic, space.[21] However, at both scales, cosmographers and cosmologists have available a graticule, mathematically articulated, similar to

---

changes. The distance of local structures within a few billion light-years of us today can be well approximated with a comoving frame.

19. Consider a flat rubber sheet that is pulled equally left to right and front to back: the comoving frame over time is the fixed relative distance along the rubber sheet among objects attached to the sheet; the peculiar velocity is any movement one or more objects may have with respect to the place on the sheet where they should be attached.

20. Weinberg 1972, 469; Hogg 1999. The FLRW metric and the comoving distance are natural metrics and natural calibrations, based on the measurable expansion of our universe.

21. See inset 3 of plate 7. How far is the Hercules cluster from our local group, which is usefully and predictably placed at the center, at coordinates {0, 0, 0}? To calculate *proper* distance, we must use Hubble's law:

recessional velocity = $H_0$ × proper distance,

where $H_0$ is Hubble's constant, which indicates the recessional velocity of any object from us, due solely to the expansion of space. The value of $H_0$ is 74 km/s per megaparsec, which is equivalent to 23 km/s per million light-years. Interestingly, this is its contemporary value, as the "constant" is actually a parameter that changes over the evolution of the universe. Since the recessional velocity (from the local group) of the Hercules cluster is approximately 7,000 km/s along every one of the three axes, we can use Hubble's Law to calculate that the proper distance along any *one* axis is approximately 303 million light-years. But now we must use the simple Euclidean Pythagorean theorem in 3-D to calculate the distance from {0, 0, 0} to the Hercules cluster:

the one used in mapping our own planet—namely, a comoving frame with proper distances, at given moments of the evolution of the universe.

Abstracting from distance, cosmography also requires measuring relative angles from Earth, the sun, or some other central location, to say where "in the sky" something is. The *galactic coordinate system* in polar coordinates imagines a sphere with the sun as its center and the plane of the Milky Way as 0° latitude, its equatorial plane. The *latitude* of any other object is taken as an angle measured from the sun to the object's position "north" or "south" of the galactic plane. The 0° *longitude* line stretches from the center of the sun toward the center of the Milky Way, connecting with the constellation Sagittarius.[22]

For scales larger than our own galaxy, a supergalactic coordinate system can be deployed. It can be a three-dimensional Cartesian "box," with axes labeled *supergalactic-X* (SGX), and so forth, as in inset 3 of plate 7. Alternatively, supergalactic latitude and longitude can be specified, with the equatorial supergalactic plane being defined (by convention) as a plane with many local galaxies and structures, such as the Virgo supercluster of galaxies.[23]

The FLRW metric; proper distance and comoving frame for distances; and galactic and supergalactic coordinate systems for Euclidean, or for latitudinal and longitudinal, 3-D locations (often with respect to the sun as relative center) provide—in toto—an overarching *cosmological graticule* for mapping large-scale structures.

## *Data Collection and Management*

If there is a clear first stage of calibration in creating the CMB and local universe maps, there is likewise a clear second stage of data abstraction. Satellites receive (collect) and process (manage) photon data for both the CMB map and the redshift mapping of the local universe.

In both cases, the economy of cosmological representation trades exclusively in photons. More recently, gravitational waves have been added as a potentially powerful future source of data to turn into representations.

## *Generalization*

Finally, according to the abstraction-ontologizing account, the third stage of mapmaking is map generalization. The satellite-based CMB mapping

---

$\sqrt{[(x - x_0)^2 + (y - y_0)^2 + (z - z_0)^2]}$. This distance is then $\sqrt{(303^2 \times 3)} = 525$ million light-years.

22. Blaauw et al. 1960; Think Astronomy, n.d.

23. De Vaucouleurs et al. 1991; Courtois et al. 2013.

projects also generalize photons according to certain representational conventions, such as added color schemes.

High-level cosmological physical theory influences and shapes the earlier, more empirical two phases of map abstraction. Ontologizing 0—representation testing—takes place, as initially explored in chapter 3. For instance, the FLRW metric is integral to spatial calibration. As a component of underlying spatial assumptions, however, it stems from general relativity. Thus, formally abstracted space becomes ontologized space, especially during calibration of units and coordinates. For instance, because we have found the universe to actually *be* flat, Friedmann's equations and the FLRW metric *are* the physical universe, though they need not have been, and though MINT still applies (see chapter 5)—at least, insofar as we remain within the domain of general relativity, and with the important proviso that this finding might need to be reinterpreted in light of developments in the dynamic and creative discipline of cosmology, as pluralistic ontologizing continues.

## Literal Cartographic Maps in Geology

In 1915, the German geophysicist, meteorologist, and polar researcher Alfred Wegener published a groundbreaking book in the history of geology. In *The Origin of Continents and Oceans*, Wegener assembled a slew of arguments in favor of the continental drift theory.[24] Continents move around on Earth's crust, according to Wegener, sometimes parting to make way for the ocean, and sometimes crashing together.

We now know, for instance, that South America and Africa began splitting some 190 million years ago, gradually allowing the South Atlantic Ocean to flow between them. And the Indian subcontinent itself virtually flew up from the southern tip of Africa, rammed into Eurasia, and crumpled the land upward (a process called *orogeny*) to form the Himalayas, the world's tallest mountains.[25]

In 1915, though, continental drift sounded like science fiction, especially to staid American geologists.[26] But Wegener's methodologically sophisticated book called for bringing together different and independent

---

24. Wegener 1966; Wegener was not the first to suggest the idea. He explicitly acknowledges, for example, a relevant 1910 paper by the American geologist Frank Bursley Taylor (see ibid., 3–4).

25. Every mountain over 7,000 meters is found within the several Central Asia mountain ranges that are part of the "Greater Himalayas," or at least the Himalayan orogenic belt.

26. Oreskes 1999.

strands of evidence from the earth sciences to reach "the truth of the matter."[27] Indeed, his book conciliated distinct strands of relevant data: geodetic, geophysical, paleontological, and paleoclimatic.[28] Each topic has its own focal chapter.

Wegener's book is suffused with map thinking. He includes sixty-three figures, almost all of them literal cartographic maps.[29] As he says in his introduction, it was in glancing at "the map of the world" and noticing the "congruence of the [South American and African] coastlines" that he registered "the first concept of continental drift."[30] Cartographic representation imagines prehistoric geographies and geologies, integrates them, and makes them concrete.

Literal maps, as we learned in chapter 2, are maps at spatial and temporal scales containing at least some geographic objects, such as continents, and are guided by both spatial and general map analogies. A literal map is, of course, the standard way of representing Earth, and it serves an important role in geology, the science of our planet. Many geological maps show straightforward distributions of physical objects or properties at particular positions; conceptualizations of geographic space are implicit in such maps. For instance, an ocean floor chart shows a wild topography of deep gorges, massive mountain ranges, and enormous plateaus.[31]

A glance at a few of Wegener's actual, literal cartographic maps illustrates the importance of map thinking to him and to the rich discipline of geology.[32] For instance, he includes maps of earthworm distribution;[33]

27. Wegener 1966, vii, 78.
28. On consilience, see E. O. Wilson 1999.
29. Wegener's actual cartographic maps include "longitudinal" cross sections of continents and crust, such as that of the "Lemurian compression" showing the Indian subcontinental sheet pushing into the Asian. See Wegener 1966, 84, fig. 22. Wegener also presents a few abstract, maplike diagrams, such as a "diagrammatic cross section through a continental margin," sketching the continents sinking deeper than the oceans into the lower "sial" layer, or mantle (36, fig. 9, see also 84, fig. 22; 212, fig. 60).
30. Wegener 1966, 1. This geometric congruity is the case Giere (1999, 128–34) explores in a section on Wegener.
31. Winther 2019b traces oceanographic mapping efforts: in particular, Marie Tharp and Bruce Heezen's mid-twentieth-century mapmaking projects.
32. For background on the history, theories, and social context of geology, see Bjornerud 2018; Frodeman 1995; Gould 1988; R. Laudan 1987.
33. See Wegener 1966, 117, figs. 30, 31.

he refuses to confine himself strictly to earth sciences in his arguments about the tectonic movements of continents. The same genera of various families of earthworms are found in the Western Hemisphere (Mexico, Florida, and northern South America) and in far-off Africa, whereas these genera are not found in closer-by locations in North America. This pattern suggests either that different worm populations were once united or that some of them somehow traveled from one place to another. But how could they have traveled? Earthworms neither fly nor survive in saltwater, so for them to have crossed the thousand-mile-wide Atlantic seems impossible.

Some researchers had suggested that ancient land bridges allowed animals such as earthworms to crisscross among the continents. But such land bridges would have had to be immense and then to have inexplicably disappeared without a trace. Wegener, who defends evolution,[34] argues that only the theory of continental drift had the resources for explaining this biogeographic pattern.

Earthworms are not the only organisms with a pattern of distribution hinting that the continents move. Many plants and animals are distributed among the continents in patterns that, independently of one another, support the theory of continental drift, as Wegener pointed out in extensive citations of others' research.[35]

Now consider Wegener's creative "reconstruction of the map of the world" maps, in which he hypothesizes the location of the continents during a particular period. Figure 6.2 shows his reconstruction of the Carboniferous period, which lasted from 360 to 300 million years ago.[36] Coal—now a notorious source of greenhouse gas emissions—formed where vast, ancient rainforests were buried and compressed over millions of years. Today, rainy regions that can support such forests exist mainly at the equator and, to a lesser extent, near 30 degrees north and south of the equator.

"Coal signifies a rainy climate. . . . Any rain belt which . . . *forms a great circle* round the globe must obviously be equatorial only," Wegener declaimed. "If we can establish in addition, as in this case, that the belt is 90° *from the centre of a large region of inland ice*, we are all the more

---

34. See Wegener 1966, 6.

35. See, for example, Wegener 1966, 101, fig. 27; 102, fig. 28. As a subsequent study pithily capturing many of these arguments stands the global biogeographic work on moss mites by the Danish zoologist Marie Hammer (see Hammer and Wallwork 1979, and references to Hammer therein).

36. See Wegener 1966, 137, fig. 35; cf. 18–19, figs. 4, 5.

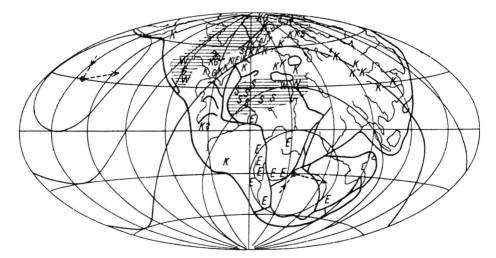

FIGURE 6.2. A geological, literal map superimposing the current-day position of the continents onto their position during the Carboniferous period, ca. 360 to 300 MYA. In this earlier era, much land was concentrated near the South Pole, while the North Pole was out to sea. The Carboniferous period poles are the physical rotational poles of Earth, not the magnetic poles. $E$ = ice traces (*Eisspuren*); $K$ = coal (*Kohlen*); $S$ = salt; $G$ = gypsum; $W$ = desert sandstone (*Wüstensandstein*); hatch marks = arid zones. (Wegener 1966, 137.)

justified in concluding that it occupied an equatorial position. . . . *The conclusion that the European Carboniferous coal beds were formed in the equatorial rain belt is therefore unavoidable*, without any reference to drift theory."[37]

The equator may have shifted over time—but from where? Any hypothesized past equator will be a great circle dividing Earth into two equal hemispheres. Wegener's postulated ancient, Carboniferous equator is a strong candidate, which is evidenced by several observations: (1) coal is a consequence of the ecology and geology of rainy regions with heavy plant growth, which is characteristic of the equator (but also possibly, and to a lesser extent, near 30 degrees north and 30 degrees south); (2) most coal today is found nowhere near our equator; (3) instead, coal is found in a belt roughly comprising a great circle, from the United States up to Europe and Eurasia and then down into China and Eastern Australia;[38] (4) traces of ice (the *E*s in fig. 6.2) from the Carboniferous are found

---

37. Wegener 1966, 138.

38. If you use a string or follow your finger along a globe, you will see this approximate great circle, especially if you imagine that the shapes of continents

in regions that are exactly 90° south from this great circle, indicating an ancient southern pole region.[39] From these facts it follows that the great circle coal belt ($K$ in fig. 6.2) must be the ancient equator.

Proponents of unmoving continents accepted these facts about the conditions likely to produce coal and the current locations of coal deposits, but not the logically inevitable conclusion. Instead, they drew maps with the continents exactly where they are today. Rather than allow the continents to move, *they permitted the equator to move*. On these steady-state Earth maps, coal deposits were inconsistently distributed, with, for instance, the equator arching down through South America, paradoxically near traces of ice in the ancient southern pole region.[40] Only Wegener's continental drift theory—map thought through, for example, his reconstructed map of the Carboniferous period—could consistently explain the enormous amounts of coal in present-day Australia, China, Russia, the United Kingdom, and the United States.

In general, Wegener's drift hypothesis required him to move dialectically between visualizations of Earth's surface today and yesterday, acting *as if* certain geological, meteorological, and ecological features had to be different in the distant past. He imagined multiple worlds—one contemporary and many other worlds in different epochs of the past—just as my multiple representations account suggests. Interestingly, though, Wegener's drift theory did not explain *how* continents move. Wegener productively, but without a clear mechanism such as plate tectonics, assumed the continents were blocks moving along the crust, like icebergs.[41] And, ironically, Wegener imagined the continents had fixed shapes.[42]

The complex spatial underpinnings of Wegener's coal map are particularly interesting. By superimposing contemporary spatial latitudinal coordinates on Carboniferous latitudinal coordinates, the map situates

---

have changed over time. For a more accurate modern map of the Carboniferous that takes into account changing continental shapes, see the geologist and cartographer Christopher R. Scotese's (n.d.) map of the Early Carboniferous, 356 million years ago (MYA). Ancient precursors to South America and Africa (the continents we typically think of when we map think the equator) were far from the Carboniferous equator.

39. The equivalent ancient Arctic is out to sea—the upper left of figure 6.2.

40. See Wegener 1966, 143, fig. 37.

41. Wegener 1966, 183, fig. 46. Wegener had made multiple expeditions to Greenland, which was then part of Denmark.

42. C. O'Neill et al. 2016. Oliver Baker and Jennie Dusheck provided constructive feedback.

the eye according to a familiar frame of reference. In essence, the complex map graticule of figure 6.2 depicts a *dual space* of the difference made by three hundred million years of continental drift. Latitudes for two different worlds are here shown. Contemporary latitudes are indicated in the standard manner, as straight or slightly curved lines, with the centers at modern South and North poles. Meanwhile, Carboniferous latitude lines are drawn as concentric circles around the South Pole three hundred million years ago, near South Africa. The Carboniferous-era equator arcs through all the *K*s etched on Wegener's map. Each of these two worlds—Carboniferous and contemporary—is treated as both an ontologizing and a merely-seeing-as. We are ontologizing pluralistically. Carboniferous ontologizations are implied, tracked, and imposed by the very existence of the coal belt and traces of ice in unlikely regions, today. Only an overarching world in which continental drift occurs could incorporate, make sense of, and fully explain these two different graticules.

Similarly to many great scientific minds willing to think outside the box, Wegener faced a scientific community resisting his theory and its underlying assumptions, his analogies (e.g., continents riding on the mantle "behave like open water and large ice floes"[43]), and his inferences from a large variety of kinds of facts.[44]

Map thinking continues to be essential to geology. Reflect on the cartographer and geologist Marie Tharp's maps of the deep, including her collaboration with fellow American Bruce Heezen.[45] Tharp's maps changed the face of the earth sciences: "This physiographic map '*is* in some ways the ocean floor,' former Heezen graduate William Ryan later mused: 'It's our only multi-dimensional picture of it . . . that map and every subsequent revision to it.'"[46] Through the *North Atlantic Physiographic Diagram* (1957), *Indian Ocean Floor* (1967; plate 9), and *World Ocean Floor Panorama* (1977), Tharp gave us the ocean floor.

---

43. Wegener 1966, 37.

44. Oreskes argues that, among Wegener's colleagues across the Atlantic, what caused dismay was less the lack of a mechanism or disagreement about facts, and more that Wegener's theory "violated deeply held methodological beliefs and valued forms of scientific practice" (1999, 6). For instance, his was a "theory-first" methodology rather than an "inductive" one (313) common among American geologists from roughly the mid-1920s to the mid-1960s. Sébastien Dutreuil provided constructive feedback.

45. On Tharp's biography, see Felt 2012 and Winther 2019b.

46. Doel, Levin, and Marker 2006, 620; ellipses original.

Tharp's representations also suggest a mechanism for explaining the ocean floor's features. Particularly striking about the 1957 map of the Mid-Atlantic Ridge is the *valley* depicted inside the ridge. According to Tharp, Heezen "initially dismissed my [rift valley and continental drift] interpretation of the profiles as 'girl talk.'"[47] Ironically, the rift valley V-shape indentation was indeed a form of girl talk, in a genuinely productive way. This all smacked of continental drift: the plates coming apart, with lava oozing out from within. Tharp's maps became the world.

In the end, Marie Tharp looked back on her remarkable life with gratitude:

> Not too many people can say this about their lives: The whole world was spread out before me (or at least, the 70 percent of it covered by oceans). I had a blank canvas to fill with extraordinary possibilities, a fascinating jigsaw puzzle to piece together: mapping the world's vast hidden seafloor. It was a once-in-a-lifetime—a once-in-the-history-of-the-world—opportunity for anyone, but especially for a woman in the 1940s. The nature of the times, the state of the science, and events large and small, logical and illogical, combined to make it all happen.[48]

While geological mapping can be a prelude to colonialism and profit—for example by fossil fuel companies—a geological map's cartopower can also be a crucible for genuine wonder, scientific humility, and curiosity for what remains to be discovered, respected, and loved.

## State-Space Maps in Physics and Physical Chemistry

Some areas of science—for instance, thermodynamics, electromagnetism, and quantum mechanics—produce maps, but not extreme-scale or literal cartographic maps.[49] In these sciences, maps are a set of abstract relationships (which can be thought of as a formal, mathematical system). Thermodynamics, for instance, is the study of the interactions of heat, work, and energy, suitably abstracted and mathematized.

State spaces are idealized and formalized approximations of the com-

---

47. Tharp 1999; cf. Tharp and Frankel 1986. Helen Longino provided constructive feedback.

48. Tharp 1999.

49. A plausible argument could be made that diagrams of electron orbitals are spatial, in that they represent probability distributions of presence in tiny spaces. Perhaps they are extreme-scale maps.

plex behavior of actual systems. Recall from chapter 2 that the state-space map analogy imagines a state or phase space as a map communicating a scientific model or theory. Each of a physical system's abstracted parts is represented as a variable in one or more equations constituting a *state-space model*. This map represents the territory—the system—that the mathematical model or theory is about. The state-space map is drawn in a formal, mathematical space, with the variables in the theory's equations as the dimensions of the space. The equations determine all the allowable points in the behavior of the system—the system's *permissible states*. The equations also express any allowable movement—that is, *possible system trajectories*.[50] To explore an example from thermodynamics and physical chemistry, we can begin with the *ideal gas law*, an important model in physics and engineering that can be shown in a state space, and that can thereby describe an actual physical system.

Applied to any specific gas, the ideal gas law helps predict how the gas will behave when pressure ($P$), volume ($V$), or temperature ($T$) changes. This information is useful in real life having to do with, for example, refrigeration, ventilation, or anything inflatable.[51] To arrive at the calculation, however, the gas is imagined a certain, simple way. As one science website puts it: "One can visualize [an ideal gas] as a collection of perfectly hard spheres which collide but which otherwise do not interact with each other. In such a gas, all the internal energy is in the form of kinetic energy and any change in internal energy is accompanied by a change in temperature."[52] The world is here ontologized as consisting of gas molecules that are, in essence, tiny, hard balls (other assumptions must also hold for this law to be true).

The ideal gas law and its equation are popular in high school, gymnasium, and university textbooks across many human languages:

$PV = nRT$.

---

50. *State* and *phase* space may be used interchangeably; see Nolte 2010. There are many kinds of state and phase spaces in physics and physical chemistry. For pithy discussions of the state-space version of the semantic view of theories, including discussion of laws of succession and coexistence, see Lloyd (1988) 1994, 11–25; Suppe 1989, 4; van Fraassen 1989, 222–25; Weisberg 2013, 26–29; Winther 2016, §3.1.1; for additional detailed characterizations of the semantic view, see Beatty 1980; López Beltrán 1987; Suppes 2002; Thompson 1989, 2007; van Fraassen 1980, 2008.

51. Aage Bisgaard Winther provided constructive feedback.

52. Nave 2017.

In this equation, $n$ expresses the number of molecules or atoms in the system and $R$ is a constant value specific to the ideal gas law.[53]

On its own, the ideal gas law is easy enough to understand in the form of this analytical equation, and could easily be rendered visually as a surface that does not fold in on itself along any of the dimensions given by the variables of the equation.[54]

However, the larger system of which gas is a part—the system of phase changes of matter—requires more complicated equations. For this level of complexity, a visual rendition that is closer to cartography becomes especially desirable.

Figure 6.3 is a state-space map of the type sometimes called a *phase diagram*. It models the three main phases of matter on Earth—gas, liquid, and solid—for a given substance under a range of conditions of pressure, volume, and temperature, which are the variables of the state space. Note that gas here, and in general, is especially prevalent under relatively high volumes and temperatures, as well as under relatively low pressures.

This representation incorporates assumptions. For example, the theory underlying figure 6.3 assumes that the substance under consideration is pure and homogeneous and that the system is closed—for example, there are no influences from external temperature differentials. And this map follows many of the same representational abstraction practices as ordinary mapmaking, such as selection, simplification, and symbolization, which chapter 3 addressed.

Studying the entire surface of figure 6.3, note that we can slice the three-dimensional surface into an infinity of two-dimensional planes given by the three pairs $\{P, V\}$, $\{P, T\}$, or $\{V, T\}$, holding the third variable (respectively, $T$, $V$, or $P$) constant at different values. It is as if we are slicing the surface, sequentially, along each of the three dimensions. In some parts of these planes, we can write out fairly simple analytical equations mathematically, as we did above with the ideal gas law. For instance, abstracting from temperature in the gas portion of the surface, Boyle's law states that the pressure and volume of an ideal gas are inversely proportional—increase one and you decrease the other.

Now, consider a $\{P, T\}$ plane, averaged across the entire volume range, as shown in the lower-right image of figure 6.3. As in the case of a

---

53. This *molar gas constant R* has been determined, by experiment, to equal $(PV)/(nT)$.

54. "The $p$-$V$-$T$ surface for a pure substance is sufficiently complex to make its complete representation by a reasonably simple analytical equation of state very unlikely" (Sprackling 1991, 214).

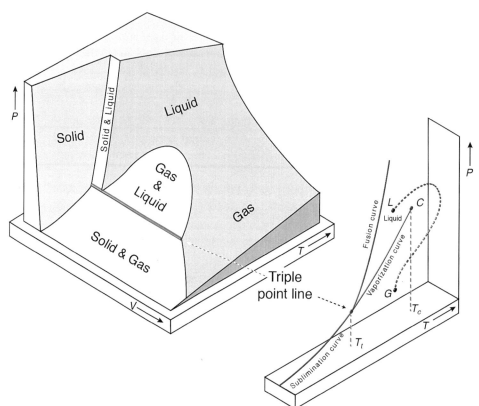

FIGURE 6.3. Two forms of state-space map—surface (*upper left*) and graph (*lower right*)—depicting the phases of matter at different combinations of pressure (*P*), volume (*V*), and temperature (*T*). The substance mapped here is a hypothetical yet typical pure substance that contracts upon freezing (unlike water). At the triple-point line, all three phases coexist. The graph is a projection of the *PVT* surface onto an averaged *PT* plane (much as the geoid of Earth is projected onto a flat map), with the triple point at temperature $T_t$ and the critical point at $T_c$. Note that a gas can "opalesce" to a liquid for temperatures and pressures above *C*, as shown by the dotted line, a movement from *G* to *L*. (Adapted by Mats Wedin from Sprackling 1991, 211, fig. 13.1, and 213, fig. 13.3.)

geographic map projection of Earth, every point on the *P–V–T* surface is associated with some point in this plane, as if a shadow were cast on the {*P, T*} plane by a spotlight shining along the *V* axis. The fusion (melting-freezing) and vaporization (boiling-condensing) curves, while found along different regions of the *V* axis, are here squished and squashed—averaged—onto the same plane.

For temperatures above the critical temperature, $C$, there is no clear distinction between gas and liquid, regardless of pressure.[55] Note the meandering "dotted path" connecting $L$ and $G$ on the map in the geography of the $\{P, T\}$ plane (fig. 6.3, lower right). Along that path, a substance can beautifully and surprisingly pass from gas to liquid without condensing. This is because above the critical temperature, the substance is simultaneously in both states, and below $C$ it returns to one or the other state.

State-space maps can be thought of instructively as topographic maps, informed by the necessary variables of the theoretical models (e.g., $P$, $V$, and $T$), and as maps organizing the data points of the empirical models presented or expressed within those state spaces.[56]

## Analogous Maps in Mathematics

Pure, formal mathematics is the study of abstract structures—symbolically or diagrammatically rendered—often removed from empirical studies of worldly processes, interactions, and objects. Mathematics works by means of analogy from one mathematical subdiscipline to another; analysis of proofs; the search for invariances and symmetry; abstraction; generalization; and other reasoning and representation practices.[57] It first establishes the form and internal relations of idealized structures and then considers how, why, and for what purposes one formal structure can be transformed into another structure.

But mathematics is also wonderfully useful in its application to the

---

55. Galison (2003) uses such a mixed gas-liquid phase of *critical opalescence* as an apt metaphor for theory-practice entanglement: rarefied conceptual abstractions from mathematics and philosophy, on the one hand (metaphor: the gas), intertwine with concrete technologies, instruments, and institutions, on the other hand (metaphor: the liquid). Just as gas and liquid cannot be differentiated above supercritical temperatures and pressures, the abstract and the concrete cannot, and perhaps *should not*, be cleanly cut in investigations of the history of science, Galison teaches. After all, just as "wildly fluctuating phase changes reflect light . . . as if from mother-of-pearl" (40), who knows which gems of surprising insights would emerge from an actively integrated history, sociology, and philosophy of science? For William James's deployment of a related air/water metaphor, see Winther 2014c.

56. See also discussion of theoretical and empirical models in chapters 3 and 5.

57. Mac Lane 1986, §§6, 7, 431–38.

natural sciences.[58] In addition, the sciences sometimes influence how mathematics itself is conceived. One avenue to understanding how mathematics is informed by cartography is through its high-level definition of *mapping* and the two related concepts of "*Abbildung*" and "function."

The concept of "mapping" is important to linguistics and computing, as well as mathematics. According to the *Oxford English Dictionary*, *mapping* in mathematics is defined as "a correspondence by which each element of a given set has associated with it one element (occasionally, one or more elements) of a second set," and the *Oxford Living Dictionaries* defines *mapping* in linguistics and mathematics as "an operation that associates each element of a given set . . . with one or more elements of a second set."[59]

A mathematical mapping is thus a translation or transformation of objects or relations in one set (or universe) into objects or relations in another set (or universe). For instance, the function $y = x^2$ is a mapping of values of the $x$-dimension onto values of the $y$-dimension, and any given object (value) in the set of all $y$ is mapped onto by precisely two objects (values) in the set of all $x$, $+$ and $-\sqrt{y}$.

Cartography, the earth sciences, and mathematics have a long-intertwined history that has influenced the mathematical comprehension of mapping. Some of that history flows through the great German mathematician Carl Friedrich Gauss, who showed how to domesticate curved surfaces. Cartographic map projections, Gauss recognized, are a special case of curved surface transformations.

Starting in 1818, Gauss directed the geodetic survey of the Kingdom of Hanover (today Lower Saxony) (fig. 6.4), following methods developed by, among others, the Cassini family, whom we met in chapter 3. The experience of making literal cartographic, geographic maps informed Gauss's studies of curved surfaces to such an extent that it has been said that the nineteenth-century mathematical understanding of curved surfaces derived from Gauss's experiences on that project.[60]

Gauss was on a par with Isaac Newton in terms of his intellect and

---

58. On the "network of mathematics," see Mac Lane 1986, 417; cf. Weyl 1940.

59. Oxford English Dictionary, online ed., s.v. "mapping," sense 2a, accessed December 12, 2018; *Oxford Living Dictionaries*, s.v. "mapping," accessed December 12, 2018, https://en.oxforddictionaries.com/definition/mapping.

60. Dunnington 1955, 163. See also A. I. Miller 1972. Such curved surfaces included his investigations of geodesics and geodesic triangles, metrics, curvature, and developable surfaces for the transformation functions discussed in chapter 3. For his 1825 and 1827 essays, see Gauss (1902) 2005. For details on these ideas, see, for example, Friedman 1983, 8–12.

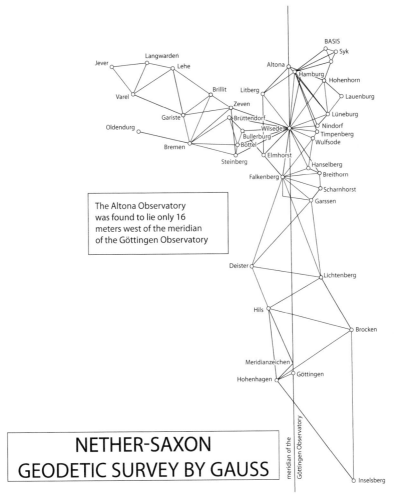

FIGURE 6.4. Cities and mountains in Lower Saxony that Carl Friedrich Gauss triangulated and surveyed with fieldwork from 1821 to 1825. The principal points were Hoher Hagen (here, Hohenhagen), Brocken, Inselsberg, and Hils. Gauss worked under the patronage of the Kingdom of Hanover and in conjunction with a Danish survey ordered by King Frederik IV. Names are as used by Gauss and are also checked against the map on the back of the old 10 Deutsch mark, which has Gauss on its front. (Dunnington 1955, 123, redrawn by Mats Wedin.)

influence on mathematics. One of his achievements was to show how to transform a curved surface by running it through a series of coordinate (graticule) changes in order to map it onto another curved surface—a surface named with the German word *Abbildung*.[61] Like *mapping* in English, *Abbildung* in German implies both the process of transforming one thing into another, and that very representation relation.[62]

Gauss's cartography and map thinking lie at the roots of both modern mathematics and cosmology. The concepts of *Abbildung* and mapping became increasingly important over time—especially in the twentieth century—in abstract algebra, analysis, and set theory. In particular, Gauss's student Bernhard Riemann used the *Abbildung* concept to explain manifolds, which are topological spaces that resemble (alternatively: *are*) classic Euclidean spaces locally, but often not globally.[63] Such abstract

---

61. Gauss 1873, 193, 212. Papadopoulos lists a number of influential mathematicians concerned with map projections and other cartographic matters, empirical and theoretical: "Lambert, Lagrange, Gauss, Beltrami and Darboux" (2016, 2). His article details Leonhard Euler's contributions to the study of map projections. Euler "was perhaps the most prolific eighteenth-century mathematician" (J. P. Snyder 1993, 70). Interestingly, and returning to themes from chapters 3 and 4, in 1777 Euler provided "the oldest proof that one can not map a sphere into a plane without distortion" (Lapaine and Divjak 2017, 270).

62. The canonization of "mapping" as a translation for *Abbildung* can clearly be seen in the English translation of Hermann Weyl's mathematical and physical texts. Weyl was intimately familiar with *Abbildung* from geometric and topological contexts harking back at least to Gauss (e.g., Weyl 1952, 1950). In discussing "a definite continuous *representation* of [a thermodynamic] surface"—in other words, the state-space map we encountered above—Weyl writes, "Geographical maps are familiar instances of such representations" (1952, 86). Moreover, the translator of Weyl's *Space—Time—Matter*, Henry L. Brose, wrote, "*Abbildung*, which signifies representation, is generally rendered equally well by transformation, inasmuch as it denotes a copy of certain elements of one space mapped out on, or expressed in terms of, another space" (quoted in Weyl 1952, xiii). Another term, *zuordnung*, has multiple senses in mathematics, including "functional coordination" and "mapping." John von Neumann's text on quantum mechanics translates various of our German terms or concepts to "mapping." For instance: "Abbildung" (1932, 85) → "mapping" ([1955] 1983, 163); "Zuordnung" (1932, 31) → "mapping" ([1955] 1983, 58). A fuller exploration of the *mapping* term and concept in physics and mathematics, in and across languages, would be fascinating.

63. For instance, see Riemann (1851) 1867, his inaugural dissertation. For discussion of Riemann's deployment of *Abbildung* and *Bild* (picture), see Ferreirós 2012. On these terms, including their relation to current concepts of model and representation—as well as to the older German terms *Darstellung* (account, rep-

spaces are critical to modern theoretical physics. Gauss's map thinking also influenced Einstein's theory of general relativity, a central tenet of modern cosmology.[64]

Although space may not be invoked specifically in the contemporary mathematical *Abbildung* mapping concept, the concept is clearly charged with map thinking, via Gauss, his contributions to geometry and topology, and his influence on Riemann and Einstein.

Mathematical mappings, which rely on functions and coordinates, are—just like geographic maps—both correspondences and transformations. As examples, consider the equation for the parabola above, or a diagrammatic proof of the Pythagorean theorem for a right triangle.

The very concept of "function" is close to "mapping," even if not cartographically inspired. The concept has a rich history, starting with the calculus of the seventeenth century, and it remains fundamental to all of mathematics.[65] The *OED* reports that in its mathematical sense, the term *function* is "due to Leibniz and his associates [who use] . . . *functiones* in a sense hardly different from its ordinary untechnical sense, to denote the various 'offices' which a straight line may fulfill in relation to a curve, viz. its tangent, normal, etc."[66] The word emphasizes activity, and in Latin is

---

resentation, or depiction) and *Vorstellung* (idea, concept, or performance)—see Janik and Toulmin 1973; Martínez 2001.

64. Einstein fully acknowledged this influence, writing in 1949 to Gauss's great-granddaughter that "the importance of C. F. Gauss for the development of modern physical theory and especially for the mathematical fundament of the theory of relativity is overwhelming indeed." Quote © Albert Einstein Archives, The Hebrew University of Jerusalem. Reprinted with permission. Indeed, in 1922 Einstein announced that he had understood in 1912 that "Gauss's theory of surfaces holds the key for unlocking this mystery" of the relation between geometry and physical laws. He continued, "I realized that Gauss's surface coordinates had a profound significance" (Einstein's December 1922 Kyoto address, cited in Pais 1982, 211). Einstein also noted: "However, I did not know at that time that Riemann had studied the foundations of geometry in an even more profound way" (211–12). Einstein, it seems, thought Gauss was the only person who could have come up with the theory of surfaces (Moszkowski [1921] 2014, 186; Dunnington 1955, 350).

65. Lützen 2003 reviews the history of the function concept. De Toffoli 2017 and De Toffoli and Giardino 2014 allude to the power of the map analogy for the structures and practices of topology (see chapter 4, note 85). Thomas Icard provided constructive feedback.

66. *Oxford English Dictionary*, s.v. "mathematics," sense 6, which is "a variable quantity regarded in its relation to one or more other variables in terms of which it may be expressed, or on the value of which its own value depends."

defined as "a performing, executing, discharging."[67] This is an important sense in which the term is used in modern mathematics.

Mapping—whether in mathematics and physics or cartography—is also an activity. It is a transformative process of establishing robust relations between representations, or between a representation and an ontologized world. Elements or relations in one area are seen as parallel to elements or relations in another area. Mathematical mappings and functions are thus *analogous maps* (fig. 2.2): they are a relation between two distinct sets (or two universes), just as a cartographic map is a relation between a territory and a public, visual representation. Mathematical maps serve many of the same purposes as cartographic maps: guidance, correspondence, discovery, and inspiration.

In this section, I have tried to map think mathematics, refracting formal methods through a cartographic lens. Mathematical structures may be too abstract and distinctive to be called maps per se, and of course they draw on many other resources besides cartography. Mathematical structures can be called maplike, however, in their general organization and use, and even in parts of their history.

## Conclusion

We have looked at examples of four map types from several intellectual disciplines of the physical sciences writ large. Cosmologists construct extreme-scale astronomical maps that highlight empirical features of the universe, and such maps help us understand the cosmic cradle that has enabled both life and consciousness to emerge.

Returning to Earth, the literal cartographic maps of geology hint at the interplay of plate tectonics and biological evolution. In thermodynamics and physical chemistry, we can cartographically interpret the phase or state space that physical chemists use to explain state changes of matter.

Mathematics provides one methodological structure of the physical sciences, and we find the general map analogy at work in the history and philosophy of mathematics.

Each kind of map can be associated with certain underlying spatial assumptions and particular strategies of space thinking: cosmological graticules, dual space, variables as dimensions, and dynamic functional transformations, respectively. Other systems of underlying assumptions figure in these disciplines, but these maps suffice to illustrate the role of map thinking in the formal sciences and mathematics. Such maps provide

67. Lewis and Short 1879.

a visual, theoretical, and conceptual space enabling the representation of processes unfolding dynamically through time, at different scales.

The theoretical physicist Steven Weinberg has written of cosmology: "Now we must begin to fill in this map with the islands of matter and the seas of radiation that make up the physical contents of the universe."[68] The same might be true, by analogy, of the map I am here creating of map thinking.

---

68. Weinberg 1972, 469.

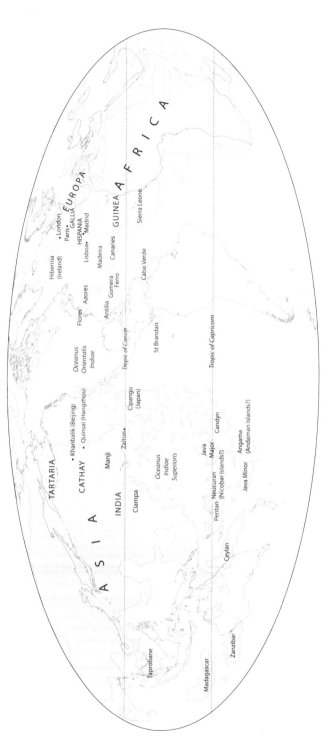

FIGURE 7.1. *Columbus's Worldview* is a consensus of various original and reconstructed maps. This best-practices consensus map shows the world as imagined by Columbus and others of his era, with the modern shapes of mapped land underneath. In Toscanelli's (which Columbus almost certainly studied; see text) and Martellus's (which he may have studied) maps, the three classical continents, Africa, Asia, and Europa, are immense. (Illustration by Mats Wedin and Rasmus Grønfeldt Winther using data from Bartholomew 1890, fig. 2, 10 "Form of the Ocean According to Toscanelli, Martin Behaim and Columbus"; PawelMM 2010; Winsor 1886, 103; Van Duzer 2012, fig. 1, 9, 2019, fig. 1.1, viii; M. A. Mitchell 2015, "1474 Paolo Toscanelli [Conjectured]" [map].)

# 7

# Mapping Ourselves

Modernity emerged with perhaps the most impactful cartographic reification ever: the European stumbling drive westward and southward into a "new world," for them. This reification was in part guided by ancient and Renaissance maps; crystallized with the landing of three Spanish ships, *La Niña*, *La Pinta*, and *La Santa María*, on what is now the Bahamas on October 12, 1492; and carried out according to a worldview of Christian religious absolutism and European superiority.

A key player was, of course, Christopher Columbus, who is alternately cast as historical hero and villain. An autodidact who fell in love with ocean exploration as a teenager, Columbus read copiously about world geography and fancied himself informed enough to declare that he had discovered a new and quicker route to Asia.[1] Given our own many wrong maps, in science and beyond, we should foster humility in all the ways we map and explore the world. Both accurate and inaccurate maps have consequences.

Rebounding from previous failures to pitch his exploration, Columbus turned to Ferdinand and Isabella of Spain, who ultimately financed his dubious venture. With four separate voyages from 1492 to 1502, Columbus instigated some of the largest and most violent geographic and demographic disruptions in the history of our species—disruptions that came to define modernity.[2]

The relentlessness with which Columbus pursued his endeavor reflects

---

1. As the Yale historian Edmund S. Morgan wrote, "[Columbus] studied *these* books, made hundreds of marginal notations in them, and came out with ideas about the world that were characteristically simple and strong and sometimes wrong, the kind of ideas that the self-educated person gains from independent reading and clings to in defiance of what anyone else tries to tell him" (E. S. Morgan 2009, 6).

2. McLean 1992; Sardar 1992; Mann 2011.

the fervor of his confidence in his calculations. You see, Columbus had prepared heavily for his round of investor pitches. Having studied Pliny, Ptolemy, and Marco Polo, among many others, and corresponded with the Florentine astronomer, astrologer, geographer, and cartographer Paolo dal Pozzo Toscanelli, Columbus optimistically calculated that Eurasia occupied nearly 80 percent of Earth's circumference; but his work was based on information so miscalculated that his generous allowance was thousands of kilometers off. In his view, the trip from Lisbon to Japan or China might be as little as 4,000 or 5,000 kilometers.[3] In reality, the trip would have been approximately five times that.[4]

To his final days, Columbus clung to an image of a planet without the Americas and the immense Pacific Ocean.[5] Columbus not only conflated map and world, he confused peoples from one part of the world with peoples from another. In his original letter announcing his discoveries to King Ferdinand, he described *Indians* ("los Indios") living in *the Indies* ("las Indias").[6] He had fallen into a trap that can ensnare any of us: placing outdated, disproven, or limited ideas above past or present evidence. This trap is pernicious reification.

While the maps that Columbus worked with are lost, a modern reconstruction using various contemporaneous maps as sources illustrates his worldview, broadly shared by many others (but not all) at the time. One source of *Columbus's Worldview* (fig. 7.1) is Toscanelli's 1474 map.

---

3. Wilford explains, "Columbus stretched the length of the Eurasian continent, from Cape St. Vincent [in Portugal] to eastern Asia, from Ptolemy's already inflated [*sic*] 177° to 225°. To that Columbus added another 28° from the discoveries of Marco Polo, plus 30° for the reputed distance from China to the east coast of Japan. Columbus had now accounted for 283° of land. Since he proposed to sail to Asia from the Canaries, which were 9° west of Cape St. Vincent, Columbus confidently concluded that he had only 68° of ocean to cross before reaching Japan" (2000, 77–78). Moreover, Columbus had diminished the radius of Earth, "scal[ing] downward the estimated length of a degree of longitude, making it 25 percent smaller than Eratosthenes had calculated and 10 percent smaller than Ptolemy had taught" (78). This gives approximately 5,000 kilometers of ocean to cross. The only change I would make to Wilford's statement is that all of Eurasia is actually approximately 180° of longitude wide.

4. Heading west, Lisbon to Shanghai (9° west to 121.5° east) is approximately $360° - 130.5° = 229.5°$ of longitude; at 30° latitude, 96.5 km/° longitude × 229.5° = 22,147 km (= 13,761.508 miles).

5. Morison 1942, 384–85. The story is a bit more complex, as Columbus did believe in an antipodal earthly paradise, but he took all his (Caribbean) discoveries to be extensions of Asia. See note 9 below.

6. Columbus 1493.

Details remain murky, but it is known that Columbus corresponded with Toscanelli, who sent him a letter and chart in 1481. It was likely onboard Columbus's flagship, *La Santa María*.[7] A globe created in 1492 by Martin Behaim shows nearly the same layout as Toscanelli's map.[8]

Drawing on assumption archaeology, four assumptions common to this late Renaissance cartographic worldview about Earth can be delineated: (1) Earth is a globe; (2) Earth is a *small* sphere (relative to correct Ancient Greek, Islamic, and modern science); (3) there is no land, sailing westward, between Europe and Asia (i.e., the Americas do not exist); (4) Eurasia is enormous, occupying much of the globe.[9] Such assumptions can be read from figure 7.1, which is superimposed on a modern map of continental shapes. Captured by this cognitive model and mental map, Columbus, Toscanelli, Martellus, and Behaim, among many others, shared the view that the bulk of the eastern parts of Asia were "the Indies"

---

7. Wilford 2000, 77–78; Crone 1978, 47. Apparently, Columbus consulted a map on his first voyage, which Bartolomé de las Casas claimed was the one Toscanelli had sent. Columbus's own calculations were based on selective choices of some sources (e.g., Alfragan, the medieval Islamic geographer) over others (e.g., Eratosthenes, the first person to calculate the circumference of Earth—and correctly, at that).

8. There is no evidence that Columbus knew about Behaim's globe, yet his model and Behaim's agree closely in their geographic imagination; both were based on inappropriate data and arguments, ignoring other much more accurate ancient wisdom (Phillips and Phillips 1992, 79).

9. Other assumptions pertain to the ontology of geography. In particular, Nicolás Wey Gómez discusses the role the tropics played in late medieval Europe: "The transition from a view of the tropics as forbidding inferno to one of the tropics as prodigal paradise was slow to dawn on the Age of Exploration" (2008, 53). A range of religiously motivated ontological questions set hearts and mouths afire. Was the world's geography "closed" (EurAsiaAfrica is all that exists) or "open" (new lands, or extended networks of EurAsiaAfrica, or both, are possible, and exist)? And what would happen if you sailed not just west, but also *south*? Columbus believed the tropics to be inhabitable, and also to hold great wealth. He and his supporters "argued that the inhabited world generally extended not just farther east and west than some believed but also farther north and south, into the allegedly inhospitable regions of the globe" (Wey Gómez 2008, 111). Interestingly, Columbus did believe there would be an *antipodes*, and that he had found, on his 1498 third voyage, the lands and river of earthly paradise (in what is now Venezuela, with the Orinoco)—an enormous ("grandísima") austral land, one of many of which there never was news ("haya otras muchas en el Austro de que jamás se hobo noticia"; Colón [1498] 1892, 288; my translation). He thus also assumed (5) an open world and (6) habitable tropics (cf. Zuber 2011).

and that the ocean there was solely the Indian Ocean.[10] The world ontologized from these maps was, by our standards, very small and incomplete.

In this chapter, I show that insisting on one best map or one overarching theory weakens map thinking and scientific thinking. I first explore *migration maps*, then *brain maps*, then *statistical causal maps*. For each, I identify simplifying, or *abstractive-averaging*, assumptions. Next I trace the potential pernicious reification of these assumptions, discussing *arrowized assumptions* in migration maps, *decompositional assumptions* in brain maps, and *linear-model assumptions* in statistical causal maps.

Finally, by identifying countermaps, I make explicit the misleading tendency of standard maps to universalize and narrow the world.

## Migration Maps

One human activity subject to rich and complex mapping practices is migration. People become migrants in many ways. They can be vulnerable and oppressed, or economically or militarily powerful. Migrants include those walking "out of Africa" in the early history of our species; arriving in Australia fifty thousand years ago; populating North America probably some ten thousand to twenty thousand years ago; and invading disparate territories as many Roman, Persian, and Mongol soldiers, for example, did over the centuries. Migrants are people who move from one geographic place to another for a significant period of time.

Migrants are subject to numerous representations. They are homogenized, classified, and normed by governments, courts, nongovernment organizations (NGOs), politicians, and academic experts such as anthropologists, geographers, and political scientists. The way people who move within or across geopolitical regions are represented by mapmakers is particularly important in our contemporary globalized and politicized life.

Map thinking about migration is a tense, pluralistic space—a socially and politically contested area of discourse and action. Considering "migrants" as a *kind* or *category* of person based on one characteristic allows us to understand how certain peoples are perniciously reified as *nothing but* migrants, both by social institutions and by themselves, internally and psychologically.[11] How might alternative classifications such as *refu-*

---

10. In figure 7.1 above: Oceanus Orientalis Indiae; Oceanus Indiae Superioris. See also Behaim's globe in Ravenstein 1908.

11. For some anthropological and political theory on migration, see Gammeltoft-Hansen and Nyberg Sørensen 2013. On kinds of people, and the looping effect (the process of individuals understanding and experiencing them-

*gee, asylum seeker, pioneer,* or *expatriate* change the identities and expectations of a person who moves? How are people escaping drug cartels different from those escaping rising seas or more generally striving for a better life? How can we show patterns of movement without dehumanizing those whose journeys are the basis of our cartographic creation?

### ARROWIZED ASSUMPTIONS

*Arrowized assumptions* are a type of abstractive-averaging assumption. Like all abstractive-averaging assumptions, they simplify a subject matter by abstracting away and averaging out differences among individual people, objects, or processes. What distinguishes *arrowized assumptions* is that they treat all people in a particular migration flow as similar. In other words, arrowized assumptions idealize and simplify human migratory patterns of movement. Representations produced according to these assumptions tend to portray people as if they lack agency, marching in a linear and automatic fashion over time. Many maps sum up migrants' journeys with monolithic arrows that seem to sweep a population from one region to another, like a weather front.

Arrowized assumptions introduce significant risks of pernicious reification. They can be challenged, however, by countermapping migrants' movements, struggles, and subjectivities. Although people are indeed moving, they are more than their movement. Some radical countermappings demonstrate this by reimagining space in a functional or fractured manner, while others use traditional Euclidean space.[12]

An assumption archaeology of standard migration maps produces the following list of simplifying arrowized assumptions:

1. *Migrant averaging.* Every migrant is roughly equivalent to every other migrant. Individuals are perceived as having similar life stories, fears, and expectations, and being subject to similar conditions. If differentiated at all, they are classed into a very small number of internally homogeneous kinds.
2. *Fixed routes or idealized patterns.* Migrants are assumed to travel along fixed routes, typically represented by arrows or other symbols of physical movement (fig. 7.2). Alternatively, idealized net movement patterns among adjacent regions are represented. Each

---

selves according to the categories normalized by power structures), see Hacking 2002, "Making Up People" chap. 6, 99–114; 2007a.

12. For the functional or fractured reimagining, see Olsson 1965; Tazzioli 2015. For countermapping within a traditional Euclidean space, see J. Walsh 2013.

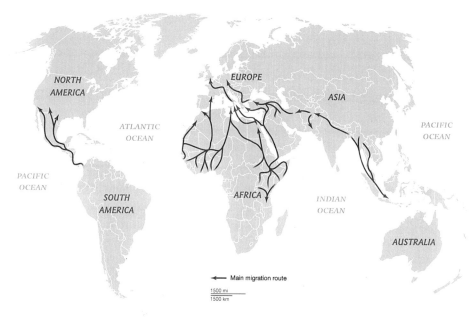

FIGURE 7.2. World migration routes. Arrows indicate averaged migrant flows. Political borders, in white, partition countries. (Matt Chwastyk / National Geographic Creative, National Geographic Image Collection.)

region (whether county, state, or department) is abstracted as a mere centroid, point, or averaged border (fig. 7.3).
3. *Lack of individual agency.* Migrants are depicted as passive. Like molecules of water in a river, they are carried along by larger forces, such as those of history and calamity.
4. *Linear temporality.* Migrant traveling can be neatly represented with arrows or other visual indicators of flow in a Euclidean, GIS space. Movement seems to happen within linear, continuous, and isotropic time.

## ARROWIZED MAPS

We can flesh out three of these four assumptions by examining the work of key historical thinkers. In the late 1800s, the renowned German-English geographer and cartographer Ernst Georg Ravenstein developed an idea that underpins modern migration theory.[13] Ravenstein inferred a set of *laws of migration* from numerous migration maps of the United

13. The cartographer Waldo Tobler (1987, 162) has referred to Ravenstein 1885 as "famous," an unusual adjective in his vocabulary. In turn, Tobler 1981 and

Kingdom of Great Britain and Ireland. He found that most migrants tend to proceed only short distances because "each main current of migration produces a compensating counter-current" (fig. 7.3), and that women are more migratory than men.[14] Ravenstein classified migrants as *temporary, local, short-journey, migration by stages,* or *long-journey*.[15] His classification system serves as the primary excavation site for my archaeology of arrowized assumptions.

Ravenstein's "temporary migrants" type illustrates arrowized assumption no. 1 from the list above, migrant averaging. Ravenstein judges temporary migrants to be a "floating element." He writes that this class "makes its presence felt very decisively at our naval and military stations, at health and pleasure resorts, in university towns, and in places abounding in boarding schools."[16] His classification thus averages people from distinct stages of life and of different socioeconomic status. With the temporary migrant class, Ravenstein lumps together adults migrating because of their work, families on vacation, and young adults or children away at school.

Ravenstein's "migration by stages" type exemplifies arrowized assumption no. 2, fixed routes. According to Ravenstein, a migrant "wanders from parish to parish, settling down at each place for a time." Ravenstein writes about distinct "streams" passing through different cities and counties on their way from Ireland to London by way of Liverpool, Cheshire, and Staffordshire or by way of Plymouth and Hampshire.[17] It is as if these streams are etched into the British landscape.

As illustrated by figure 7.3, standardized maps use abstraction to show idealized patterns of movement rather than showing the myriad, meandering routes actually taken. Twentieth-century strategies for describing traveling and movement include idealized patterns with symbols and colorful arrows to indicate quantitative features of migrations.[18] Many modern migration maps portray fixed, abstracted, and idealized routes, even though many of the individuals traveling the territory do not follow them.

Finally, arrowized assumption no. 3—lack of individual agency— seems conjoined with assumption no. 1, migrant averaging. Ravenstein presents some evidence that the main motive for migrants on the British

---

1987 are featured in the "Novel Methods for Flow Mapping" section of a standard textbook; see Slocum et al. 2005, 364–65.

14. Ravenstein 1885, 199, 196.
15. Ravenstein 1885, 181–84.
16. Ravenstein 1885, 183.
17. Ravenstein 1885, 183.
18. Tobler 1981, 1987; Phan et al. 2005.

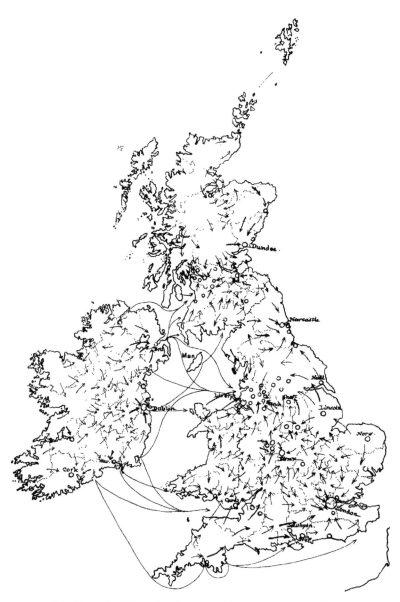

FIGURE 7.3. *Currents of Migration* shows spatial movement of people across counties of the United Kingdom of Great Britain and Ireland. Although Ernest G. Ravenstein used two censuses as data sources for other maps in his article, no information is given for this map. Cities are represented by circles; counties by stippled lines; movement of people seems to be indicated by arrows. (Ravenstein 1885, map 5. Courtesy of the Royal Statistical Society.)

Isles to leave home is the search for work.¹⁹ He assumes that more remunerative work is the primary, if not sole, reason for leaving home. And yet people also move out of other desires and dreams. Ravenstein's reasoning highlights labor as a paramount motivator for migration, and generalizes the decisions of individuals by providing only a broad-strokes context for their struggles.

### COUNTERMAPPING MIGRATION

Countermapping is a way to challenge standard arrowized assumptions. It can reimagine who and what migrants are; whose purposes migration serves; and how governments, NGOs, researchers, and others can make representational decisions that are more inclusive and responsible. Following my multiple representations account, countermapping provides a new way to see things allowing us to make new sets of worlds and realms.

For instance, consider the "death maps" presented by Humane Borders / Fronteras Compasivas.²⁰ These maps indicate the location at which each of approximately three thousand Mexican and Central American migrants slogging through the Arizona desert have died, usually of dehydration and exposure. The organization has also distributed these death maps as warning posters with arcs indicating the distance, in number of walking days, from several cities or border crossings. These countermaps might dissuade some people from crossing into the United States through the Arizona desert, but they also show the locations of critical emergency water stations placed by the organization to reduce the number of deaths.²¹

Another migration countermap is the *Spaces in Migration* map, which instead of tracing routes of migration marks some impulses and impediments of migration (fig. 7.4). This map, by Robby Habans, represents the experiences of Tunisians escaping revolution in their home country by way of Italy, including locations of departures, shipwrecks, and "struggles, resistances, escapes." *Spaces in Migration*—both the map and the multiauthor book that it illustrates²²—challenges cartographic narratives that average or lump together all peoples escaping economic

---

19. Ravenstein 1885, 181.

20. See Humane Borders / Fronteras Compasivas and Puma County Office of the Medical Examiner, n.d.

21. James Walsh writes: "By approaching technology as an ally ... spatial tracking and geosurveillance can be ethically designed and employed to promote democratic practices and address injustice" (J. Walsh 2013, 982).

22. See Garelli, Sossi, and Tazzioli 2013.

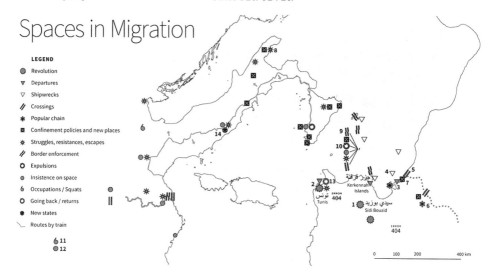

FIGURE 7.4. *Spaces in Migration* captures the nonlinear voyages taken by Tunisians escaping revolution in 2010, moving through Italy, and eventually ending up in Paris. The map represents hotspots of revolution in Tunisia, multiple points of border enforcement, and deadly shipwrecks. Map orientation seems purposefully unconventional. I read this as a countermap. (Garelli, Sossi, and Tazzioli 2013, 170–71. Courtesy of Pavement Books and Martina Tazzioli.)

misfortune or sociopolitical upheaval into a few kinds, such as migrant or refugee. This map works hard to respect the complexity and uniqueness of multiple migrant narratives.

In line with the broader conceptualization of *representation* explored in part 1, countermaps open up possibilities for imagining new kinds of space. For instance, Martina Tazzioli speaks of "spatial turbulences," "scattered spaces" and of migration maps that "fracture in the face of migrant[s'] unmappable spaces." In a classic formalist study, Gunnar Olsson calls for "physical distance" to be "replaced by a new concept vaguely termed 'real,' 'functional' . . . or even 'emotional' distance." In his study of migration in Sweden, Olsson constructs "nodal regions" unequal in physical size but equivalent in "population, economy, etc."; agential and social factors, rather than mere physical factors, anchor the space of migration maps in his study. Olsson's cartographic suggestions are intended less for wayfinding than for honoring migrant subjectivity and agency, and thereby raise political consciousness.[23]

Countermaps of migration invite humanization by explicitly repre-

23. Tazzioli 2015, 3, 17, 16; Olsson 1965, 8–10.

senting personal narratives and choices, as in figure 7.4, or by making room for intention—for example, by distinguishing refugees from retirees. Countermaps add first-person experience and narrative back into the map, depicting migration as contingent and complex, agential and intentional.[24] Such maps need not assume a constant timeline or a fixed route. In contrast to arrowized assumption no. 4, linear temporality, time can be extended as people stay for a while or partly return. As a result, migration countermaps are much more than Euclidean spaces with absolute Newtonian clocks ticking in the background. They include a personal dimension calibration: a sense of *place* in addition to space.[25] Contrary to implicit graphical assumptions in standard migration maps such as *National Geographic's* (fig. 7.2), migration countermaps presume neither that arrows of migration already exist in the world (arrowized assumption no. 2) nor that arrow lengths correspond neatly to determined times (arrowized assumption no. 4). By expanding our conception of what constitutes space and time, we expand our means of representation.

## Brain Maps

In the first half of the nineteenth century, many educated Europeans became convinced that the outside shape of the skull conformed to the brain within, whose specific structures corresponded to "brain functions" such as love of children, need for approval, and wit. The Scottish lawyer and phrenologist George Combe argued that "the organs of the mind can be seen and felt, and their size estimated—and the mental manifestations also that accompany them can be observed, in an unlimited number of instances."[26] This purported science is known as *phrenology*.

Today, brain scientists officially reject phrenology, but not the idea that different parts of the brain perform assigned functions, at least to an extent. Indeed, debates about relations between brain structure and mental function are a driving force in the mind and brain sciences. But what does it mean to say, for instance, that the prefrontal cortex is associated with executive functions such as problem solving, or that the amygdala stores emotionally laden memories?

To achieve contextual objectivity in cognitive science and neuroscience, we must start by observing certain simplifying assumptions that underlie research in these areas—namely, *decompositional* assumptions,

---

24. Line Richter provided constructive feedback.
25. For a philosophical history of the concept "place," see Casey 1997.
26. Combe 1847, 56.

which imply that a system is really a set of distinct and preprogrammed, yet interacting parts.²⁷

### DECOMPOSITIONAL ASSUMPTIONS

In his book *Incognito*, the neuroscientist David Eagleman writes: "The three-pound organ in your skull . . . is composed of miniaturized, self-configuring parts, and it vastly outstrips anything we've dreamt of building."²⁸ The idea is that the brain, like a computer or a machine, is a set of parts, each specifiable and comprehensible independently of the others. And the complexity of the mind depends on the complexity of the brain. As depicted in *The Turing Phrenunculus* (fig. 7.5), such parts are typically understood to read and interpret environmental input (D-A-T-A) according to strong internal programming (e.g., sex drives). That is, the brain is divided into distinct modules with intrinsic functions, running according to their own unique programming.²⁹ The tendency in this broadly Turing-phrenological decompositional model of the brain is to understand the system by decomposing it into its parts and mapping out the workings of each part.³⁰ Yet, I believe, paraphrasing Shakespeare,

---

27. Because my focus is on general relationships between brain structure and cognitive function, other important mapping efforts in the brain sciences are set aside. For instance, classic experiments on retinal maps in Hubel and Wiesel 2004, or topographic maps in the visual cortex (see Dehaene and Brannon 2011), or research on grid cells or place cells (see Montemayor and Winther 2015), or even on how time is cognized and mapped (Montemayor 2013).

28. Eagleman 2011, 2; cf. Carter 2010, chap. 1.

29. My concerns here overlap with Robert A. Wilson's, who sees "smallism" as ubiquitous across the sciences: "Small things and their properties are seen to be ontologically prior to the larger things that they constitute, and this metaphysics drives both explanatory ideal and methodological perspective" (R. A. Wilson 2005, 22; cf. Winther 2009d). Somehow, the small carries causal or constitutive power, or both, of the large. But holists disagree (see Levins and Lewontin 1985; Oyama [1985] 2000; Winther 2006b, 2011b; McGranahan 2017), as do those focused on interaction (see Winther, Wade, and Dimond 2013; Longino, forthcoming; see also below). Brian Cantwell Smith, Helen Longino, Lucas McGranahan, and Alina Shron provided constructive feedback.

30. Although this section focuses on the brain, the fields of neuroscience, computer science, and philosophy also shed light on the ways the *mind* might operate. In particular, *the computational theory of mind* remains strong, inspired and shaped by the pathbreaking work of Alan Turing (1937, 1938, 1950), reflected in figure 7.5. Some less-traditional computationalists argue that the mind can contain semantic representations (and not just pure syntax; see Rescorla 2017)

FIGURE 7.5. *The Turing Phrenunculus.* What is on a person's phrenological brain? This brain is surrounded by a Penfield homunculus, and its mind operates according to Turing computational theory. The *Turing machine*, a kind of universal computing machine (here with gears, valves, and dial indicators), operates according to certain built-in algorithms or programs, while reading D-A-T-A from the real world, interpreted by many to be binary information (0/1). Freudian parts with actual phrenological names are depicted. (Illustration by Larisa DePalma; concept by Rasmus Grønfeldt Winther; inspired by Penfield and Rasmussen 1950 and Hermans 2013.)

that there is more between heaven and Earth than can be dreamt of in our philosophy of the brain and the mind.

Phrenology helps us understand the abstractive-averaging assump-

---

and could be operating according to statistical Bayesian patterns (and not just according to purely deductively closed rules; see Tenenbaum et al. 2011). Dissenters to computationalism include Gessell and De Brigard (2018); Horgan (2016); Horst (2007); and Brian Cantwell Smith (1996, 2015).

tions underlying significant areas of mainstream cognitive neuroscience, and certainly of public understandings of neuroscience. An assumption archeology of concepts and principles underlying phrenology identifies five decompositional assumptions about how the brain is structured and how the mind functions. These decompositional assumptions remain common in contemporary strategies for studying the brain and mind, and possibly date as far back as Aristotle's anatomical investigations:[31]

1. *Distinct structures.* Distinct brain structures or regions exist, such as the amygdala or hippocampus. An ongoing goal of neuroscience is to accurately partition and map such structures, at various scales. These structures can exist at different levels, from neuronal cells and tissues to larger regions. And while they do vary somewhat in size or location among individuals, the regions are fairly constant across individuals.
2. *Unique functions.* Each of these regions or structures has a unique role. This could be either one main function, such as memory; or a consistent set of functions, potentially at various levels of processing, such as short- or long-term memory subfunctions. While functions—or subfunctions—might occasionally overlap across brain regions, the overall functional profile of each brain structure is distinctive.
3. *Clear measurability.* Brain regions or structures are measurable using well-defined empirical procedures and calibration strategies such as cell morphology, tissue structure, or location within a spatial grid. Measurements improve with the development of new scientific technologies and methodologies. Mental function, unlike physical structure, is only *indirectly* measurable. Since the late nineteenth century, researchers have assessed function with a variety of protocols, including but not limited to: cognitive tests, measurements of skull form and shape (à la phrenology), and, today, functional magnetic resonance imaging (fMRI).
4. *Strong structure-function mapability.* A central aim is to get as close to a one-to-one structure-function mapping as possible. The number of one-to-many or many-to-many correspondences between structure and function should be reduced to the extent possible. Each structure or region would ideally map to one function or a small, unique set of functions. For instance, the amygdala could map to emotional states such as fear and anxiety.
5. *Explanation of abnormality.* Cognitive dysfunctions or abnormalities are primarily caused by structures at various scales failing to

---

31. In *On the Parts of Animals,* for instance.

fulfill their standard function. What we label as mental disorders or illnesses originate solely and simply in brain regions. Somatic states, social interactions, and environmental features are mostly irrelevant.

I shall use these five basic decompositional assumptions to explore some standard ways of mapping the brain. I will also introduce countermaps that resist the pernicious reification of these assumptions.

### PHRENOLOGICAL MAPS

It is not difficult to trace these decompositional assumptions to nineteenth-century phrenology. The neuroanatomist Franz Joseph Gall, founder of phrenology, argued that innate mental faculties in humans and animals are grounded in distinct and independent brain structures, or "brain organs."[32] In his work, Gall reports discovering these brain organs by various means, including mapping skull protuberances from plaster models of skulls and mapping brain lesions in people with brain disease. Gall's research was and remains controversial, and some of his publications were criticized even in the nineteenth century, including for plagiarism and for neglecting relevant work by other researchers.[33]

*The Turing Phrenunculus* (fig. 7.5) resonates with a classic nineteenth-century phrenological map of the skull entirely consistent with decompositional assumptions nos. 1 and 2 above. This map portrays clearly demarcated areas of the skull alleged to correspond to brain organs that are in turn mapped to mental functions, such as self-esteem and secretiveness. Also consistent with decompositional assumption no. 3, such cognitive faculties are taken to be measurable. With assumption no. 4, structure-function mapping is taken to be ideally one-to-one.

Finally, regarding decompositional assumption no. 5, the phrenologist Combe argued that humans have the "*possibility* of health, vigour, and organic enjoyment," provided that various "organic laws" are met. Examples of such laws or conditions, he said, include an infant having a complete complement of brain and body parts, and the satisfaction of mental exercise.[34] But he held pathology to be caused primarily by intrinsic factors.[35]

---

32. F. J. Gall 1798, 1835; Goyder 1857, 143–52.

33. Gall replied that he "never failed to exhibit literally the opinions of" a wide variety of thinkers (F. J. Gall 1835, 303).

34. Combe 1847, 138, 133–41.

35. See also Hacking 1990, chap. 14, "Society Prepares the Crimes," which illuminates the social and political context of phrenology.

Phrenology presented a simplistic and indeed reductionist worldview of the mind-brain relationship. It continues to serve as a canonical example of pseudoscience. After all, it ontologized—and then universalized and narrowed—assumptions about mental faculties across all members of our species. Yet despite its flaws, notes the historian of science Sherrie Lyons, "phrenology played a crucial role in furthering naturalistic approaches to the study of mind and behavior."[36] Hacking refers to phrenologists as drawing "maps of the head."[37] Phrenology's powerful decompositional assumptions endure today in the form of partitioning our brain regions and associating them with aspects of mind and behavior, even beyond our species.

## THE SOMATOSENSORY AND MOTOR HOMUNCULI

Some of phrenology's "organs of the brain" model can be traced forward to the work of the renowned twentieth-century neuroscientist Wilder Penfield. Penfield was a Montreal neurosurgeon and brain "cartographer."[38] He is most famous (perhaps infamous) for experiments on patients who were awake and awaiting surgery for epilepsy. Penfield and collaborators electrically stimulated different parts of the patients' cerebral cortex, which is considered the "thinking part" of the brain, responsible for functions such as attention, memory, and thought. While different parts of the cortex were stimulated, Penfield's conscious patients told him what they felt, and the researchers observed which body parts moved.

Penfield used the responses of the patients to map locations in the cortex for both sensory and motor function. He created a distorted drawing of somatosensory and motor homunculi to convey the basic spatial arrangement of brain function.[39] An artistic rendition of such homunculi is shown wrapped around the large phrenologically structured head of the subject of figure 7.5.

Penfield and his Canadian neurologist colleague Theodore Brown

---

36. Lyons 2009, 51. Chap. 3, ibid., provides further details regarding phrenology and its social, cultural, and political impact.

37. Hacking 1990, 122.

38. P. J. Snyder and Whitaker (2013, 279) inform us that Penfield studied with Sir Charles Sherrington, an influential contributor to contemporaneous neuroscience, while at Oxford, and trained under Harvey Cushing, a leader of neurosurgery, while at Johns Hopkins. See also Costandi 2008.

39. For an image with at least four homunculi distributed across highly schematized sensory brain regions, see P. J. Snyder and Whitaker 2013, 286, fig. 7, taken from Penfield and Jasper 1954, 105.

Rasmussen were aware of the potential for pernicious reification of their images:

> But physicians have always felt the urge to seek an explanation before adequate information was at hand from which to draw conclusions. An excellent illustration of this failing, at a time when Galvani and Volta were just initiating the study of nerve conduction, was the appearance of a treatise on the *Anatomy and Physiology of the Nervous System* by Gall and Spurzheim (1810). In this pretentious vehicle the authors presented to the public the so-called science of the mind, which they termed *phrenology*.[40]

Despite this critique, Penfield and Rasmussen adopted phrenology's decompositional assumptions for mapping brain functions.[41] Their resulting homunculus, or "cortex person," maps have been immensely influential, long used by neuroscience students and researchers across the globe as heuristics and inspiration for memorizing the putative topographic organization of the sensory and motor parts of the cortex (fig. 7.5).[42]

Penfield and Rasmussen's seductive artwork easily distracts the viewer's attention from Penfield's actual patient experiments. Even if neuroanatomists are able to contextualize the representations and avoid pernicious reification, the same may not be true of beginning students and lay readers exposed to the homunculus images.[43]

Penfield's classic homunculi studies mapped cognitive function to brain location, and *his homunculus images became the brain itself*. Although Penfield employed careful vocabulary to characterize his representations, his figures contributed to pernicious reification. The concretization of Penfield's visualizations, especially as they have disseminated across the intellectual and lay public landscape, deserves further study.

---

40. Penfield and Rasmussen 1950, 1. The "treatise" referred to is probably Spurzheim's 1815 text *The Physiognomical System of Drs. Gall and Spurzheim*. Gall and Spurzheim do not seem to have written a text together. Penfield and Rasmussen called their own images "figurines." See Penfield and Rasmussen 1950, 44, 57, 214, 215.

41. For instance, multiple chapters of the book have subheadings with "localization" in the title.

42. P. J. Snyder and Whitaker 2013, 289.

43. P. J. Snyder and Whitaker 2013, 277, 289–90. Anderson (2014, 205) critiques this image. For a philosophically rigorous study of the diffusion of scientific knowledge, see Longino 2013, chap. 10, "The Social Life of Behavioral Science"; cf. Longino 1990.

## FUNCTIONAL MAGNETIC RESONANCE IMAGING (FMRI)

With modern fMRI technology, we can experimentally test Gall's and Penfield's assumptions about the structure and function of the brain, and explore the dangers of ontologizing—and then universalizing or narrowing, or both—decompositional assumptions. A medical imaging method commonly used on the brain, fMRI has had a profound impact on neuroscience.[44] In a 2008 review in the journal *Nature*, the neuroscientist Nikos Logothetis noted that approximately eight papers were being published *per day* that mentioned fMRI; 43 percent of all such papers published through 2007 investigated the functional or structural partitioning "associated with some cognitive task or stimulus."[45]

In most cases, fMRI works by detecting blood-oxygen levels. These levels are taken to be a surrogate for brain activity, because the molecule hemoglobin picks up oxygen in the lungs and delivers it to the brain, releasing it to needy neurons. The ratio between oxygenated and deoxygenated blood in different parts of the brain therefore indicates how active those structures are at any given moment. This kind of mapping has some of the same intentions as phrenology and Penfield's homunculus maps in that it also partitions the brain into regions and seeks to measure their independent function.

Maps built on fMRI data are as subject to pernicious reification as any other representation. Despite the high-tech nature of fMRI, inferences from the resulting images are hardly automatic and transparent. According to Logothetis: "The fMRI signal cannot easily differentiate between function-specific processing and neuromodulation [i.e., changes in nerve action caused by the fMRI signal itself] ... and it may potentially confuse excitation and inhibition."[46] Attempting to draw robust conclusions about causality from fMRI structure-function mapping (decompositional as-

---

44. Logothetis states: "Magnetic resonance imaging (MRI) is the most important imaging advance since the introduction of X-rays by Conrad Röntgen in 1895" (2008, 869). fMRI is a variation on MRI; whereas MRI focuses on structure, fMRI focuses more on function over time.

45. Logothetis 2008, 869. The number of PubMed abstracts mentioning fMRI has increased exponentially since the early 1990s (Poldrack, pers. comm., June 2017).

46. Logothetis 2008, 877. See also "the epidemiology of visualization" discussion in Rose and Abi-Rached 2013, 74–80; cf E. G. Jones 2008, which is a review of an atlas of cortical cytoarchitectonics.

sumption no. 4) demands critical thought, assumption archaeology, and integration platforms.[47]

COUNTERMAPPING THE BRAIN

Contextual objectivity in the brain sciences requires identifying and exploring countermaps. Let us turn now to two research programs countermapping the brain, that of the neuroscientist Russell Poldrack and the neuroscientist and philosopher Michael Anderson.

*Cognitive Ontologies: Poldrack*

One scientific, specialized use of the term *ontology*, derived from computer science, refers to controlled vocabularies predefining the words and phrases to be used in tagging information, for easier communication and more reliable information retrieval. *Ontology*, in this usage, means a set vocabulary and its definitions.[48] The map analogy provides one way to understand how scientific concepts are organized into such ontologies and how they can be overhauled.

Consider how Thomas Kuhn uses map thinking, and analogy 8 from chapter 2, to explore how a historian translates the scientific language of paradigms no longer operative (e.g., Galen's humoral theory) to contemporary paradigms.[49] A paradigm's conceptual ontology, Kuhn observes, is analogous to a coordinate system specifying latitude and longitude. It is thus a kind of partitioning frame. The same world can be mapped, in science, using different lexica or ontologies, and, in cartography, using distinct coordinate systems. The historian thus catalogs and systematizes a paradigm-specific scientific vocabulary, acting as an "interpreter and

---

47. For instance, brain electrical fields provide another way to match structure and function, when measured, e.g., with electroencephalograms (EEGs) or, more invasively, by electrode probing of local field potential (LFP) (Agarwal et al. 2014).

48. As Barry Smith, a philosopher who has been instrumental in establishing biomedical and other ontologies, has written, "Ontologies created by scientists must, of course, be associated with implementations satisfying the requirements of software engineering. But the ontologies are not themselves engineering artifacts, and to conceive them as such brings grievous consequences. Rather, ontologies ... are in different respects comparable to scientific theories, to scientific databases, and to scientific journal publications" (Barry Smith 2008, 21).

49. See Kuhn 2000a, 50–53; 2000b, 63.

language teacher."[50] For instance, Kuhn compares notions of "water" before and after the chemical revolution of the 1780s.[51] Prior to the work of the French chemist Antoine Lavoisier, water "was an elementary body of which liquidity was an essential property." Lavoisier and peers altered "the taxonomy of chemistry" such that a "chemical species" could exist in three states of matter: solid, liquid, and gas.[52] The ontology changed. Ice was now water, whereas before it was not.

In neuroscience, a *cognitive ontology* classifies cognitive functions and specifies the theoretical, experimental, and operational relationships among these functions and other concepts. According to Poldrack, underlying functional assumptions about mental tasks and activities are often folk ontologies—that is, commonsense ontologies—rather than more precise and empirically evaluated scientific ontologies.

Consider Poldrack's ongoing development of a cognitive ontology in a project at Stanford called the Cognitive Atlas, whose website declares:

> Cognitive neuroscience aims to map mental processes onto brain function, which begs the question of what "mental processes" exist and how they relate to the tasks that are used to manipulate and measure them.[53]

Much like Kuhn's historian, Poldrack and his colleagues are creating an integration platform, translating distinct conceptual ontologies from different neuroscience research programs and publications.[54] Thus, through a collaborative effort spanning institutions, the Cognitive Atlas has been populated with a basic vocabulary of mental processes and tasks used to calibrate and test those processes experimentally. These efforts integrate pluralistic critical thinking in disciplines as varied as neuroscience, medicine, computer science, statistics, and philosophy.[55]

---

50. Kuhn 2000a, 43–47.
51. See Kuhn 2000b, 81–83.
52. Kuhn 2000b, 82.
53. http://www.cognitiveatlas.org/about (accessed May 5, 2016).
54. Cf. Poldrack 2010.
55. In one study, a Poldrack team employed machine learning, statistical algorithms, and a 200,000-dimension data set to create a countermap that reversed the usual model of fMRI research: "Whereas previous imaging studies have nearly always focused on determining the neural basis of a particular cognitive process using specific task comparisons to isolate that process . . . our classifier analysis answered a very different question: What [mental] task is the subject most likely engaged in, given the observed pattern of brain activity?" (Poldrack, Halchenko, and Hanson 2009, 1370–71).

Poldrack's drive to replace a commonsense view of cognition with a scientific one resonates with a critique of the reification of intelligence and IQ by the evolutionary biologist Stephen Jay Gould. According to Gould, "Reification—[is] in this case, the notion that such a nebulous, socially defined concept as intelligence might be identified as a 'thing' with a locus in the brain and a definite degree of heritability—and that it might be measured as a single number, thus permitting a unilinear ranking of people according to the amount of it they possess."[56] For Gould, reification in general is "the propensity to convert an abstract concept (like intelligence) into a hard entity (like an amount of quantifiable brain stuff)."[57] These might be reasonable concerns to have about fMRI studies as well, and about cognitive ontologies that have not been rigorously tested and integrated.[58]

## Functional Fingerprints: Anderson

In another example of countermapping, the neuroscientist Michael Anderson operates a contextually objective research program moving beyond decompositional assumptions. In explicitly resisting decompositional assumptions nos. 1–4, Anderson's work stands as a testament against what he calls "grand unified theories of brain function."[59] Furthermore, Anderson (among others) vehemently resists the computationalist metaphor of thinking and feeling as software on the brain's hardware,[60] which—recalling David Eagleman's quote above—pervades contemporary cognitive science (fig. 7.5).[61]

Anderson's dynamic interactive-brain model builds on the twin concepts of "neural reuse" and "affordance." Neural reuse is the notion that "individual neural elements (at multiple spatial scales) are used and reused for multiple cognitive and behavioral ends."[62] Affordance is the concept of the "relationships between an organism's abilities and objects in the world

---

56. Gould (1981) 1996, 269.
57. Gould (1981) 1996, 27.
58. For a balanced view of what "neuroimaging can and cannot reveal," see Poldrack 2018.
59. Anderson 2014, 77.
60. Anderson 2014, xx.
61. Anderson 2014, xx. Recall Eagleman writing, "The three-pound organ in your skull . . . is composed of miniaturized, self-configuring parts, and it vastly outstrips anything we've dreamt of building" (2011, 2).
62. Anderson 2016, 1; Stanley, Gessell, and De Brigard 2019.

[198]   CHAPTER SEVEN

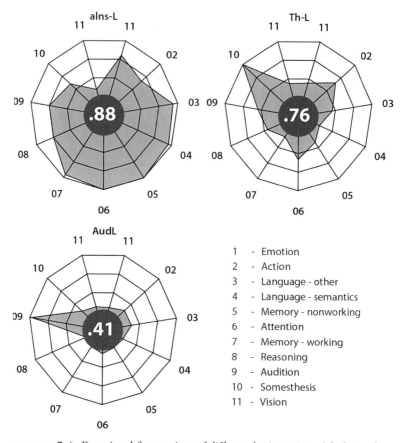

FIGURE 7.6. Functional fingerprints of different brain regions (*clockwise from top left*: left anterior insula, left thalamus, left auditory cortex) show the functional overlap across brain regions, as well as neural reuse for functional tasks—that is, many-to-many mappings between structure and function. (Anderson 2016, 3. Courtesy of Cambridge University Press and Michael L. Anderson.)

that indicate opportunities for action."[63] Ultimately, Anderson's empirically grounded model captures complex feedback between brain and environment, as well as neuroplasticity for sensorimotor functions. His model begins with the assumption that brains evolved to control action, and thus brain dynamics are ultimately embedded in physical space.[64]

To distinguish a decompositional framing of cognition from Anderson's, consider his method of "functional fingerprints" (fig. 7.6). Individ-

63. Anderson 2016, 7. This concept can be traced to Gibson 1979.
64. Anderson 2016, 7.

ual regions of the brain do not have single functions (as in phrenology or Penfield's homunculus), Anderson argues. Instead, like individual people, different regions have "personalities" that can be involved in many functions.[65] Imagine a group of people at work who go home to separate families or attend separate sporting or musical events. In each environment, they play different roles and coordinate with different groups. In the same way, each brain region can dynamically and flexibly cooperate with different regions.

For instance, Broca's area—a region on the frontal lobe of the dominant hemisphere of the brain—has classically been linked to speech production and language. Anderson documents how it is also involved in action recognition, imagery of human motion, action imitation, and other mental functions.[66] In short, Anderson develops a countermapping, anti–*Turing Phrenunculus*, neural-reuse perspective on the brain.

In fMRI and other neuroscience methods, checks on pernicious reification exist. This is particularly so within pluralistic, interdisciplinary research communities trained to be cautious about inferring cause from correlation. Working within an integration platform of diverse methods, models, and disciplines can forestall pernicious reification.[67]

Poldrack's cognitive ontologies and Anderson's overlapping functional fingerprints of distinct brain regions provide provocative ways of reimagining and countermapping the brain.[68]

## Statistical Causal Maps

The magic of statistics is the ability to represent the *frequencies* or *distribution* of particular traits of objects or processes; these objects or processes are organized into *populations* (e.g., the age distribution of all inhabitants

---

65. Anderson 2014, 114.

66. Anderson 2010, 245.

67. On EEG and magnetoencephalography (MEG) see, for example, Suppes 2012. For philosophical discussion about localization, see Bechtel 1982; Bechtel and Richardson 1993; Winther 2011b.

68. Anderson 2014, 46. Much more could be said about Poldrack's and Anderson's new spatial conceptualizations, ontologies, and maps, and related research programs; see, for example, Noë 2010. Or consider evidence that "the mapping from cortex to muscles is not fixed, as was once thought, but instead is fluid, changing continuously on the basis of feedback in a manner that could support the control of higher-order movement parameters" (Graziano 2006, 105); or recent work on functional segregation and integration in the brain (see Shine et al. 2018).

of a country). Guided by probability theory, statistics can also data-mine broad *descriptive patterns* "bottom-up" from the measured data, including linear relationships between different properties of an object (e.g., rates of incidence of various kinds of cancer at different ages). Mapping ourselves, indeed.

Such descriptive maps can then be used to test "top-down" probabilistic *causal models* of a system, including predictions made by probability models of, for instance, how the global economy will react to a change in the price of crude oil, or patterns of neural activation given a change in the organism's environment.[69] To make such models, researchers erect probability models containing theoretical random variables—with well-defined features such as a mean (average) and variance (spread around the average, over time, or over different objects of the same kind)—using the mathematics of what is called a *probability distribution function*, unique to each random variable.[70] As we shall see below, some probability models can also be described via the causal maps called *path diagrams*.

Statistics came of age from roughly the late nineteenth century to World War II, when the field was driven largely by a handful of mathematically minded biologists obsessed with the relative causal influence of genes and environment in evolutionary theory, psychology, medicine, and agriculture. Key statistical ideas were created by Francis Galton, Raphael Weldon, Karl Pearson, R. A. Fisher, Sewall Wright, and Charles Spearman, to name the most influential, in their investigations of intelligence, the milk yield or fat percentage of cattle, and disease.

As of today, statistics and probability theory may be the most widely used branch of mathematics across the sciences, thanks in part to the modeling innovations made possible by the rise of computer technology in the late twentieth century.

## LINEAR MODEL ASSUMPTIONS

Are internal factors (such as genes, biology, and ancestry) or external ones (such as environment, culture, and technology) more important in explaining the etiology, development, and population distribution of mental illness? What about differences in cognitive ability? If you have a mole on your right foot in the same place that your mother has one,

---

69. In contemporary philosophy and psychology, probability theory and statistics are also used in the attempt to articulate a "logic of induction" or a "formal epistemology."

70. For conceptual introductions to probability theory and statistics, see Hacking 1990; Daston 1995; Diaconis and Skyrms 2017.

should we explain this with an appeal to shared genes, shared diet, or shared weather, or some interaction of such factors? Does smoking cause lung cancer, or does poverty—or maybe both do? How would we know?

A fundamental method in statistics is to find correlations between one or more *independent* factors and a *dependent* factor, and then to tease out which of these correlations might imply a causal interaction. Setting up randomized controlled experiments allows us to test the abstractive-averaging assumptions relevant to what in statistics are called *linear model assumptions*. Only when we can make certain abstractive-averaging assumptions about errors, the shapes of the mathematical functions mapping independent and dependent variables, and so on, is linear statistical inference powerful.[71]

These assumptions ground the fundamental statistical inference machinery surrounding the basic idea of a *regression line*. The mathematics are a bit tricky here, but in much statistics all correlations among factors are assumed to be linear—that is, they are assumed to follow some kind of simple regression line of the form

$Y = mX + b + e$.

Here $X$ is an explanatory, causal variable (factor), $Y$ is the dependent variable, $m$ the slope of the line, $b$ the $y$-intercept, and $e$ the error term or residual that relates to the spread of the measurements around the line.

The linear model assumptions are then as follows:

1. *Linear relation.* Explanatory ($X$) and dependent ($Y$) variables are linearly related, much like the mass of an object and its weight, or voltage and current across a resistor. While threshold phenomena are important in biology (e.g., digital nerve signals being sent only once a certain charge has accumulated), linear correlations are also common (e.g., the gradients of various chemicals produced differentially by cells during early development).
2. *Uncorrelated residuals.* The variance or spread around the regression line does not correlate with various levels of measured traits or properties. The error or residual variance remains the same. (Moreover, the mean or average of residuals at every point along the line is 0.)
3. *Multivariate normal distribution.* All explanatory variables are multivariate normal (i.e., a Gaussian bell curve).
4. *Longitudinal error independence.* The errors or residuals of a given

---

71. Freedman 2009; Winther 2014a, 5–8; Edge 2019.

property for a given object or process at different times remains the same.
5. *Cross-variable explanatory impotence*. We cannot predict the behavior of any explanatory variable from any of the others. That is why we *need* the different variables in our theoretical causal structure. The probabilistic causal model is an enormous Gaussian machine, with multiple independent variables.
6. *Ignore outliers*. When any data point seems particularly odd or not part of the sample, ignore it, or at least give it minimal weight.

## CORRELATION AND CAUSATION

When a single causal factor is extremely strong and an underlying mechanism can be plausibly and clearly identified, correlation can, for all practical purposes, imply causation. You drop a rock in a lecture hall; it falls and thumps as it hits the floor. We could do an experiment varying temperature, air quality, loudness, magnetic field, and so on, and perceive that none of these makes any difference to whether or how the rock falls. Similarly, an astronaut in the International Space Station might drop a rock at the same set of variables and report that the rock remained immobile, floating in front of her hand, each time. Because the single systematic factor that differed between the two experimental contexts is gravity, we can infer that gravity is what causes the rock to fall on Earth. Conversely, the lack of a net downward gravitational force explains why the rock does not fall inside a space capsule in free fall around Earth.[72]

## "GENETIC" AND "ENVIRONMENTAL" DISEASES

Finally, compare the etiology of two diseases, one of which is "genetic," the other largely "environmental."

Huntington's disease is a late-onset neurodegenerative disease. Its expression follows an inheritance pattern consistent with the existence of a single dominant and autosomal (non–sex linked) gene in affected individuals. Solid evidence for genetics as an explanation for Huntington's disease emerged in the early 1980s from villagers living along the shores of Lake Maracaibo in Venezuela. A family-tree analysis of more than three thousand interrelated Venezuelans confirmed the dominant, autosomal allele (gene variant) pattern. Furthermore, a molecular study using DNA from lymph cell lines taken from this large family helped map the allele

---

72. As we saw in chapter 5, note 66, another example of a powerful, single explanatory factor in a system is greenhouse gas emission and global temperatures.

responsible for Huntington's disease to a particular location on the fourth human chromosome.[73]

This 1983 study was a watershed moment in the development of risk analysis and diagnostic methods for genetic diseases, as well as of biomedical ethics.[74] While there remains no cure for Huntington's disease and actual clinical diagnostic tests were not available until the 1990s, this was one of the first times researchers had identified the etiology of a disease caused largely by genetic differences, mapping it onto a specific and identifiable place on a chromosome.

A more environmentally sensitive disease is lung cancer. Two influential statistical studies investigating risk factors for lung cancer were Doll and Hill's 1950 preliminary report on the role of smoking and Cornfield and colleagues' careful 1959 study of relative risks.[75] Drawing on enormous amounts of survey and medical data, these studies showed that various kinds of carcinomas had much higher relative incidence in smokers than in nonsmokers, and that lung cancer incidence had increased dramatically in the twentieth century alongside an increase in smoking behavior.[76]

These studies received serious pushback from the tobacco industry and from the eminent statistician R. A. Fisher, who appealed to the possibility of a *spurious correlation* between smoking and lung cancer, owing to a separate independent factor—for instance, a genetic basis—that was explanatory of both smoking and lung cancer.[77] However, the robust and well-argued large-scale statistical studies rightly won the day. Doubt can be useful in scientific inquiry, but when that doubt is poorly thought out, it can be severely harmful in cases where the consequences are high and money is on the side of those wishing to preserve the status quo, whether

---

73. Gusella et al. 1983. I shall return to these genetic terms in the next chapter.

74. On the relation between the history of Huntington's disease and systematic social discrimination, see Wexler 2010.

75. See Doll and Hill 1950; Hill 1965; Doll 2002; Cornfield et al. (1959) 2009.

76. Nonstatistical evidence included "additional confirmations . . . on the induction of cancer of the skin in mice painted with tobacco-smoke condensates" (Cornfield et al. [1959] 2009, 1176).

77. Strictly speaking, this does not rule out the smoking as a possible "cofactor" of lung cancer, something Fisher apparently admitted as his death approached (Doll 2002, 505), even though he was much more skeptical of this while writing his interventions collected in Fisher 1959. Stolley 1991 discusses the incoherence of Fisher's arguments against the causal links, and Oreskes and Conway 2011 addresses the general strategy of moneyed corporate interests in defending their bottom line.

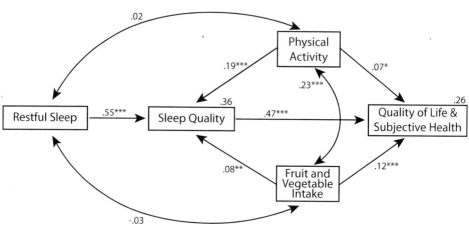

FIGURE 7.7. A causal model from a study of how sleep quality and other variables correlate with quality of life and subjective health. The path diagram is based on data from 790 Dutch and German participants who filled out a survey about their eating, exercise, and sleeping habits. (Tan et al. 2018, fig. 1.)

in the context of individual health and disease (e.g., the smoking industry in the mid-twentieth century) or of environment and ecology (e.g., fossil fuel producers).

## PATH DIAGRAMS AS STATISTICAL CAUSAL MAPS

Statistics and probability theory are important for disentangling causal factors when there is *causal complexity*—that is, many powerful factors are involved in the production of a phenomenon. A key tool for statistical disentanglement is the causal map known as a *path diagram*. Path diagrams are derived according to the statistical abstractive-averaging linear methodological assumptions described above. They show hypothesized causal influences among different factors (fig. 7.7). The numbers above the arrows indicate the strength of the correlation between the independent (explanatory) and the dependent variable—the higher the number, the stronger the correlation; the arrow direction specifies which variable is dependent (it could possibly be both).[78]

Figure 7.7 calculates the direct and indirect correlations among five

---

[78]. The *correlation coefficients* are between 0 and 1 because they are normalized against the standard deviation (i.e., the square root of the variance) of each variable. Asterisks indicate levels of *statistical significance*.

potentially explanatory or causal factors, and with the final independent variable ("Quality of Life & Subjective Health").

Some key lessons are that, while physical activity and fruit and vegetable intake do not contribute to restful sleep (as shown by the tiny correlation coefficients), physical activity in particular directly explains some of the variance of sleep quality. Moreover, sleep quality, physical activity, and fruit and vegetable intake taken together explain 26 percent of the total variance of quality of life and subjective health.[79] In cases where we have randomized and controlled variables; have sought to root out spurious correlations and confounding variables; and have followed other good statistical methodology, such path diagrams are strong causal maps.

The first path diagram was the evolutionary biologist and statistician Sewall Wright's 1921 causal map explicating the relative influence of different factors in the development of guinea pigs (fig. 7.8).[80] Here the lowercase letters are variables representing the path correlation coefficients, as explained in the figure caption.

Wright's path diagram illustrates the orthodox paradigm and standard causal map for unthreading the multiple, interacting causes during individual development of many cognitive and physical objects and processes, such as intelligence or disease. Is genetics or environment more important, and how? Might the factors interact (individuals of a given genotype have different phenotypes depending on the environment in which they find themselves) or even just co-vary (individuals with certain genotypes tend be more prevalent in certain environments)?

A standard method for dealing with these questions is to cross all genotypes against all environments to see which single genes, gene combinations, single environmental variables, or environmental combinations have the largest explanatory effects on the organism's traits. Individuals of known genotypes are standardized by crossing them in a variety of ways, often inbreeding them to create lines with fixed genetic characteristics. Environments with known conditions are produced by controlling temperature or humidity settings or by randomizing unknown conditions (e.g., placing different cages of guinea pigs in randomly assigned different parts of the lab, in order to exclude loudness or potential magnetic fields).

---

79. See figure 7.7, the value above the box "Quality of Life & Subjective Health."

80. Pearl and Mackenzie 2018 highlight the role of Wright's intellect and path diagrams in their history of the development of causal inference in statistics and probability theory. See J. Woodward 2003 for one philosophically detailed discussion of causation.

[206]　　　　　　　　　　CHAPTER SEVEN

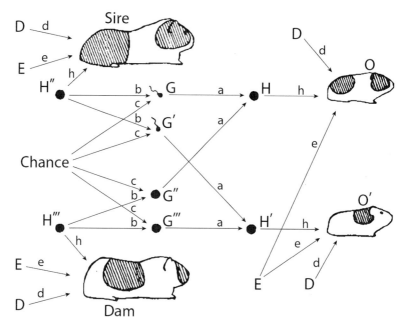

FIGURE 7.8. An article by Sewall Wright (1920, 328) includes this "diagram illustrating the causal relations between litter mates (O, O′) and between each of them and their parents. H, H′, H″, and H‴ represent the genetic constitutions of the four individuals, G, G′, G″, and G‴ that of four germ cells. E represents such environmental factors as are common to litter mates. D represents other factors, largely ontogenetic irregularity. The small letters stand for the various coefficients."

In mixing and matching many genotypes against many environments, we identify statistical variance and causal explanations of the phenotype.

### WHEN CAUSAL MAPS BECOME THE WORLD

In accordance with the multiple representations account of chapter 5, we first use the map to ontologize the processes occurring in the system. We believe our map to truly and utterly capture the causal structure of the world.

But recall the next stage: we take the causal model to be merely one way to view the world and its descriptive patterns.

What then? We hypothesize many causal model maps, each carrying a certain perspective, associated with a set of partitioning frames, and identifying particular kinds of causal factors. For instance, we could postulate a different causal structure than that of figure 7.7, bringing in family structure or psychological profile as potential causal variables. (Even maps with the same perspective and partitioning frame can postulate different statistical correlation coefficients among the factors—depending, for instance, on age, as was also the case with the study of fig. 7.7.)

Countermaps are possible. Consider statistical countermaps, which take issue with the linear assumptions made by standard statistical causal maps. It is important to note—as Sewall Wright also understood—that there are numerous causal interactions within and without a developing individual, each of which is somewhat explanatory. This fact was also grasped by the American psychologist and behavioral geneticist Eric Turkheimer, who has provocatively discussed what he calls "the three laws of behavior genetics":

1. All human behavioral traits are heritable.
2. The effect of being raised in the same family is smaller than the effect of genes.
3. A substantial portion of the variation in complex human behavioral traits is not accounted for by the effects of genes or families.

Somewhat in line with Susan Oyama, Richard Levins and Richard Lewontin, and Evelyn Fox Keller, Turkheimer riffs on the third law, providing an alternative statistical causal countermap based on a "gloomy prospect" in which neither environment nor genes, whether alone or together, can really explain much that is of interest in *mapping ourselves* as cognitive and biological agents. According to Turkheimer, "The additive effect of genes may constitute what is predictable about human development, but what is predictable about human development is also what is least interesting about it."[81]

Helen Longino wonders what exactly we are explaining with statistical causal maps when we try to explain human behavior such as aggression or sexual activity. Are we explaining individual behaviors in terms of individual properties, as "scaled-down" methodological individualism (based on linear model assumptions) would have it? Or do we accept genuinely emergent "scaled-up" group cultural properties, explainable only at the group level? This is a classic methodological debate in the human and social sciences, with many interlocutors of a statistical bent favoring methodological individualism. Although Longino accepts scaled-up and scaled-down perspectives as acceptable maps, she introduces a third, surprising countermap:

---

81. Turkheimer 2000, 160, 164, and 161, fig. 1. For Oyama as well as Levins and Lewontin, see note 29 above. For Keller, see chapter 8.

TABLE 7.1 The assumptions, standard maps, and countermaps probed in this chapter

| Kind of map | Abstractive-averaging assumption | Standard map | Countermap |
| --- | --- | --- | --- |
| Migration | Arrowized | *National Geographic* map of migrant flows across continents (fig. 7.2); Ravenstein's *Currents of Migration* (fig. 7.3) | Humane Borders / Fronteras Compasivas (n.d.) migrant death maps; *Spaces in Migration* (fig. 7.4) |
| Brain | Decompositional | *The Turing Phrenunculus* (fig. 7.5); Penfield's homunculus | Poldrack's cognitive ontologies; Anderson's functional fingerprints (fig. 7.6) |
| Causal | Linear model | Path diagrams (figs. 7.7, 7.8) | "Gloomy Prospect"; Longino on interaction |

We might want to know about the distribution, spatial or temporal, of some interaction or about the conditions that facilitate or inhibit particular kinds of interaction, again regardless of the properties of the individual participants. Not only is the differential distribution of these interactions not explainable in terms of similar distributions of causal factors operating on individual members of the population, the interactions themselves cannot be decomposed into the individual behavior of the interacting individuals. The interactive behaviors are phenomena in their own right.[82]

According to this perspective, *interactions*—sometimes statistically characterized, but not necessarily so—are themselves phenomena to be explained, and to do the explaining, in new kinds of causal (statistical) countermaps yet to be developed.

If standard, linear statistical models can become the world, so can countermaps emphasizing complexity and interaction.

## Conclusion

The variety of ways to map *Homo sapiens* is fascinating and dizzying, as well as politically and culturally charged. The three ways of representing ourselves that we have examined—migration maps, brain maps, and statistical causal maps—are powerful loci for identifying ontological and methodological assumptions about ourselves (table 7.1).

82. Longino, forthcoming, 16.

Conventional assumptions partition moving peoples, brain structures and mental functions, and causal influences in simple, clear, and distinct ways. These simplifying assumptions sweep away personal history, indeterminate geographic paths and boundaries, and unpredictable interactions among individuals. Questions emerge from these standard maps: How can we block or overcome the pernicious reification of assumptions and achieve contextual objectivity? Whose agenda does a given map or countermap further?

For map thinking generally, assumption archaeology, together with my multiple representations account, permits us to open up to ontological pluralism. The contextual objectivity of maps and emergent countermaps requires exploring many representations. Integration platforms provide methods and tools we can use to carry out this interrogation and to organize the results.

A tendency to linearize and flatten concepts and causes seems to drive abstractive-averaging standard assumptions. This tendency pervades science and is perhaps deeply human. Yet scientific progress requires both cartographic modes, map and countermap. Without abstractive averaged maps, science could not effectively partition a system or offer explanations or predictions. Without countermaps, we would not be able to re-imagine our world and identify the vastly dangerous incompleteness of standard assumptions. Countermaps also draw attention to flow, fuzzy boundaries, and dynamic feedback and interaction, all of which are essential features of most systems studied by science.

Now let us turn to the rich context of genetics to put integration platforms fully to work.

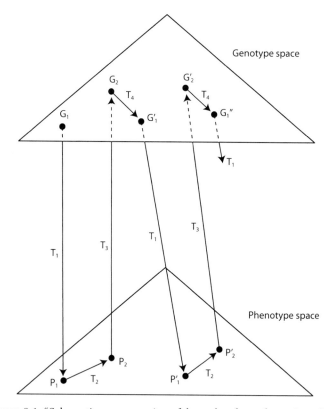

FIGURE 8.1. "Schematic representation of the paths of transformation of population genotype from one generation to the next" (Lewontin 1974, 14). Two formal state spaces are represented—genotype and phenotype—along with the processes within and between these. (Courtesy of Richard Lewontin, formatted by Mats Wedin.)

# 8

# Mapping Genetics

In the middle of the nineteenth century, a monk growing peas in the garden of an Augustinian monastery in Central Europe worked out the first rules of inheritance and quietly founded the field of study that would become genetics. Gregor Mendel was the son of struggling Czech farmers so poor that Mendel's younger sister donated her dowry so he could finish school. In college, he studied philosophy and physics, then joined the Augustinian friars so he could earn a living while continuing to study and teach science. Most of his publications were on meteorology, and he also had a keen interest in astronomy.[1]

During Mendel's life, his groundbreaking 1866 paper on inheritance went unnoticed. Although Charles Darwin was then developing his theory of evolution by natural selection—and knew that inheritance was the missing link for completing it—there is no evidence Darwin knew about Mendel's work, or would have understood or cared for its quantitative methodology.[2] Not until 1900, sixteen years after Mendel's death, did other scientists belatedly discover his work and recognize its importance.[3]

In his paper, Mendel described a set of experiments, the culmination of years of study.[4] He chose seven traits of garden pea plants that did not blend: a pea plant was either tall or short (not medium), and the seeds were either smoothly round or wrinkled (not in between). He also chose traits he knew were "true-breeding," meaning that a short plant that was self-pollinated would always produce short offspring.

1. Masaryk University Mendel Museum, n.d.
2. Galton 2009. On Darwin on inheritance, see Winther 2000.
3. Dunn 1965; Provine 2001.
4. "Experimentally, therefore, the theory is confirmed that the pea hybrids form egg and pollen cells which, in their constitution, represent in equal numbers all constant forms which result from the combination of the characters united in fertilization" (Mendel 1996, 24; italics suppressed).

TABLE 8.1 Punnett squares for the two crosses described in the text

| Granddad \ Grandmom | p | p | Dad \ Mom | P | p |
|---|---|---|---|---|---|
| P | Pp | Pp | P | PP | Pp |
| P | Pp | Pp | p | Pp | pp |

*Note*: In the first-generation cross (left), a white-flowered Grandmom (*pp*) is crossed with a (true-breeding) purple-flowered Granddad (*PP*). This produces all hybrid offspring (*Pp*). In the second-generation cross (right), the purple-flowered *Pp* female and male offspring produce offspring of which 25 percent are white-flowered. Punnett squares were developed in the early twentieth century; Mendel himself used another formalism.

Using these traits, he performed a series of crossbreeding experiments. For example, crossing true-breeding, white-flowered peas with true-breeding, purple-flowered peas produced offspring that all had purple flowers (table 8.1, left). Mendel then allowed these offspring hybrids to self-pollinate. This time, instead of getting offspring that all had purple flowers, he discovered a generation with approximately 75 percent purple flowers and 25 percent white flowers (table 8.1, right).

Based on these and other experimental crosses, Mendel concluded that each of his seven traits (plant height, seed form and color, pod shape and color, and flower position and color) was determined by a pair of a single kind of *factor*, one version from each parent. Those factors are what we now know as *genes*. Each version of the gene, or factor—whether it led to a purple flower or white flower, or a tall plant or a short plant—is what we now call an *allele*, or variant. Furthermore, we now describe the purple trait as *dominant*, because it dominates the white-flower factor in a cross, and we describe the white-flower trait as *recessive*. In a bit, we will see how Mendel used these genetic experiments to arrive at two important principles of inheritance.[5]

Mendel was a countermapper whose ideas did not fit with the dominant nineteenth-century views about development, inheritance, and evolution. Many influential biologists (including Darwin) believed in the inheritance of acquired characters, or "Lamarckism"—the idea that environmentally induced changes during an individual's life (e.g., a weightlifter's strong arms) could be inherited, since these changes affect the individual's hereditary germinal material. For Mendel, factors were preexisting and precoded, basically invulnerable to change. Although he lacked the social

---

5. The story of what exactly Mendel thought is perhaps more complicated; see Olby 1979. On the important relation between teaching and research in Mendel's case, see Jungck 1985.

status that would have helped give his ideas credibility, Mendel's radical theory would be accepted and even perniciously reified in the twentieth century, just as Gerardus Mercator's 1569 countermap became the ontologized, universalized, and narrowed map projection for centuries to come.

In this chapter, we will create an integration platform that helps us understand how scientific advances are replacing this pernicious reification with contextually objective pluralistic ontologizing, as per my multiple representations account. The research discipline, technologies, and mechanistic processes of genetics are richly varied. One way to produce an overarching integration platform for mapping genetics from such complexity is to introduce a set of outrageously different genetic maps (seven of them, in this chapter) as visual heuristics into distinct domains of genetics. An integration platform that allows us to systematically compare genetic maps will also fulfill a second important function: we can check the limits of each map by comparing it with countermaps within its domain, and even contrasting it with maps and countermaps from *other* domains.

Each of the seven maps of our mapping-genetics integration platform is illustrated with case studies of the laboratory organism *Drosophila melanogaster* (fruit flies), with suggested applications to *Homo sapiens*.

## Building a Mapping-Genetics Integration Platform

To build our integration platform, we will consider two types of major assumptions, pause to define some basic terminology of genetics (see also the glossary at the end of this chapter), and then list the map types that will facilitate our integration.

### ASSUMPTIONS

The field of genetics is replete with maps. We have already seen that every map carries assumptions and builds worlds. By laying bare two assumptions behind genetic maps—the existence of a partitioning frame, and reflections on space and time—we can generate contextual objectivity and avoid pernicious reification.

1. A *partitioning frame* is like the viewfinder in a camera: it carves out the parts, at the chosen scale, of systems that have been perspectivized. The frame delimits what will be represented. Partitioning frame assumptions—reflected in the scale, colors, and symbols chosen for a map—might include the decision of which objects, processes, features, and relations are important to include for a given map's

purpose. Partitioning frames for each of the seven genetic maps are importantly distinct, as you shall see.[6]
2. Genetic maps track and represent implicit and explicit events happening over *time* and in different kinds and scales of *space*. (Recall fig. 3.2.) For example, a linear genetic map represents a particular type of DNA molecule making up genes. Different scales of time are also implicit in genetic maps, including but not limited to: the fractions of a second of molecular reactions; the months of embryonic development; or the millions of years of the course of evolution.

These two map assumptions—a partitioning frame and how space and time are defined—suggest the mapmaker's abstraction and ontologizing processes. Assumption archaeology brings these two assumptions into relief, helping us understand the structure and function of genetic maps. As we explore each map type, keep in mind assumptions about space and time and those related to partitioning frames.

TERMINOLOGY

Some additional terminology comes up as we first encounter the map types.

A *gene* is either the length of a single strand of double-helix DNA or different parts along a strand of DNA that are read as encoding a single message. A *genotype* may be either all the genes in a cell or parts of a DNA strand's nucleotide sequence. DNA strands are, in turn and in reality, folded up and wrapped with proteins to form *chromosomes*—there are as many long DNA strands as there are chromosomes.

The *phenotype* of an organism comprises all its observable traits, including all aspects of its metabolism, behavior, and structural features (*morphology*). Geneticists tend to think of such features—the phenotype—as coded by the genotype. The word *phenotype* is also sometimes used for an individual trait in relationship to the gene that produces it.

MAP TYPES

Here are the seven kinds of genetic maps examined below:

1. The *linear genetic map*, which can show the arrangement of gene names along a chromosome, or indicate the sequence of nucleotides
2. The *gene expression map*, which can show a gene's effect by color-coding the gene (along the genome) and the body area affected

6. Winther 2003, 2006c, 2009a, 2011b; cf. Wimsatt 2007.

3. The *genotype-phenotype map*, relevant to evolution, including *reaction norms* that account for environmental variation
4. The *literal cartographic genetic map*, such as that of *Drosophila* populations across northern Mexico and the Southwest USA
5. The *comparative genetic map*, comparing the gene sequences of different species
6. The *adaptive landscape map* of evolution, where gene combinations are mapped against a fitness dimension of reproductive success
7. An *analogous map*, such as the *Tree of Life analogy* first introduced by Charles Darwin, and later assembled with data gleaned from specific genes

Let us now turn to these maps.

## The Linear Genetic Map

A central fact about genes is that they are arranged linearly. Genes are ordered along the one-dimensional space of DNA, similar to a linear musical score. Much as a musical score contains codes such as key and time signatures to guide its reading for the production of sounds, a biological system contains a genetic code for translating DNA nucleotide components into amino acids, which in turn form proteins that carry out specific functions as, for example, hormones or receptors. The sounds produced by musicians reading a score are analogous to the proteins and the eventual phenotype "read" from the DNA. Different orchestras with different musicians might play a given score with a different inflection, just as individual organisms might read a given segment of genetic information a bit differently—but in both cases there is a linear order and logic. Recall that different instruments may play different sounds based on the same overall musical score, both because the score is written for different instruments, and because particular musicians interpret tempo and timbre distinctly. Similarly, a given DNA strand has a certain specified order constraining how that strand might be read by different protein-making orchestras of distinct species, and the exact song performance will vary significantly from species to species. The cellular orchestra reading DNA in order to make proteins takes the linear DNA, reads from it a complementary RNA strand, and either takes that RNA strand as such, or, much more often, cuts and pastes and transforms that RNA strand, using it to make a curated linear amino acid strand. This is called *gene expression*.[7]

Linear genetic maps show genes arranged with respect to one another

---

7. A video elucidating gene expression can be found at Mandal, n.d.

on a line. That the DNA strand is packaged into chromosomes was a mid-twentieth-century discovery—influenced perhaps most powerfully by Crick and Watson's 1953 discovery of the biochemical structure of DNA (and their metal-assemblage 3-D map of it, examined in chapter 5). Phenotypic linkage maps show the relative arrangement of factors or genes that have been observed statistically, through breeding experiments, to correspond to phenotypic variation—by analogy, a world map. In contrast, nucleotide maps are like city maps, homing in on the specific genetic structure that helps explain, at a biochemical level, *why* phenotypic variation occurs.

Linkage and nucleotide genetic maps are at two different scales, with the latter providing a detailed picture that the former only begins to show at its grosser level of resolution. To look more closely at these two map types, let us proceed historically, and start with linkage maps.

## LINEAR GENETIC MAPS OF PHENOTYPIC LINKAGE

Mendel's work established two principles of inheritance. First, according to Mendel's principle of segregation, a sexually reproducing individual has two alleles (gene variants) for a given factor in each of their various cells, but those two alleles separate in the production of a *gamete* (an egg or sperm). For instance, in table 8.1 above, alleles $P$ and $p$ separated in both Mom's eggs and Dad's sperm. Second, according to his principle of independent assortment, the alleles for one gene separate independently from the alleles of another gene. This means that Dad's contribution of $P$ instead of $p$ is not made any more or less likely by him contributing a particular allele for another factor. Mendel derived these principles from careful mathematical analysis of the large amounts of data he accumulated by planting many generations of sweet pea plants.

Mendel's independent assortment principle had worked in his limited selection of pea plant traits. The suggestion by the American physician and geneticist Walter Sutton in about 1903 that genes were on chromosomes complicated Mendel's picture, as genes will not sort independently if they are located close together on the same chromosome. Alfred Sturtevant, one of Thomas Hunt Morgan's students (see below), saw in a flash of inspiration how this principle of linkage could be used to map genes: *we can use numerical deviations from independent assortment to calculate the relative locations of multiple genes situated in the same linkage group (i.e., chromosome).* Drosophila have just four pairs of chromosomes, and the undergraduate Sturtevant pulled a typical all-nighter to map six genes on one of those, the X sex chromosome. Sturtevant's mapping logic, published in 1913, influenced all subsequent gene mapping inferred from crossbreeding experiments (fig. 8.2, panel 1).[8]

What is the logic of mapping a phenotypic linkage group? Every offspring receives two sets of chromosomes, one from each parent via egg or sperm. (Across a population, more than two allele types might be possible for a single gene, but an individual will receive from its two parents either the same allele or at the very most only two kinds of alleles.) Alleles of a gene can be of the same type on both chromosomes, but are also sometimes different.

Consider, then, a simple situation of two genes with two alleles: *A* and *a*, and *B* and *b*. The capitalized letter for each gene represents the dominant allele, and the lowercase letter, the recessive. Let us concentrate on three cases:

> *Independent.* Imagine a mother's alleles are *Aa* and *Bb*, while the father is *aa* and *bb*. When the two genes are on different linkage groups (chromosomes), they segregate independently, one allele from each parent gene going into each gamete. The eggs will be, in equal numbers, *AB*, *Ab*, *aB*, and *ab*; sperm will all be of *ab* formation. The offspring are respectively *AaBb*, *Aabb*, *aaBb*, and *aabb*, again in equal quantities. It is as if you did a single Punnett square similar to table 8.1, twice: the two offspring types of Mom's *Aa* and Dad's *aa*, that is, *Aa* and *aa*, combine with equal probability with the two offspring types for the other gene, *Bb* and *bb*. In table 8.2, the frequencies of the four columns are equal.
>
> *Co-traveling.* Now visualize a parent's alleles of two different genes being inherited as a single unit. This can in principle happen when the genes are immediately next to each other on a chromosome (*AB* next to each other on one chromosome in the pair, alleles *ab* together on the other chromosome in the pair). Mom's eggs will be all *AB* or *ab*, and Dad's sperm all *ab*. Thus, half of the resulting offspring will be *AaBb* (*A* and *B*, which are dominant, determining both traits in the individual phenotype), while the other half will be *aabb*, with the phenotype displaying the recessive traits. This is the mirror image of independent, as if *AB* and *ab* in Mom were the two alleles of the "same" gene. Columns 2 and 3 of table 8.2 disappear.
>
> *Linked.* Now imagine the genes to be on the same chromosome but not immediately adjacent. This very common and important scenario is logically in between independent and co-traveling. Alleles of

---

8. Sturtevant 1913.

TABLE 8.2  Two-gene Punnett square

| Dad \ Mom | 1<br>AB | 2<br>Ab | 3<br>aB | 4<br>ab |
|---|---|---|---|---|
| ab | AaBb | Aabb | aaBb | aabb |
| ab | | | | |
| ab | | | | |
| ab | | | | |

*Note*: This two-gene Punnett square shows the four egg types of a mother heterozygous for both genes, and the sole sperm type of a father homozygous for the recessive allele for both genes. A full Punnett square would indicate all rows; but because the offspring genotypic rows are identical and do not change in proportion, the bottom three are omitted for ease of reading. See text for further details.

different genes will now sometimes be inherited together—but not every time. This is because, during gamete formation, the pairs of chromosomes line up together. A given pair can then sometimes break apart, with the parts of the two different chromosomes combining with each other. The process is called *crossover*; the resulting genetic arrangement is called the *recombinant genotype*. If, again, Mom has *AB* on one chromosome in the pair and *ab* on the other (the *homologous* chromosome), then *AaBb* and *aabb* will be the regular genotypes of the offspring (recall Dad is *aabb*), and *Aabb* and *aaBb* the recombinant genotypes (columns 2 and 3 of table 8.2). The higher the relative number of recombinant genotypes for any two genes, the further apart the two genes are on that chromosome in all individuals of that species. In table 8.2, this corresponds to cases where the frequencies of columns 1 and 4 are equal, and both of these frequencies are from just above 25 percent up to 50 percent, and conversely for columns 2 and 3—they are also equal, but go from 25 percent down to 0 percent. (The case where each column is 25 percent is the independent scenario; the case of 50 percent for columns 1 and 4, and 0 percent for columns 2 and 3 is co-traveling.)

Counting offspring with simple and observable phenotypic effects— the effects of these gene combinations—therefore gives an indirect measure of *map distance on the linkage group* between any two genes. In particular, we wish to know the relative number of recombinant genotypes for a given cross. Map distance is expressed in *map units*, also known as *centimorgans*. This measure is indirect and approximate because what it actually expresses is a frequency of observed traits as a result of recombination, which only *implies* the distance: the lower the relative number (frequency) of recombinant types, the closer together the genes likely are. At zero map units (0 percent recombination), the genes are co-traveling. If two genes are fifty or more map units apart, they assort independently, even if they are on the same chromosome. This is because so many crossover events occur between them that they are randomly shuffled with respect to each other. Finally, just above zero and up to fifty map units, the genes are linked. In short, for this broad range of scenarios, tallying offspring with simple and observable phenotypic effects gives a *map distance* for any two genes (e.g., between *yellow* and *vermilion*: 33 percent recombination = 33 map units; fig. 8.2, panel 2).[9]

Thomas Hunt Morgan's lab at Columbia University, famously known as the Fly Group, flourished during the 1910s and 1920s and produced a steady flow of publications. Morgan's students included Sturtevant, whom we met above, and Calvin Bridges and Hermann Joseph Muller. The Fly Group used this mapping logic to produce linear genetic maps of phenotypic linkage for *Drosophila*. Because all the genes on a chromosome are ultimately inherited together, as many phenotypic linkage group maps exist as there are chromosomes. In the case of the fruit fly, that number of linkage groups or chromosomes is four. In the case of humans, the number is twenty-three, and today we know that the human genome—its complete DNA code—averages about 3,600 map units.[10]

In 1931, the American geneticists Harriet Creighton and Barbara Mc-

---

9. Strictly speaking, the situation for linear genetic maps of phenotypic linkage is more complex. First, recombination events are not randomly distributed along the chromosomes; chromosomes often have particular "hot spot" locations with higher recombination. Second, when recombination happens at a given location, it can sometimes suppress another recombination event happening close by on the chromosome (this is called *interference*). Finally, genes far apart (but less than 50 centimorgans) can sometimes have two recombination events happen between them (more generally, any even number of recombination events), which would be equivalent to no recombination having happened; hence, counting those offspring as non-recombined types will underestimate the map distance between those two genes. There are at least two ways to address all this complexity. First, map distance is fairly additive when we add multiple pairs of genes that are each very close together. Second, there are a variety of mathematical "map functions" that estimate map distance by factoring in nonrandom recombination, interference, and double crossovers (see, for example, Kosambi 1944; Casares 2007; Vinod 2011).

10. Morton 1991.

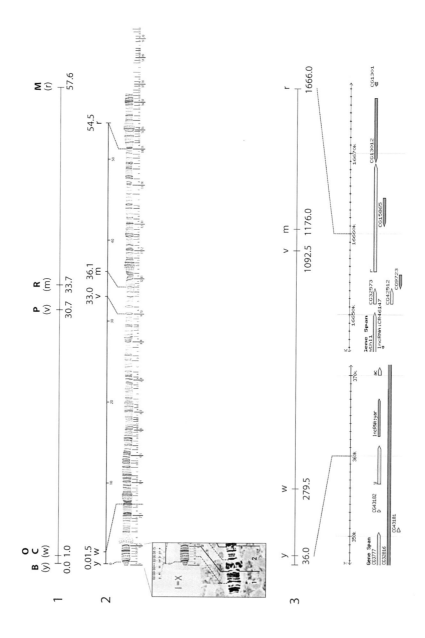

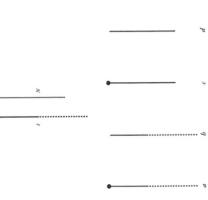

FIGURE 8.2. *The Evolution of Linear Genetic Maps in Four Movements.*

(1) Sturtevant (1913, 49) mapped what he thought were six genes: black (body; B); color (eye; C); eosin (eye; O); vermilion (P); rudimentary (wings; R); miniature (wings; M). The nomenclature, later adopted in Bridges 1935, is written under Sturtevant's letters.

(2) The same five genes are shown with revised map units, as is Bridges's correspondence of linkage group to microscopic salivary chromosome bands (these are X sex-linked genes). An inset indicates a cutout from Bridges's original pictures.

(3) *Yellow, white, vermilion, miniature,* and *rudimentary* are represented according to contemporary base distances (numbers indicate location in 10-kilobase distances).

(4) Creighton and McClintock's (1931) work on corn provided strong evidence for the hypothesis that genetic crossovers take place because chromosomes ($I$ = interchange chromosome; $N$ = normal chromosome) pair up during meiosis. In their words: "$a$ = knobbed, interchanged chromosome; $b$ = knobless, interchanged chromosome; $c$ = knobbed, normal chromosome; $d$ = knobless, normal chromosome; $a$ and $d$ are non-crossover types; $b$ and $c$ are crossover types" (493).

(Adapted by Mats Wedin and Rasmus Grønfeldt Winther. Additional sources: FlyBase, http://flybase.org/reports/FBgn0004034 and http://flybase.org/reports/FBgn0003189.)

Clintock reported a set of experiments with corn showing that heritable factors (genes) coding for observable traits could be tracked in the microscopic appearance of corn chromosomes (fig. 8.2, panel 4).[11]

A few years later, Thomas Painter and Calvin Bridges published papers showing the same thing in fruit flies (fig. 8.2, panel 2).[12] By 1935, the Fly Group labeled genes by the recessive (rather than the dominant) allele; *C* and *O* turned out to be two alleles of one gene (rather than two co-traveling genes, as Sturtevant 1913 seems to have thought); and interactions among genes such as rudimentary and miniature, were still accepted but were downplayed in gene naming. Thus, in both corn and flies, it was possible to map genes via breeding experiments; to infer linear genetic maps of phenotypic linkage; and to actually see corresponding physical changes in the chromosomes under a microscope. Only somewhat metaphorically speaking, *the physical chromosomes are their own phenotypic linkage group map.*

## LINEAR GENETIC MAPS OF NUCLEOTIDES

As we know now, genes exist on strands of DNA. Every strand of DNA consists of varied sequences of four kinds of *nucleotides*. Each nucleotide comprises a sugar molecule, a phosphate molecule, and one of four molecules containing nitrogen: adenine, thymine, cytosine, or guanine (A, T, C, or G). The sugar and phosphate molecules line up in alternating order to form the two sides of a ladder, with internal rungs that may be either AT pairs or CG pairs. The ladder twists to form a spiral, the famous "double helix" of DNA. The sequence of A, T, C, and G along the DNA strand is the *genetic information* (implied by fig. 8.2, panel 3). Sequences of three rung halves are called *codons*, each of which can be translated into one of twenty kinds of amino acids, according to the *genetic code*. Linear strands of amino acids then fold up according to specific thermodynamic and biochemical rules to form proteins, crucial to all life.

In the 1970s and 1980s, researchers developed methods that permitted the absolutely detailed mapping of the sequence of every single DNA nucleotide, in linear order. Mapping entire genomes of different species

11. Sutton 1903; Creighton and McClintock 1931.

12. Although Painter 1933 has historical priority on the idea of conceptually anchoring phenotypic linkage maps in chromosome images, it was Bridges who creatively and steadfastly developed lenses, staining techniques, classification schemes (e.g., "the Totem Pole," T. H. Morgan 1939, 356, fig. 1), and breeding machinery. Furthermore, he provided a detailed "key to the banding of the chromosomes" in Bridges 1935. See Kohler's (1994, 73–77) discussion of "constructional map" and "working and valuation map," vis-à-vis the Totem Pole.

eventually became possible. A draft linear genetic map of nucleotides of *Homo sapiens*, created by the Human Genome Project and publicly funded across the globe, was published in a special issue on the human genome in *Science* magazine in 2001. Included was a large foldout gene map poster. Opposite the immediately classic article's first page was an advertisement for New England Biolabs.[13] The spread emanated the world-making cartopower defined in chapter 5, and the hype, hope, and promise of the mapped genome as the dawn of a new era in a new millennium.

The advertisement's image superimposes a west-coast-of-Africa portion of a colonial-era map—complete with clipper ships—on an otherwise naked woman. The map is draped across her body as gauze swept back by a sea wind, her arms stretched alluringly above her head. Her forward-arched stance suggests a ship's figurehead. She peers toward an unseen horizon, the unexplored territory of genetic engineering. The image gives the ad sexualized (woman; Eve) and racialized overtones.

In the ad, New England Biolabs offers custom restriction enzymes for the selective amplification of nucleotide sequences. With such amplified DNA, genetic interventions in a brave new world could be birthed. The ontologizing cartopower of the advertisement's image involves the possibilities of gene therapy and genetic engineering potentially curing diseases and controlling cognitive capacities, establishing new realms of humanity, and reaping new levels of corporate profits.

Today, the genomes of hundreds of species have been substantially sequenced. However, only a dozen or so species still act as *model organisms* (plate 10). We increasingly have a treasure trove of linear genetic maps of nucleotides. It is important to know, however, that the exact sequence of all the DNA in an organism's genome does not necessarily reveal very much about the function of individual genes.

## ASSUMPTIONS OF THE LINEAR GENETIC MAP

In linear genetic maps of the phenotypic linkage kind, the partitioning frame is premised on heritable and observable features, such as the height of pea plants, or the color of their seeds. When chromosomes themselves become a kind of material stand-in for phenotypic linkage maps, the partitioning frame includes physical bands and other visible, structural chromosomal features, as depicted in panel 2 of figure 8.2. Space is at the chromosomal scale. Time is the time of chromosomal crossover.

For linear genetic maps of nucleotides, the partitioning frame involves molecular criteria, including gene parts such as stop codons and concepts

13. Venter et al. 2001; see also Dupré 2004; Gannett 2014.

such as the *reading frame*.[14] The first full nucleotide map of a free-living organism was Craig Venter and collaborators' 1995 rendition of the 1.8 million nucleotide genome of the bacterium *Haemophilus influenza*, replete with annotations of the functions of many of its genes.[15] In such a map, space is nucleotide space, and the genes, composed of nucleotide sequences, are indicated as colored lines, with the colors reflecting gene function. Time is implicitly coded as possible gene expression (that is, when a cell uses a gene, as it were, to create a protein) and, following the abstraction-ontologizing account, time is also the time of abstraction of the behind-the-scenes experimental and visualization protocols. Nucleotide linear genetic maps are what most people think of when they hear the term *genetic map*.

## The Gene Expression Map

Symmetry and pattern are fundamental to biological form. Consider the symmetries and repeating patterns in bodies: our rows of ribs, or the segmentation patterns of earthworms and caterpillars. Genes can be fundamental to understanding and explaining such spatial organization. There is a cascade of genetic effects in fertilized eggs and early embryos, discovered in the late 1970s and early 1980s, that tell the very early developing embryo which way is up and which down, left and right, front and back. This is the case in organisms as different as fruit flies, zebrafish, and humans, and it is one kind of gene expression.[16] From before we even leave the womb, we are segmented and oriented, as if consulting a map in utero.

The *gene expression map* uses color-coding to match the gene nucleotide sequence on the chromosome with the tissue or body part that that gene affects. Take, for example, the *homeotic*, or *HOX*, genes, which control the body plan and the body segment identity—such as wing or leg or eye—of different parts of the body. A color-coded gene expression map of HOX genes maps the genes' chromosomal location to the related body segment in the developing embryo or larva, and in the adult. The HOX genes are clustered along chromosomes, and their chromosomal order corresponds to the order of their spatial expression in the embryo or larva during development, as well as to the spatial order of segment identity in the adult.[17] (Apologies for reifying life to animals here.)

---

14. Gerstein et al. 2007; Griffiths and Stotz 2013.
15. Fleischmann et al. 1995.
16. Nüsslein-Volhardt 1995 provides a synoptic description of such genetic cascades.
17. Khan Academy, n.d.; BIOPRO Baden-Württemberg 2012.

Other gene expression maps also visualize with color: for example, global microarray maps identify all genes expressed at a myriad of points, cells, or regions, of the brain, say. Color indicates which genes are expressed where in the body.[18]

In general, the partitioning frames for such extreme-scale maps involve the particular genes whose expression is being mapped, and the body regions in which the genes are expressed. The linearly arranged genes that are the origin of the expression are depicted in a linear map that amplifies, as we saw immediately above and in chapter 3. A range of scales of territorial space might appear together on gene expression maps. We can either representationally amplify actual, tiny cells onto paper or screen, showing with colors, symbols, etc. the spatial organization of gene expression, or we can *reduce*, as with a regular geographic map, the entire body of a human adult or whale onto paper or screen. Time is the inherent time of development, whether that is hours or years. Gene expression maps as here discussed are thus extreme-scaled spatial maps, portraying where in the body genes are expressed; they can also be causal maps when they depict the organization and order of gene expression processes in a regulatory network.[19] The gene expression map plays a crucial role in summarizing experimental results, providing explanations of gene effects, and guiding further research.

## The Genotype-Phenotype Map

If the map is not the territory, so the genotype is not the phenotype. Gene expression takes place in the context of a variable environment. And the phenotype of an individual—its morphological, metabolic, behavioral, and other observable traits—is influenced by interactions among the individual's genes, and by interactions of its genotype with the changing environment.

How different can identical twins of a given species become? *Norms of reaction* describe the different patterns of phenotypes that result from a single genotype's interaction with different environmental factors, at distinct magnitudes. Norms of reaction depict the organism's flexibility and responsiveness to environments across both time and space. Sometimes genes hold phenotypic variation on a short leash; sometimes the leash

---

18. Shen, Overly, and Jones 2012; Lukk et al. 2010.

19. Gene regulatory network maps can also portray the causal structure of gene expression, as discussed in Davidson 2006. For more general network maps of metabolism and cell activity, see Kohn 1999; Ito et al. 2000; Collins and Barker 2007. On networks in general, see Winther 2010 and references therein.

is long. But these reactions are so complex and varied that identifying specific, clear-cut norms of reaction is difficult.

Particularly vexing is the case where the organism under consideration is a human being, with an environment constituted partly by human culture. Is our biological heredity or our cultural upbringing more important in determining our looks, health, memory, and charm? The relative causal influence of biology versus culture became a pervasive object of study in Western science with the rise of European colonialism and the study of natural history vis-à-vis the oppression of peoples across the globe, alongside the emergence of liberal, individualistic Enlightenment thinking. We started exploring these questions in the last chapter.

In the twentieth century, two key tools breathed new scientific life into these topics and questions. The first is the statistical method of analysis of variance (ANOVA), described in chapter 3, which helped clarify thinking about the relation between genotypic, environmental, and phenotypic variances, and about concepts such as heritability and genotype-environment interaction. Of concern in the present section, however, is the second tool, *genotype-phenotype mapping*, first powerfully presented by the biologist Richard Lewontin in *The Genetic Basis of Evolutionary Change* (1974).

Lewontin's map (fig. 8.1) depicts two state spaces, the genotypic (G) and phenotypic (P). The laws of transformation (T) proceed within and between these spaces, from one prime-indexed generation to the next. In particular:

$T_1$ transforms elements in the genotypic domain into phenotypic elements, dependent on the environment. Think of these as the laws of gene expression and development.

$T_2$ mathematical laws apply to evolutionary change within generations. In $T_2$, evolutionary processes—such as selection, migration, and genetic drift—change the frequency distribution of phenotypes during the passage from juveniles to reproductive adults.

$T_3$ laws transform the (surviving and reproducing) phenotypic elements back into genotypic elements. Organisms with surviving phenotypes reproduce, but they rarely do so randomly. For instance, sexual selection can lead to the pairing of individuals of similar phenotype more often than randomly.

$T_4$ laws predict gene frequencies of genotypes in the next generation based on those in the previous, adult one (according to principles such as segregation, independent assortment, and linkage, seen above).

Lewontin's genotype-phenotype map provides a highly simplified and abstracted map of the entire developmental and evolutionary process, tracking the units relevant to the process.[20] The partitioning frame of this state-space map identifies dimensions of two formal state spaces suitably idealized for visualization purposes, reducing to simple points the multiple dimensions of genotypic features (e.g., gene combinations) and multiple dimensions of phenotypic features (e.g., different morphological features). Time is the time of molecular, developmental, and evolutionary processes.

One particular countermap to the genotype-phenotype map that seems especially important to illuminate is the theoretical biologist Mary Jane West-Eberhard's overarching vision of the evolutionary process. West-Eberhard teaches: "Some people think of ontogeny in terms of how genes are 'mapped on' to the phenotype, but this is too flat an image for the dynamic events of development. It would better evoke the events of development to ask, 'How do environmental supplies, partially ordered by the genome, affect the highly reactive phenotype that exists before they arrive?'"[21]

## The Literal Cartographic Genetic Map

The literal cartographic genetic map is a powerful source of empirical information, useful for testing evolutionary and ecological mathematical models. Among the first and most influential literal genetic maps were those of the Russian-American geneticist Theodosius Dobzhansky and the Italian Luigi Luca Cavalli-Sforza.

One of the architects of the 1930s-era synthesis of genetics and evolutionary theory, Dobzhansky mapped the geographic gradients of genetic variation in fruit flies (fig. 8.3). He also mapped ecological conditions such as temperature, humidity, latitude, and altitude, exploring how such factors exerted selection pressure on particular genes and gene combinations of *Drosophila*. Dobzhansky and many others also modeled such ecological conditions in the lab to directly demonstrate selection over short periods of time.[22] One of the architects of statistical inference about an organism's or species' evolutionary history, or *phylogeny*, Cavalli-Sforza

---

20. For different and complementary analyses of the genotype-phenotype map, see Alberch 1991; Fontana 2002.
21. West-Eberhard 2003, 98.
22. Winther, Giordano, Edge, and Nielsen 2015.

[228] CHAPTER EIGHT

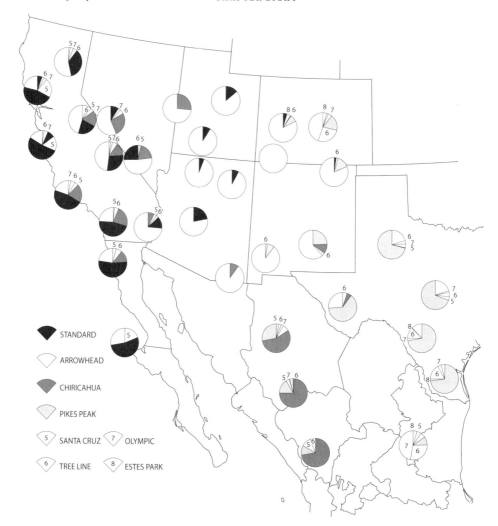

FIGURE 8.3. "Relative frequencies of various gene arrangements in the third chromosomes of different populations of *Drosophila pseudoobscura* in the western United States and in northern Mexico." (Dobzhansky [1937] 1951, 138, fig. 10; adapted by Mats Wedin.)

and collaborators drew on the geography of human genetic diversity to infer past human migration routes.[23]

Bearing in mind that both migration and random genetic drift can structure geographic variation, Dobzhansky argues that we should con-

---

23. See Cavalli-Sforza, Menozzi, and Piazza 1994; Anthony W. F. Edwards and Cavalli-Sforza's eight coauthored papers appear in Winther 2018a.

sider the full range of genetic variation in populations, whether in fruit flies or in humans, rather than some grand average. Random genetic drift played an important theoretical role in the development of collaborators Cavalli-Sforza and Edwards's statistical phylogenetics.[24]

The partitioning frame of literal genetic maps includes two general kinds of parts: geographic parts, such as oceans, mountain ranges, and urban areas, and genetic parts, including genes and their frequencies.

Space in these maps is literal and geographic. The maps show the range and location of genetic variation available for natural selection. These literal maps also sometimes portray the ecological conditions (e.g., temperature and humidity, predator species) determining the extent and kind of natural selection. All this information hones evolutionary explanations, and helps judge the accuracy of evolutionary predictions. Time is either *synchronic*, at a given time slice, or *diachronic*, showing changes in gene frequency over time in a given region.

## The Comparative Genetic Map

Comparative genetic maps guide our understanding of evolutionary changes in the regulation, order, number, and nature of genes across species. Considering our overall DNA, humans are about 99 percent identical in nucleotide sequence to chimpanzees. For genes that encode proteins, we are roughly 85 percent identical to mice and 60 percent identical to bananas. This tells us how much the cells of nearly all organisms share. Basic metabolic and reproductive "housekeeping" functions work well enough to have barely changed over hundreds of millions of years of evolution, across an impressive range of species.

Evolution is conservative and opportunistic, adapting what already works to new situations, rather than building from scratch. This modus operandi constrains later change. Because evolution is limited by which genetic information is accessible, and because phenotypic parts and properties were already available in ancestors, comparing the genetic maps of groups of related species can explain the structure and function of genes in lesser-known species.

The *comparative genetic map* is a set of linear genetic maps of different species allowing us to draw inferences about gene expression in different species, and the evolutionary function of genes. We can use these maps to make inferences about shared ancestry, gene rearrangements and tinkering, the evolution of protein structure and function, and possible

---

24. Cavalli-Sforza, Barrai, and Edwards 1964; Cavalli-Sforza and Edwards 1967; Winther 2018a.

cross-species *homologies* (similarities in structure as a result of sharing a common ancestor). Such maps may also shed light on the evolution of higher-level phenotypic features such as wings; the ability to withstand extreme temperatures; and behavioral and cognitive processes.

Starting in the 1990s, genomic information for fruit flies became widely available through the online tool FlyBase. Today, the genomes of twelve different species of *Drosophila* are available to researchers.[25] Databases for other organisms are also available, including *E. coli* bacteria, yeast, zebrafish, and the plant *Arabidopsis* (plate 10).

Consider the iron-centered heme proteins hemoglobin and myoglobin. Hemoglobin in the blood picks up oxygen from the lungs and delivers it throughout the body; myoglobin supplies oxygen to the muscles. These were the first two proteins whose three-dimensional structures were mapped. Both kinds of heme proteins are found in almost all vertebrates. Comparative studies prove that, hundreds of millions of years ago, a single gene coding for a heme protein existed in a distant ancestor.[26] After a series of gene duplications, the various duplicates began to evolve separately into different kinds of heme proteins. One such gene became the ancestral line for myoglobin and another for hemoglobin. The genes for myoglobin and hemoglobin—and, later, hemoglobin's two subunits, alpha and beta globin—could then evolve independently.

Interestingly, studies of nucleotide maps in mammals demonstrate that a human alpha globin subunit is much more similar to a mouse or elephant alpha globin subunit than to a human beta globin subunit.[27] As we just saw, these two homologies have different origins: alpha globin genes across different species are *orthologous*, sharing functions and traceable to a single ancestral gene in a direct common ancestor, whereas alpha and beta globin genes within single species are *paralogous*, having arisen from an ancestral duplication event in a distant ancestor, and typically taking on new functions before each of the two genes specialized across species as orthologous genes.[28]

The partitioning frame of comparative genetic maps includes that of linear genetic nucleotide maps, in which the genetic parts are identified

---

25. *Drosophila* 12 Genomes Consortium 2007; Stark et al. 2007.

26. Dating these deep events exactly is difficult, because molecular clock signals typically reach back to six hundred to eight hundred MYA, conflicting with the upper bound of multicellular animal evolution of approximately 540 MYA established by the geological record of the Cambrian explosion. Alberts et al. 2014, 229, fig. 4-76; Hardison 2012, 2, fig. 1, 4, fig. 2.

27. Hardison 2012, 7, fig. 3, 8, fig. 4.

28. Pevsner 2015, 70–79.

across many species, rather than just one. Space is the extreme-scale space of linear genetic nucleotide maps, and potentially also the various spaces of phylogenetic trees, discussed below. Time is the evolutionary time of the formation of species (*speciation*), which tends to be measured in millions of years.

## The Adaptive Landscape Map

Literal genetic maps show us the terrestrial distribution of genes and illuminate the history of populations, typically in a single species, while comparative genetic maps shed light on the origin of multiple species. To further understand the evolution of organismal complexity, however, we need to consider selection—and this key evolutionary force can be cartographically captured by the *adaptive landscape map*.

Sewall Wright gave us the first adaptive landscape map in a 1932 paper.[29] Wright's visualization offers students of evolution an intuitive and abstract way to think about the key evolutionary concept of fitness. Like Carl Friedrich Gauss and Alfred Wegener (discussed in chapter 6), Wright had surveying experience—a college course in surveying from his father, and a year as a railroad engineer in South Dakota.[30] His understanding of mathematics and cartography generated an imagined abstract space resembling a landscape.

On his state-space map, each specific combination of many alleles at many loci genes is depicted by a point on a flat $XY$ plane. Wright means to include about a thousand loci, each with ten alleles. Figure 8.4 is a "very inadequate representation" of the field of gene combinations, and yet it visually aggregates individuals with increasingly different genotypes.[31] That is, individuals who are relatively farther apart along the $x$-dimension or $y$-dimension, or both, have fewer alleles in common, and those that are closer together share relatively more.[32] The third dimension of this partitioned state space is the *relative fitness* value, analogous to height and depth as depicted on a topographic map. The relative fitness of a certain

---

29. Strictly speaking, he sketched a one-dimensional map in a 1931 letter to R. A. Fisher (Provine 1986, 272).

30. Provine 1986, 19–23.

31. S. Wright 1932, 357–59.

32. To crudely put it: an individual at the leftmost edge of the map might have the recessive allele (or have five of ten alleles) for five hundred loci, and an individual at the rightmost edge might have the dominant allele (or the other five alleles) for those same five hundred loci. Conversely for the $y$-dimension summarizing the five hundred other loci.

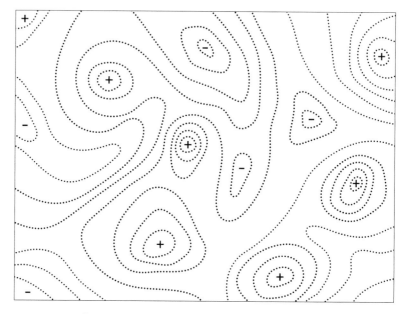

FIGURE 8.4. "Diagrammatic representation of the field of gene combinations in two dimensions instead of many thousands." See text for details. (Wright 1932, 358, fig. 2; formatted by Mats Wedin.)

genotype is how many offspring individuals with that genotype leave, relative to individuals of other genotypes, in a given environment, and given population. In this sense, fitness, called *adaptive value*, is abstract and inherently comparative.

The adaptive landscape map's contour lines (*isolines*) represent levels of equal relative fitness, with peaks (+) and valleys (−).[33] Wright ex-

---

33. Provine claims that this map is incoherent because it represents discontinuous genetic combinations as a continuous surface (1986, 310). There is merit to this worry, but note that each of the two dimensions is already a visual summary of an enormous amount of gene combinations, at many gene loci (five hundred, say, in the case of Wright's thousand total loci). Thus discontinuity is almost continuity. Furthermore, the philosopher of biology Jonathan Kaplan (pers. comm., November 2015) defends this map: "Each location on the landscape is a genotype, with an associated fitness given that particular developmental environment. [The 1932] Wright landscapes are explanatory—organisms in high-fitness spaces reproduce more than those that do not, and leave a 'cloud' of related organisms around them, some of which are in even higher fitness regions, and some in lower." More generally, there is another kind of adaptive landscape map that refers to entire populations. The $x$ and $y$ axes are no longer the particular alleles of an individual

plains: "In a rugged field of this character, selection will easily carry the species to the nearest peak, but there may be innumerable other peaks which are higher but which are separated by 'valleys.' The problem of evolution as I see it is that of a mechanism by which the species may continually find its way from lower to higher peaks in such a field."[34] Time here is evolutionary time, often measured in generations or many years. It can be depicted as zigzagging arrows moving across the landscape, where each small, straight section of an arrow represents one or several generations.

Wright also uses this metaphorically compelling map to imagine how genetic drift might facilitate evolution: "The species moves down from the extreme peak but continually wanders in the vicinity."[35] The adaptive landscape map can also be used to show evolutionary opportunities—paths over the adaptive landscape.[36] Just as for regular travel in a landscape with cliffs or narrow gulches, evolutionary movement in the gene space of individuals and populations due to natural selection is constrained by contour lines of adaptive value.[37]

---

at a variety of genetic loci; each dimension is now a *population frequency* of a gene, or set of genes. Moreover, fitness is not individual fitness, but now the mean fitness of a population for a certain *interaction system* of gene frequencies. This association between gene frequencies and fitness produces an adaptive landscape surface of "mean selective values" (S. Wright 1977, 3, 445–46). This map need not have the same topography as the adaptive landscape surface of individual gene combinations described in the main text above.

34. S. Wright 1932, 358–59. See also S. Wright 1988.

35. S. Wright 1932, 362.

36. Refer to the oft-reprinted figure 4 from S. Wright 1932, 361, showing movement between two neighboring peaks under six sets of conditions of population size, mutation rate, and selection differentials. Here and elsewhere, Wright relies extensively on adaptive landscape maps to develop and communicate his Shifting Balance Theory (SBT) of evolution (see Winther, Wade, and Dimond 2013), thus demonstrating the versatility of maps as both research and communication tools. The maps are neither necessary nor sufficient for SBT, and have been used for other purposes.

37. These maps do have their critics. In order to avoid "talk[ing] as if natural populations lived on fitness surfaces rather than on the earth's surface in ecological settings," we should remain cognizant of the threat of pernicious reification (Provine 1986, 308). Despite such dangers, adaptive landscape maps have served immensely useful purposes in evolutionary biology (Carneiro and Hartl 2010; Plutynski 2008; Svensson and Calsbeek 2012; Wade 2012). For instance, these maps play important inferential and heuristic roles for Gavrilets's (2004) development of "holey" landscapes.

Is the map the territory?[38] One reason to say not is that a multidimensional state space of genetic combinations (e.g., Wright's thousand dimensions) cannot be faithfully reproduced in maps of low dimensionality. Another issue is that evolutionary forces besides random genetic drift, migration, and natural selection—such as developmental and phylogenetic constraints—cannot be clearly captured by the adaptive landscape map. Moreover, it is a representation or model of the theory, not of complex empirical reality. It represents a simplified, abstract landscape, as given by mathematical genetic evolutionary theory. It is not the same as the world—or even the theory—that it helps us think about.

## An Analogous Genetic Map: The Tree of Life

Let us explore the representation of the evolutionary process rendered via the general map analogy: the *analogous genetic map*. Darwin's *On the Origin of Species* contained a single diagram, popularly called "the Tree of Life," which depicts the divergence of taxonomic groups of organisms. That visual, as well as more-contemporary tree and network phylogenies, lend themselves to multiple map-thinking interpretations—literal, state-space, and analogous.[39]

### DARWIN'S HYPOTHESIS

In *On the Origin of Species*, Darwin bequeathed us two ideas about the evolutionary process: natural selection and common descent. Although these are logically independent, Darwin integrated them: natural selection over time and space leads to the accumulation of small, heritable changes, which—especially with geographic isolation—leads to the formation of new species. Similarly to Wegener in the twentieth century, Darwin built his theory of natural selection on facts and data from a wide range of fields and processes: artificial selection, variation and inheritance, development, demography, and ecology.

His treelike diagram embodies the causal narrative of branching-species diversification from common ancestors, driven by natural se-

---

38. While Wright probably did not refer to these landscapes as maps, many others have, including Simpson (1944) 1984, 89–96; Lewontin and White 1960, 126; Gould 1997, 1022–23.

39. Darwin also employed literal maps to illustrate the distribution of coral reefs in his 1842 monograph; see his plate 3 world map. Michael Ghiselin provided constructive feedback.

lection.⁴⁰ It has provided an image for understanding all phylogenies generally, making it a "dominant metaphor."⁴¹ Darwin's image can also be countermapped, as I have tried to do with Heidi Svenningsen Kajita in our *Pondering Darwin's Forms of Life* (fig. 8.5).

Darwin intended his Tree of Life diagram to be a new representation of an evolutionary pattern he was trying to persuade his reader to accept. It was a metaphor resonating with ancient traditions. Thus he writes: "The affinities of all the beings of the same class have sometimes been represented by a great tree. I believe this simile largely speaks the truth."⁴²

This tree figure is foundational to Darwin's theorizing practices, exemplifying his exhortation: "All true classification is genealogical."⁴³ Consider its parts. Symbols A through L "represent the species of a genus large in its own country." The $X$-axis represents feature differences in "characters," which we might now understand as a range of phenotypes.⁴⁴ The farther apart two species are, the more different they are.⁴⁵

The $Y$-axis represents time at various scales. The horizontal lines, he writes, "represent a thousand generations, but each may represent a million or hundred million generations, and likewise a section of the successive strata of the earth's crust including extinct remains."⁴⁶ Generation time depended on the species, and, famously, Darwin and other scientists at the time did not know how old Earth was.

The tendency to vary, which Darwin took to be hereditary,⁴⁷ is represented by small lines forking off each node. Only a few of these lineages persist through time.

Darwin draws on the diagram to bolster his argument that natural se-

---

40. Bouzat 2014.

41. According to Gerald Holton (1977, 40) a "central image," such as Newton's "clockwork" universe, provides a "single, almost hypnotic image" that serves as a "fruitful oversimplification."

42. Darwin (1859) 1964, 129.

43. Darwin (1859) 1964, 420.

44. Strictly speaking, Darwin did not strongly distinguish genotype and phenotype. That distinction was drawn early in the twentieth century, in the wake of August Weismann's work, and with the rise of a variety of experimental breeding programs (Winther 2001). Thus, the $x$-dimension of his Tree of Life could also be a genetic phase space, as with Wright's map above (where fitness rather than time is the last dimension), and as is, in essence, the case with contemporary statistical phylogenetics below, first developed by Cavalli-Sforza and Edwards (Winther 2018a).

45. Darwin (1859) 1964, 116–17.

46. Darwin (1859) 1964, 124; cf. 117, 331.

47. See Winther 2000.

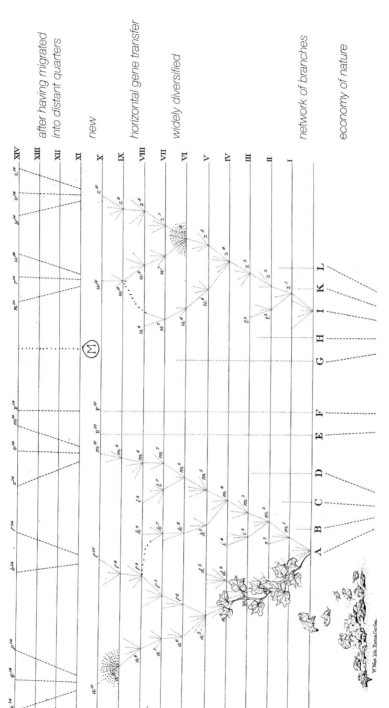

FIGURE 8.5. *Pondering Darwin's Forms of Life.* Following concepts both from Darwin's *On the Origin of Species* (e.g., widely diversified) and outside of that book (e.g., horizontal gene transfer), this figure rethinks the only illustration in *Origin*, commonly referred to as the Tree of Life. Could a new form of life be introduced de novo? How might different branches meet up via horizontal gene transfer? Is variation random in all directions? (Darwin [1859] 1964, foldout between pages 116 and 117; adapted by Heidi Svenningsen Kajita and Rasmus Grønfeldt Winther.)

lection explains the diversity of life.[48] This branching topology, in turn, explains why related species look similar, exhibiting homologous parts. The tree diagram also explains why taxonomy and classification are hierarchical (as we saw in chapter 2, where we investigated the Tree of Life as an analogy). The tree diagram's hierarchy is fractal, capturing the divergence of single species giving rise to genera, families, and so forth, all the way up to phyla.[49]

Darwin's image represents an idea, not an actual evolutionary history of any organism in particular. A tree showing the full set of evolutionary relationships among all the organisms of an actual family would require a zoomable tree (and network) with far more detail. Yet the map packs significant information. The nodes from which multiple varieties fork represent "divergence of character."[50] Diverging groups are able to "seize" and "fill" "new," "many," and "widely diversified" "places" in the "polity of nature" and "economy of nature."[51] Moreover, "forms of life" often change "after having migrated into distant quarters."[52] Darwin's language suggests movement into new functional or spatial niches and new geographic areas. From this spatial rendering and from the facts of individual death and species "extinction," we also get the fossil record buried in Earth.[53] Darwin might also have been referring, more abstractly (or even spatially), to movement into new ecological niches.[54]

Darwin held his diagram to be too simple.[55] In simplifying, it follows certain abstractive-averaging idealizing assumptions, as we met in chapters 3 and 7. For instance, it assumes bifurcating branching, and omits hybridization and reticulation. Network topologies provide an alternative, more general metaphor and visualization.[56]

48. See Darwin (1859) 1964, chap. 4, "Natural Selection," 80–130, and chap. 10, "On the Geological Succession of Organic Beings," 312–45.
49. Darwin (1859) 1964, 331–36.
50. Darwin (1859) 1964, 111–26.
51. Darwin (1859) 1964, 112, 121, 331.
52. Darwin (1859) 1964, 410.
53. Darwin (1859) 1964, 317–22.
54. For further discussion of Darwin and maps, see Camerini 1993, including nn21–23 delineating Darwin's use of maps in his work on coral reefs and geology.
55. Darwin (1859) 1964, 331.
56. A network view emphasizes the intertwinement of many branches, as different species exchange genetic material, or as one cell of one organism goes into a larger cell of another species not closely related, eventually becoming an *organelle*—analogous to an organ and performing a specific function, as does a cell nucleus or a plant's chloroplast (which carries out photosynthesis). See

## CONTEMPORARY PHYLOGENIES

Contemporary, statistical methods of representing species' evolutionary history treat frequencies of distinct genes as different dimensions of a state space, with time being an additional dimension.[57] Unique populations are points in this space, and one way of inferring a tree in such a diagram is by statistically pairing the closest points, recursively.[58] Many other statistical approaches to phylogenetic inference exist.[59]

Today, tracing mutations in HOX genes offers a beautiful illustration of how genetic trees for actual organisms can be drawn and what they say about the relative distance and variation of species at different parts of the Tree of Life's many branches. Such maps represent the interconnection of life.

Darwin's tree diagram illustrates phylogeny by depicting "divergence of character" of species over time. Today's tree diagrams often use genetic data for phylogenetic inference.[60] These diagrams rely on assumptions similar to Darwin's, even if the context is different. The aim of contemporary phylogenetic trees is to hypothesize specific evolutionary histories and the order of branching events. Both approaches—phenotypic and genetic, each contextually objective for its purpose—should be welcome in contemporary discussions; they provide map pluralism, together with network views.

The partitioning frame of phylogenetic analogous maps includes identifying features or properties at the phenotypic or nucleotide level. Such abstracted parts are compared *across* species—how and to what extent do features differ, such as wing features among *Drosophila* species, and what does that tell us about how species are related? Extant species are also part of Tree of Life analogous maps, as they are tips or leaves of phylogenetic branches. Space can be inscribed geographically and literally (spatial ecological isolation and genealogical splitting at branch nodes) or as a state space (a tree structure is statistically inferred in a formal space with dimensions of genotypic or phenotypic properties). Time is not necessarily represented one-to-one in terms of years, but branching order respects temporal order.

---

Doolittle 1999; Doolittle and Bapteste 2007; see also figure 8.5 and pages 144–45. Interestingly, Darwin was inspired by alternative topologies too, such as that of a bushy "coral of life" (Bredekamp 2019; Maderspacher 2006).

57. Winther 2018a, 29, fig. 1.
58. Winther 2018a, 429–31.
59. Felsenstein 2004.
60. See Winther 2009a, 2018a.

TABLE 8.3 Mapping-genetics integration platform

| Genetic map type | Partitioning frame | Space and time | Map type |
|---|---|---|---|
| Linear | Morphological or disease features; chromosome bands; nucleotides | Chromosomal and nucleotide space; time of crossover, and experimental protocols | Extreme-scale |
| Gene expression | Expressed genes; body regions of gene expression | Nucleotide, cellular, and body space; developmental time | • Extreme-scale<br>• Causal |
| Genotype-phenotype | Genotypic and phenotypic features captured formally | Genotypic and phenotypic state spaces; molecular, developmental, and evolutionary time | State-space |
| Literal | Relevant geographic regions; genes represented | Geographic space; synchronic or evolutionary time | Literal |
| Comparative | Same as linear genetic maps, but with different species as the source of parts | Nucleotide space; potentially, various spaces of phylogenetic trees; evolutionary time | Extreme-scale |
| Adaptive landscape | Genotypic and phenotypic features captured formally | State space, with an abstract fitness dimension; evolutionary time- | State-space |
| Analogous (Tree of Life) | Nucleotide and morphological parts; different species are the source of parts | Geographic space or state space; implicit evolutionary time; branching order respects temporal order | • General map analogy<br>• State-space<br>• Literal |

## Future Extensions: Mapping Genetics as a Paradigmatic Integration Platform

The mapping-genetics integration platform developed in this chapter, summarized in table 8.3, presents a plurality of maps. Each one of the seven map families can be usefully imagined as a complementary countermap to every other map family. Moreover, each family contains countermaps within it (e.g., network phylogenies contrasted with bifurcating tree phylogenies).[61] This integration platform thus provides a cognitive and troubleshooting activity for scientists; philosophers, historians, and

---

61. Turnbull 2004 provides another understanding of mapping and the map analogy in genetics.

sociologists of science; students; and the general public. It permits pluralistic ontologizing. Moreover, countermapping strategies to genetics as such should also be considered in the integration platform, lest we perniciously reify genetics.[62] For instance, in analyzing human health and development, pointing to genes as factors is insufficient. In addition to tracking multiple interacting environmental factors, we must also consider that the individual is a *holobiont*, a confederation of many organisms, including bacteria in the gut microbiome, whose aggregate genome can be described as a *hologenome*.[63]

Recall Ontologizing III from chapter 3. A mapping-genetics integration platform is a useful tool for abstraction and communication in the classroom. Just as it aids understanding and progress to teach a variety of map projections, not just the Mercator, so it would be beneficial to teach genetics not according to just one or two maps, but as a whole set of partial and limited maps, perhaps even using the list of seven above. Teaching students at all levels that genetics (and science generally) comprises many theories and maps blocks the pernicious reification of universalized and narrowed models.

Returning to my account of multiple representations from chapter 5, think of each gene map as one perspective on, or abstraction of, the complex and fuzzy world of genetics-related life processes. Maps collide and combine. It is by the complex interrelation, multiple cross-referencing, and illumination of distinct maps that we cast a floodlight, rather than a series of spotlights, on the phenomena of life.

Maps and mapping, space and time, and map thinking are elemental to the sciences, from cosmology to psychology. Great scientists have been inspired by map thinking: Gauss, Mendel, Darwin, Wegener, and Wright, to mention just a few discussed in these pages. Whatever the fate of my philosophical framework as presented in part 1, I would like to imagine that in the future you might be sitting at a party or talking to friends or even to strangers you have just met, and will discuss with them how powerful and deep our cognitive and social need is to *map space*—and continue from there to develop a useful problem-solving "philosophy out of mapping."

62. Critiques of an exclusively genetic point of view for understanding biological systems include West-Eberhard 2003; Levins and Lewontin 1985; Keller 2002a, 2002b; Oyama (1985) 2000; Winther 2008; Longino 2013.

63. D. Walsh 2015; E. Rosenberg and Zilber-Rosenberg 2018; Van de Guchte, Blottière, and Doré 2018.

TABLE 8.4  Glossary for chapter 8

| Term | Definition |
| --- | --- |
| adaptive value | relative fitness of a trait |
| chromosome | a DNA double helix folded up and wrapped with proteins |
| coded by | phenotypic features are coded by the genotype; more properly, the linear sequence of genes is coded onto the linear sequence of amino acids of proteins; genotypic information is only partly mapped onto the phenotype (see norm of reaction; heritability) |
| codon | a sequence of three DNA ladder-rung halves, i.e., three nucleotides |
| crossover | when parts of two paired chromosomes break apart and combine with each other during gamete formation |
| diachronic | occurring or changing over time |
| gamete | egg or sperm |
| gene expression | the process of mapping a DNA strand onto an RNA strand, and then cutting, pasting, and transforming that RNA strand before it maps onto an amino acid strand |
| genome | the complete DNA information of an organism |
| genotype | all the genes in a cell or some part of a DNA strand's sequence |
| heme | a type of protein carrying oxygen—hemoglobin and myoglobin |
| heritability | the amount of the phenotypic variance correlated with (explained by) the additive genotypic variance |
| homology | a similarity in structure as a result of evolving from the same part in a common ancestor |
| HOX gene | a plot and map body design along the head-to-tail axis |
| isoline | a contour line on an adaptive landscape |
| linkage group | a set of genes across a single chromosome that tend to act as a single correlated group of genes when gametes are produced |
| map distance | distance between two genes on a chromosome |
| map unit / centi-morgan | At 0 map units (0 percent recombination), the genes are co-traveling. In between (just above) 0 and up to 50 map units, the genes are linked. If two genes are 50 map units or more apart, they assort independently, even if they are on the same chromosome. |
| morphology | the form and structure of an organism |
| norms of reaction | charts or maps showing phenotype values at different environmental values, with each genotype drawn with its own line or curve |
| nucleotide | for DNA, a unit combining a sugar molecule, a phosphate molecule, and one of four molecules containing nitrogen: adenine, thymine, cytosine, or guanine (A, T, C, and G) |
| orthologous | genes sharing functions and traceable to a single ancestral gene in a direct common ancestor |
| paralogous | genes having arisen from an ancestral duplication event in a distant ancestor, and typically taking on new functions before each of the two genes specialize across species as orthologous genes |
| phenotype | an organism's observable traits (including cognitive and behavioral features) |
| phylogeny | evolutionary history of an organism or species in relation to others |
| principle of independent assortment | the principle that an allele for one gene can separate independently from the alleles of another gene |
| principle of segregation | the principle that a sexually reproducing individual has two alleles for a given trait, and that those alleles separate when a parent creates a gamete |
| relative fitness | how many offspring individuals with that genotype leave relative to individuals with other genotypes, in a given environment |
| RNA folding | RNA assuming various shapes to perform different functions, determined mainly by the laws of thermodynamics and the information in the linear strand of RNA |
| RNA | ribonucleic acid; has one-half the DNA double helix ladder-and-rung structure, with uracil (U) substituting for thymine |

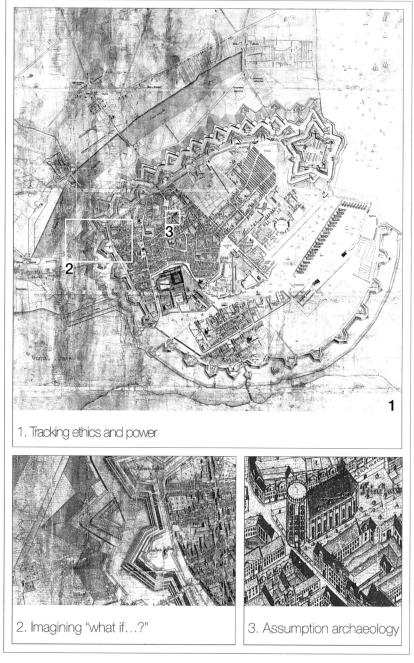

FIGURE 9.1. Christian Gedde's 1761 map of Copenhagen, with two cutouts, illustrating three philosophical methodologies from *When Maps Become the World*, as explained in the text. (Copenhagen City Archive map, adapted by Heidi Svenningsen Kajita.)

# 9

# Map Thinking Science and Philosophy

Recall the maps showing California backpackers places to trek. These backpackers, along with the climbers of Mount Everest and Cape Town's famous Table Mountain, forever change the landscape beneath their feet. As a result, the map needs to change. The original map bears the burden of having changed the world.

In this same way, the sciences are able to re-create the world, and have even created new worlds. For example, think about technologies such as the gene-editing tool CRISPR. This mapping technology, partly developed through scientific theorizing, may yet change our very genomic nature. After all, genomes can now be easily edited by a wide variety of people, and not just by biopharmaceutical companies or the medical or research communities. Or consider the abstract mathematical models that have been built, tested, and confirmed regarding the population sizes, population structure, and migration patterns of *Homo sapiens* today, and throughout our history. These theoretical models affect the public understanding of race and ethnicity, particularly as the research results are communicated into popular culture and to broader audiences by journalists, movie directors, and ideologues.[1]

Maps provide an intuitive window into how humans think and act, and steer their hopes and fears. I believe that, across the humanities and the sciences, cartographic objects—and representations that can be analogized to such objects—bring into sharp relief the wishes and anxieties we feed into our worlds. Our representations of reality shape how we understand ourselves and our universe. They influence our desire and our angst, and our imagination and our dreams, shaping the possible futures of our children and their children. Thus, barring the influence of unfortunate "antiscience" movements (e.g., antivaxxers and climate change deniers), scientific representations will inform the strategies we

---

1. Winther 2018a, 2018b.

as a society design and implement to redress social injustices and forestall ecological catastrophe.

*When Maps Become the World* has explored how maps, theories, and models—not to mention paradigms and discourses—can be used for good (contextual objectivity) or ill (pernicious reification). Central maps featured in this book have brought such points home: Martin Waldseemüller's 1507 map baptized America, thereby exuding momentous cartopower; Gerardus Mercator's 1569 map gave birth to the projection that became a key weapon of European colonization and world domination. Yet both maps also embedded sincere concern with empirical facts and mathematical correctness. Guaman Poma's *mappa mundi* countermap foretold possibilities of either an apocalyptic or a redemptive future for indigenous Americans, while simultaneously respecting Andean and European mapping practices, and Arno Peters's theatrical countermap attempted to counteract the colonialist excesses of Mercator's projection, while satisfying metric map properties distinct from those of Mercator.

This concluding chapter of *When Maps Become the World* provides three other applications of map thinking and the map analogy to areas of ongoing interest to scientists, humanists, and the general public:

- What exists (and how and why), and what responsibility do we have for that which exists, now and in the future?
- What is scientific methodology?
- What is philosophical methodology?

## Existence, World Making, and Responsibility

How do we understand the reality of the objects and processes postulated by science? Did Galen's four humors exist in some sense, despite the fact they were disproven? Were atoms or electrons or genes ever *not* real? Do social classes or the unconscious exist, and in what sense? What role do researchers or the lay public or university science students play, if any, in establishing and stabilizing the existence and reality of such objects and processes?

Differing intuitions regarding what exists are deeply rooted in the philosophical traditions of diverse cultures. Do you or I exist, or are we a dream, an emergence, a fabrication? Are we and our thoughts an integral and essential part of reality, or does reality exist separately, whether as pure matter, or with its own soul(s), logic(s), or god(s)? If the apparatuses of human thought and sensation are indeed cleaved from reality, then is reality hidden behind a translucent—or, worse, opaque—veil of appearance? Are you and I free in our will and in our actions, and can our aspirations, imaginations, and activities significantly affect the world? Our attitude toward these questions impacts what we believe and imagine we can change with our knowledge and actions; how we imagine our role and place in the universe, and in society; and which sorts of futures we might consider possible and desirable.

Philosophers and other thinkers and doers have delved extensively into such questions and concerns. Consider three families of philosophical approach to questions about what exists, and how and why:

- *Constructivism* highlights the roles that humans play in co-determining the world via scientific interventions and representations.
- *Empiricism* is concerned with how inferences can be legitimately drawn from data, measurement, and observation.
- *Realism* interprets scientific theories as mirrors of a nature independent of human will, mind, and sense, and it interprets the theories as becoming increasingly certain and proven over history, through ongoing correction of error.

Drawing on the map analogy, each of these positions can be likened to a philosophical projection onto the actual, in vivo representational practices of science. Each of the three projections yearns for completeness, in the style of Mercator's projection, and philosophically universalizes itself. Each comes with attendant fears, and asserts itself in dialectical opposition to the others. Each has a key cartographic counterpart, and each will be illustrated, briefly, with the question of the existence of genes.

Constructivism finds ever more nuanced ways in which cognitive principles, communal actions, and technology can change the material world.[2] The constructivist emphasizes human intervention and human agency. The constructivist is concerned that scientistic, neopositivist apologists—those who believe natural science can, eventually, explain everything—will downplay the historical contingency of scientific knowledge and progress. We saw Denis Wood's cartographic constructivist suggestion that mapmakers are "extraordinarily selective creators of a world."[3] Bonnie Spanier, in her analysis of scientific journals and textbooks on molecular biology, finds that there is a close connection between the gendered language of sexual reproduction and the view that DNA is the master (male) molecule of life, controlling development,

---

2. On the last, see, for example, Pickering 1984. For a general, "a priori" Kantian constructivism, see Friedman 2001, 2008; cf. Hacking 2002; Winther 2012a.

3. Wood 2010, 51; see chapter 5.

mind, and behavior within hierarchically organized life processes. It is through sexualized language that the molecular biologist and neuroscientist Francis Crick's central dogma, describing a linear flow of genetic information, has been constructed—and even perniciously reified—as the single, dominant map of the causal agency of DNA across generations. Alternative maps such as "metabolism, energy-conversion, [and] autopoiesis (self-maintenance with or without reproduction)" are by and large ignored, thereby distorting knowledge, and—I would add—universalizing and narrowing dominant maps focused on genetic information.[4] One must squint to see past the central dogma, just as it is hard to see the Peters or Robinson projection through the glare of Mercator's.

Empiricism insists on the importance of data, whether historical or contemporary. Empiricism accepts that, because the evidence consists only of the best data available at any given historical moment, there will always be a very large number of representations (in fact, an infinite number) consistent with the evidence gathered.[5] The empiricist dreads that the limitations of data will be disrespected—perhaps by merry ontologists drawing strong, unwarranted inferences about unobservable entities. This map-thinking comment by Leslie Curry quoted in chapter 4 can be seen as empiricist: "Geographical studies are not descriptions of the real world but rather perceptions passed through the double filter of the author's mind and his available tools of argument and representation."[6] The philosopher of science Kyle Stanford argues that the history of science has always contained "unconceived alternatives" to every extant representation.[7] Drawing on Charles Darwin's, Francis Galton's, and August Weismann's various influential nineteenth-century theories of heredity and development, he shows how the ontology of each was always limited, and that none of them imagined certain alternatives that would also have been consistent with the evidence available at the time.

Realism draws on sophisticated philosophical accounts of truth, reference, and natural kinds (classifications purported to correspond to the actual structure of nature) to appeal to the existence of an independent reality.[8] The realist worries that the independence of the real (that they

4. Spanier 1995, 92.

5. Longino 2002a; van Fraassen 2002; Stanford 2006; Cartwright 2007; Godfrey-Smith 2008.

6. Curry 1962, 21.

7. Stanford 2006, chap. 2, "Chasing Duhem: The Problem of Unconceived Alternatives," 27–50.

8. Kitcher 1993; Psillos 1999; Chakravartty 2007, 2016; for an epistemological attempt to inflect realism through an empirical "antirealism" lens, see Roush 2005.

have postulated) and the progress of science (that they believe in) will be unfairly diminished, either by constructivists inappropriately supposing that human conceptual construction changes the very fabric of the universe or by empiricists unjustifiably hiding reality behind a veil of appearances. As a case of map thinking in realism, consider the geographer Michael Goodchild's matter-of-fact differentiation of "other spaces, such as the space of the cosmos, of the human body, of the human brain, or of the genome" from "the geographic world," which is "defined by its extent, its geometry, and its range of useful resolutions."[9] In *The Advancement of Science*, the philosopher of science Philip Kitcher defends realism by employing a complex conceptual apparatus making a two-pronged appeal. First, Kitcher argues that certain natural kinds remain stable across historically changing scientific practices. Second, he declares that particular theory-dependent explanations are prevalent and important.[10] Specifically, Kitcher insists that the best suggestion for the predictive and explanatory success of genetics is simply "that there are genes."[11] These nuanced discussions in the philosophy of science regarding what exists have an intricate conceptual history, which has been recounted elsewhere.[12]

Turn again to map thinking. Cartographic objects simultaneously require (1) processes of social and cognitive construction, (2) empirical information, and (3) reference to a world. These three requirements align with the three philosophical families just discussed. Why, then, do philosophers so bluntly contrast constructivism, empiricism, and realism, as if they were mutually exclusive and competing in a zero-sum intellectual game? Why do we seem obligated to take sides and defend just one of these families? Perhaps it is because masculinist hierarchies continue to triumph. Furthermore, although their arguments differ significantly, their ongoing work seems to suggest that cartographers Wood and Goodchild would resist the simplistic distinctions and positionings associated with the mutual exclusivity of the three views above.

I believe that one reason these families of philosophical views have

9. Goodchild 2004, 302.

10. Kitcher's argument here is also inspired by the philosopher Hilary Putnam's statement of the "no miracles" argument for realism—that the best scientific theories are true statements about the world. "The positive argument for realism," he says, "is that it is the only philosophy that doesn't make the success of science a miracle" (Putnam 1975, 73). In other words, short of a miracle, there must be a real world to which science refers. What do *you* make of this argument?

11. Kitcher 1993, 157.

12. See, for example, Hacking 1983; Okasha 2002; Godfrey-Smith 2003; Winther 2009b.

been interpreted as mutually exclusive is the use of the term *antirealism*. Following my abstraction-ontologizing account, this is a kind of gross conceptual simplification and exaggeration, a pernicious philosophical averaging of many diverse and astute points of view. All arguments that do not have a putatively mind-independent reality as the bedrock of the world, and of our knowledge and actions, are lumped into a single category. But the descriptor *antirealism* covers too many disparate positions, including those of philosophers of science Arthur Fine, Larry Laudan, Kyle Stanford, and Bas van Fraassen, and various types of constructivism analyzed by Ian Hacking.[13] Force-fitting these views into a single category seems intellectually dishonest, because it ignores many important conceptual differences.[14] It also makes realism the default philosophical projection, similar to the Mercator projection, subsuming every alternative to it into an "anti-" category.[15]

I resist this strategy of classification. Realism should not be regarded as the default for framing the questions with which we began this section, let alone as the only game in town. Rather, by imagining each of the three families as a partial, purposive, and even political philosophical map projection, we can use assumption archaeology to identify the assumptions they share, and use integration platforms to negotiate tensions among them.

Undeniably, map thinking invites us to entertain a plurality of philosophical projections. When we try to understand the underlying assumptions of an opposing view, we might empathize with the other perspective. Doing so could even illuminate our own perspective. Constructivism,

---

13. See van Fraassen 1980; Putnam 1984; Fine 1986; L. Laudan 1990; Stanford 2006; Hacking 1999.

14. In his classic *The Structure of Scientific Revolutions*, first published in 1962, Kuhn claims that scientists "live" and "work" in a different world after a paradigm shift (Kuhn 2012: "lived," 117, "live," 133; "worked," 118, "work," 134). Is Kuhn a constructivist (paradigms make worlds) or a realist (the world exists independently of paradigms)? For a kinds-based articulation of Kuhn's constructivism, see Hacking 1993, 306. In a later discussion of Kuhn's chapter 10, Hacking sets forth fewer prescriptions and more descriptions; see Hacking 2012, xxvii–xxix. For a more realist interpretation, see Godfrey-Smith 2003. Perhaps, as with the map analogy's multiple representations account, constructivism, empiricism, and realism are merely different perspectives, each with its own set of partitioning frames.

15. This is not unlike how analytic philosophy defines *continental philosophy* or *post-structuralism*. Philosophy on the European continent, or even just in France, is unjustly and ignorantly reified to be a garbage bin full of disparate refuse.

empiricism, and realism are partial, fallible, and abstract philosophical maps (or, more accurately, each is a family of maps). Each opens a window from a particular set of assumptions through which we could intentionally ontologize pluralistically—unmapping and remapping—answers to the above philosophical questions.

For instance, imagine simultaneously taking seriously Spanier's, Stanford's, and Kitcher's respective suggestions about the existence of genes. Interestingly, each itself seems a call for an integration platform: Spanier undermines centralized control by relating and contrasting it with alternative modes of imagining life; Stanford explicitly appeals to the importance of a slew of unconceived alternatives, and of finding ways of identifying these, even when we might seem blind to these alternatives; and Kitcher's analysis implies comparing the successes of many kinds of science, including the success of various parts of genetics. Now envisage placing *all* of these within a single philosophical integration platform.

Continuing with the map analogy and tools of map thinking, I suggest that there is no complete description of science. *There is no ultimate, single whole of scientific practice on which we shine philosophical spotlights from multiple perspectives.* Empirical and theoretical scientific inquiry is too complex a set of social and cognitive processes to permit absolute and unrevisable articulation. Only perhaps as a Kantian regulative ideal—that is, an ideal state toward which we aim and move, but which we never attain—may we posit that the different philosophical perspectives on scientific practice are together getting at a monistic whole, a description and explanation for the single thing that science at its very heart and core is. My position about the open-endedness of science does not imply a rejection of realism but instead is compatible with multiple philosophical views on the status of the reality that science describes.

Furthermore, the distinction between science and philosophy can itself be tenuous and fuzzy. Very likely there is a constrained yet dynamic and unfinished pluralism all the way up, from scientific representations and scientific methodologies (constrained by data and robust methodologies) to philosophical metarepresentations and philosophical methodologies (constrained by the requirements of being conceptually consistent and surviving critique). Let us now turn to such methodologies.

## Map Thinking Scientific Methodology

Scientific inquiry is a family of methods for following curiosity's lead and guessing about how things hang together. Applying map thinking and the map analogy enriches our understanding of general scientific methodology and inference. Studying cartography has invited us to ask a series of questions: How, by whom, and for what purpose is scientific knowledge made? Which assumptions seep into different scientific representations of nature or society? How and why are there so many perspectives—philosophical, scientific, and vernacular alike—on scientific phenomena and practice?

Scientific methodology as commonly taught is a linear sequence of logically related activities:

1. researching existing knowledge;
2. hypothesizing or theorizing;
3. formulating predictions from hypotheses or theories;
4. experimenting and testing; and
5. comparing predictions with experimental results to confirm or discard hypotheses or theories.

Similarly, philosophers as distinct as the logical positivist Rudolf Carnap, the pragmatist John Dewey, the critical rationalist Karl Popper, and the analytic philosopher W. V. O. Quine have regarded science as a robust and reliable inquiry with two prongs: abstract theorizing and concrete experimenting. However, this *central dogma of scientific inquiry*, as I will call it, is belied by the astonishing complexity of real, live science. Map thinking on scientific methodology invites us to question the universality and absolute validity of this methodological central dogma. Just as experimental science in the lab may miss out on conditions produced in the field, the armchair "lab" of theorizing about scientific methodology may miss what happens in the "field" of scientific practice.

As a teacher of philosophy of science for many years, I can attest that the stepwise hypothesis-prediction-experiment model of scientific inquiry is ubiquitous among university students and the general public. The cracks start showing in this central dogma when we look for the myriad assumptions, the births of pluralism, and the twin potentials of contextual objectivity and pernicious reification, in scientific theory and practice.[16]

Consider two ways of countermapping scientific methodology: as a logical form of inference, and as a style of practice. First, as we saw in chapter 2, the nineteenth-century philosopher Charles Sanders Peirce provided a useful classification of modes of inference that remains influential to this day. As we have seen with his methods, we can distinguish deductive, inductive, abductive, and analogical inference. We can use the

---

16. See, for example, discussion on how claims about genetics and gene causality are treated by the general public in Keller 2002a, introduction and conclusion; Lewontin 1991, chap. 4 and 5.

full panoply of these forms of inference within abstraction and ontologizing mapping practices, and we do so in science as well, often applying the forms within the same field. We cycle among the modes, guided—in cartography and science alike—by our interests to choose one inferential procedure or another at different times, for use in different ways.

Second, consider the plurality of styles of scientific reasoning defended by the historian of science A. C. Crombie, Ian Hacking, and many others.[17] These include *axiomatic, probabilistic, taxonomic*, and *genealogical* styles. They exist across the sciences, even if they are occasionally concentrated in some areas of science rather than others.

This meta-analysis of my multiple perspectives on scientific methodology becomes an exercise in map thinking, where our maps are the philosophical maps of how the science itself is conducted. Just as there is no single correct map or map projection, or even a single correct and universal map abstraction and ontologizing practice, so there is no single way of interpreting or of practicing scientific methodology. Imagining scientific theory as a map, and scientific practice as mapping, opens a window through which we can see science for what it is: *an error-prone yet error-correcting, imperfect yet knowledge-producing, variegated and plural empirical, value- and power-laden enterprise.*

## Map Thinking Philosophical Methodology

Philosophy involves the systematic practice of *questioning*, especially when undertaken through dialogue and from a highly general point of view.[18] Truly, the Socratic method is an archetypal philosophical method. Another is *conceptual analysis*:[19] a representation (for example, a concept or a definition) is broken up into its necessary and sufficient theoretical components; then, each component is fully characterized and appropriate conditions for applying it are identified.[20] I complement such important methodologies and aims of philosophy with three others. One, already discussed, is assumption archaeology. Two further techniques, introduced here, are *tracking ethics and power* and *imagining*

---

17. See Crombie 1994, 1996; Vicedo 1995; Hacking 2002, 2009; Winther 2012a.
18. See, for example, Rescher 2001; Williamson 2008.
19. A. J. Ayer has effectively identified philosophy with conceptual analysis: "The propositions of philosophy are not factual, but linguistic in character—that is, they do not describe the behaviour of physical, or even mental, objects; they express definitions, or the formal consequences of definitions. Accordingly, we may say that philosophy is a department of logic" ([1946] 1952, 57).
20. Block and Stalnaker 1999; Chalmers and Jackson 2001.

"*What if . . . ?*" Each of these, while perhaps as old as philosophy itself, also emerges from map thinking.

## ASSUMPTION ARCHAEOLOGY

What is the structure of a complex set of assumptions? Under which circumstances are they operative, and who uses them? Scientific theory contains a multitude of implicit and powerful assumptions. *Homo cartograficus* collates these into operative structures, even when they partially conflict or are in tension with one another.

We have seen the concept *assumption* used broadly, to mean that which is believed to be implicit. Chapter 2 presented *assumption archaeology*. The philosopher can help bring assumptions to light by studying the unconscious content under and behind a theory—the *theoretical unconscious*. Assumptions include what Thomas Kuhn has called *ontological assumptions* and *values*.[21] In chapter 5, they included assumptions about map representational elements—for example, metric properties, symbols, and ontological or world-making elements. More generally, they also include distinctions and classifications. Even though assumptions often hide in the recesses and weeds of scientific theory and practice they cannot and should not be overlooked or ignored.

## TRACKING ETHICS AND POWER

Map thinking gives additional clarity to standard philosophical accounts of ethical thinking and overarching perspectives on ethics, such as virtue ethics, utilitarianism, or Kantian deontological ethics.

Which worlds are reified or changed according to a given representation? How, and for whom? Who wins under a given representational regime, and who loses? Is it possible for many agents to simultaneously win, or for us to even forgo the metaphor of winning—and, if so, how? Countermapping, critical cartography, and map art provide forums to think about the concrete way that cartopower—representational power—makes the world.[22]

Map thinking also complements feminist ethics, which can emphasize nurturing in interpersonal relationships or can argue that marginalized groups are in a position to notice things that nonmarginalized groups do

---

21. Kuhn 2012, 181–86. This discussion was part of the 1969 postscript to his original 1962 book.

22. On map art, see Harmon 2010; on maps and place, see Leslie 2016.

not. In this context, what we learn from cartography is the possibility of attending to subjectivity, context, and even to emotions and passions.[23] Map thinking about ideology and power helps us track and meld together "'male' ways of moral reasoning that emphasize rules, rights, universality, and impartiality [and] 'female' ways of moral reasoning that emphasize relationships, responsibilities, particularity, and partiality."[24] Moreover, maps are not abstractly male, and the territory is not concretely female. Map thinking challenges many dichotomies, distinctions, and binaries.

In part 1, we began to see ethics and power tracking via the possibilities opened up by alternatives to Waldseemüller's map, the Mercator projection, and their respective cartopower. Other potential subjects of ethics and power tracking include individual subjective perceptions and how they are influenced by bias based in feelings and social situations, or why nonmainstream and unorthodox yet valid scientific theories are frequently despised, and often hyperbolically so.

### IMAGINING "WHAT IF . . . ?"

To ask "What if . . . ?" implies imagining many possible worlds.[25] In posing this most capacious question, philosophy opens up a space for memories, feelings, hopes, and imagination. When we ask "What if . . . ?," we swap one set of assumptions for another and follow the world-making consequences of each, whether in the future or in potential existence more generally.[26] Perhaps this is a kind of future-oriented pluralistic ontologizing. We might imagine developing a cluster of worlds for which we could ask: What if we used different scientific maps—for example, replacing the self-interested *Homo economicus* (rational human)[27] with an altruistic agent or even an anarchist agent? What if any one philosophical map—including the one communicated in these pages of *When Maps Become the World*—were not centered imperialistically on a world navel? What if social relations were structured with institutions, values, and behaviors dramatically different from those in place here, today? Posing a number and variety of "What if . . . ?" questions can make room for imagining new kinds of worlds. As the scholar Bertrand Russell wrote in his chapter "The Value of Philosophy," philosophical questions "enrich our intellec-

23. Harding 1986; Tong and Williams 2014.
24. Tong and Williams 2014.
25. On conceptions, semantics, and logics of possible worlds, see Menzel 2017.
26. Frase 2016.
27. As described, for instance, in M. Morgan 2006, 2012; cf. Sen 1977.

tual imagination and diminish the dogmatic assurance which closes the mind against speculation."[28]

Figure 9.1 provides a cartographic example of the importance of each of the three philosophical methodologies I have added. This figure shows Christian Gedde's 1761 map, inspired by Turgot's 1739 map of Paris, which portrays details of Copenhagen, down to its last buildings and streets, circa 1758, after the big fire of 1728. We can track ethics and power (methodology 1). Gedde's bird's-eye-view map reflects an intra-European power differential, with the Danish Royal House of Frederik V attempting to imitate the Royal House of France embodied in Louis XV. Frederik V founded the Royal Danish Academy of Art and purchased the Danish West Indies. Slotsholmen island is labeled "Christiansborg" (in the solid white square under 3), a royal palace and set of government buildings from which power oozed. Gedde also imagined "What if...?" (methodology 2). He drew walls and fortresses around Copenhagen that never existed. It remains unclear whether he was suggesting they be built, or whether Frederik V had plans to build them, or whether this might just have been artistic license. Finally, we can engage in assumption archaeology (methodology 3). As stated by the Danish historian and geographer Peder Dam, the Round Tower is shown with approximately the right number of windows and height, but is rotated to show its entrance toward the viewer—this is perhaps as much a painting as a map, aimed at pleasing Frederik V and other influential viewers.

## An Invitation to Dream

To borrow from philosopher Wilfrid Sellars, if the *manifest* image is the world we all see (for example, the table or screen or book in front of you), and the *scientific* image is the world science tells us is there (for example, the table or screen or the book as a collection of atoms and molecules),[29] then I would say that the *cartographic* image is a metaperspective on science and its philosophical underpinnings (for example, which of many possible perspectives should you take regarding the table, and why, including an inquiry into what assumptions each perspective embeds and how we could contextually apply and integrate the perspectives).

The cartographic image is not a view on the world; that claim would be too abstractionist and pernicious. Rather, it is an approach for assessing the value and dangers of scientific theory and technology. The map thinking of the cartographic image is, I believe, but one way of instantiating

28. Russell (1912) 1997, 161.
29. Sellars 1962.

FIGURE 9.2. *The Dead Speak Tomorrow*. Four species that are extinct or nearly extinct: the white rhinoceros; the vaquita from the Gulf of California; the New Guinean tree-kangaroo; and the Amazonian Spix's macaw, also depicted in Disney's *Rio*. Do these gorgeous organisms not deserve a voice? And what of entire ecosystems that are under threat, such as coral reefs? I refuse to let them just die. (Rhino, vaquita, and tree-kangaroo illustrated by Gertrud Bohnstedt Winther; macaw and coral reef by Larisa DePalma.)

what I take to be the mother of all philosophical methodologies: *critique*. Critique means, on the one hand, grounding and justifying, and, on the other hand, criticizing our pictures and perspectives of the world, and imagining a different world.

The cartographic image invites us to reflect on and dream of new places, new territories to explore, new worlds to make, and better worlds to emancipate and care for, and to be cared by. Yet many future map-thinking tasks remain beyond those introduced in this book—particularly in the social sciences, medicine, architecture and design, and engineering. We must represent our own worlds and the world at large in the ways we intensely dream about, lest the worlds and world get represented *for* us. But we must also imagine and hold ourselves accountable to how *others* dream, and how our dreams affect others. And those others include many other voices, even nonhuman. There are many more than 7.7 billion agents on this planet (fig. 9.2).

Imagination and dreaming kindle our travels in the sciences and philosophy. And the sciences and philosophy, in turn, kindle our imagination and dreaming.

# Appendix: Cognitive Map Exercise

Here are two blank maps of continent outlines, in Mercator (fig. A.1) and Peters (fig. A.2) projections, and a blank page for an additional map you could draw.

## World Map Instructions

On each of the blank world map projections, draw in features from memory as best you can. Take however long you need.

After you draw the map, reflect on what you did. What did you remember, and what did you choose to fill in? Did you add mountains, rivers, forests, or other physical or natural features? Were you perhaps more focused on capital cities, country borders, population density, and other political or cultural features? What about the oceans and seas? Did you use color, icons, and symbols? Was there a projection you preferred? And how did you feel as you filled in the continental blanks? Did biases or preconceptions about certain parts of the world enter your mind? Did you dream of traveling?

## Home Map Instructions

On the third page (fig. A.3), try drawing a map of your home location or favorite spot. It can be a favorite city, beach, or forest, maybe even your home—any space you would like to map. How might your embodied experience of the mapping process alter your understanding of mapmaking and map use, and of your own personal and meaningful spaces?

FIGURE A.1. Mercator projection, redrawn by Mats Wedin with data from naturalearthdata.com.

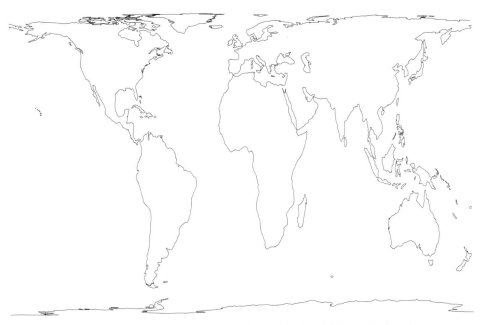

FIGURE A.2. Gall–Peters projection, redrawn by Mats Wedin with data from naturalearthdata.com.

# YOU

# References

Adorno, Rolena. 2000. *Guaman Poma: Writing and Resistance in Colonial Peru.* 2nd ed. Austin: University of Texas Press.

———. 2011. "Andean Empire." In *Mapping Latin America: A Cartographic Reader*, edited by Jordana Dym and Karl Offen, 74–78. Chicago: University of Chicago Press.

Agamben, Giorgio. 2009. *The Signature of All Things: On Method.* Translated by Luca D'Isanto with Kevin Attell. New York: Zone Books.

Agarwal, Gautam, Ian H. Stevenson, Antal Berényi, Kenji Mizuseki, György Buzsáki, and Friedrich T. Sommer. 2014. "Spatially Distributed Local Fields in the Hippocampus Encode Rat Position." *Science* 344 (6184): 626–30. https://doi.org/10.1126/science.1250444.

Alberch, Pere. 1991. "From Genes to Phenotypes: Dynamical Systems and Evolvability." *Genetica* 84 (1): 5–11. https://doi.org/10.1007/BF00123979.

Alberts, Bruce, Alexander D. Johnson, Julian Lewis, David Morgan, Martin Raff, Keith Roberts, and Peter Walter. 2014. *Molecular Biology of the Cell*, 6th ed. New York: W. W. Norton.

Alembert, Jean Le Rond d'. (1751) 1995. *Preliminary Discourse* to the "Encyclopedia of Diderot." Translated by Richard N. Schwab, with the collaboration of Walter E. Rex, 1:i–xlv. Chicago: University of Chicago Press. Available at http://hdl.handle.net/2027/spo.did2222.0001.083.

Allison, Henry and Peter Heath. 2002. *Immanuel Kant: Theoretical Philosophy after 1781.* Translated by Gary Hatfield, Michael Friedman, Henry Allison, and Peter Heath. Cambridge: Cambridge University Press.

Alpers, Svetlana. 1987. "The Mapping Impulse in Dutch Art." In *Art and Cartography: Six Historical Essays*, edited by David Woodward, 51–96. Chicago: University of Chicago Press.

*American Cartographer.* 1989. "Geographers and Cartographers Urge End to Popular Use of Rectangular Maps." 16 (3): 222–23.

Anderson, Michael L. 2010. "Neural Reuse: A Fundamental Organizational Principle of the Brain." *Behavioral and Brain Sciences* 33 (4): 245–66. https://doi.org/10.1017/S0140525X10000853.

———. 2014. *After Phrenology: Neural Reuse and the Interactive Brain*. Cambridge, MA: MIT Press.
———. 2016. "Précis of *After Phrenology: Neural Reuse and the Interactive Brain*." *Behavioral and Brain Sciences* 39:e120. https://doi.org/10.1017/S0140525X150 00631; https://pdfs.semanticscholar.org/c7c6/a8cbe16c5f836033f8f71d6244f 44b6e521d.pdf.
Andrews, John H. 1996. "What Was a Map? The Lexicographers Reply." *Cartographica* 33 (4): 1–11.
Archibald, J. David, and D. E. Fastovsky. 2004. "Dinosaur Extinction." In *The Dinosauria*, 2nd ed., edited by David B. Weishampel, Peter Dodson, and Halszka Osmólska, 672–84. Berkeley: University of California Press.
Armstrong, David M. 1968. *A Materialist Theory of the Mind*. London: Routledge.
Ayer, Alfred Jules. (1946) 1952. *Language, Truth, and Logic*. New York: Dover.
Bakhtin, Mikhail M. 1981. "Forms of Time and of the Chronotope in the Novel." In *The Dialogical Imagination*, by Bakhtin, edited by Michael Holquist, 84–258. Austin: University of Texas Press.
Balfour, Arthur James. 1917. Letter to Lionel Walter Rothschild, November 2. Available at https://en.wikipedia.org/wiki/Balfour_Declaration#/media/File :Balfour_declaration_unmarked.jpg.
Bartha, Paul F. A. 2010. *By Parallel Reasoning: The Construction and Evaluation of Analogical Arguments*. New York: Oxford University Press.
Bartholomew, J. G. 1911. *A Literary and Historical Atlas of America*. London: Dent.
Bateson, Gregory. 1972. *Steps to an Ecology of Mind: Collected Essays in Anthropology, Psychiatry, Evolution, and Epistemology*. Chicago: University of Chicago Press.
Battersby, Sarah E., and Daniel R. Montello. 2009. "Area Estimation of World Regions and the Projection of the Global-Scale Cognitive Map." *Annals of the Association of American Geographers* 99 (2): 273–91.
Battersby, Sarah E., Michael P. Finn, E. Lynn Usery, and Kristina H. Yamamoto. 2014. "Implications of the Web Mercator and Its Use in Online Mapping." *Cartographica* 49 (2): 85–101.
Baudrillard, Jean. 1994. *Simulacra and Simulation*. Translated by Sheila F. Glaser. Ann Arbor: University of Michigan Press. Originally published as *Simulacres et Simulation*, 1981.
Beatty, John. 1980. "What's Wrong with the Received View of Evolutionary Theory?" *PSA: Proceedings of the Biennial Meeting of the Philosophy of Science Association 1980* 2:397–426.
Bechtel, William. 1982. "Two Common Errors in Explaining Biological and Psychological Phenomena." *Philosophy of Science* 49 (4): 549–74.
Bechtel, William, and Robert C. Richardson. 1993. *Discovering Complexity: Decomposition and Localization as Strategies in Scientific Research*. Princeton, NJ: Princeton University Press.
Beckwith, Jonathan R. 1970. "Gene Expression in Bacteria and Some Concerns about the Misuse of Science." *Bacteriological Reviews* 34 (3): 222–27.

Bennett, C. L., D. Larson, J. L. Weiland, N. Jarosik, G. Hinshaw, N. Odegard, et al. 2013. "Nine-Year *Wilkinson Microwave Anisotropy Probe* (*WMAP*) Observations: Final Maps and Results." *Astrophysical Journal Supplement Series* 208 (2). https://doi.org/10.1088/0067-0049/208/2/20.

Berry, Mary E. 2006. *Japan in Print: Information and Nation in the Early Modern Period*. Berkeley: University of California Press.

Bertin, Jacques. (1983) 2011. *Semiology of Graphics: Diagrams, Networks, Maps*. Translated by W. J. Berg. Madison: University of Wisconsin Press. Originally published as *Sémiologie Graphique: Les diagrammes, les réseaux, les cartes*, 1967.

Bertram, Anton. 2011. *The Colonial Service*. Cambridge: Cambridge University Press.

BICEP2/Keck and Planck Collaborations. 2015. "Joint Analysis of BICEP2/*Keck Array* and *Planck* Data." *Physical Review Letters* 114 (10): 1–17. https://doi.org/10.1103/PhysRevLett.114.101301.

BIOPRO Baden-Württemberg. 2012. "The Discovery of Homeotic Genes." *Healthcare Industry* BW. https://www.gesundheitsindustrie-bw.de/en/article/news/the-discovery-of-homeotic-genes/.

Bird, Alexander, and Emma Tobin. 2017. "Natural Kinds." *Stanford Encyclopedia of Philosophy*, Spring 2017 ed., edited by Edward N. Zalta. https://plato.stanford.edu/archives/spr2017/entries/natural-kinds.

Bjornerud, Marcia. 2018. *Timefulness: How Thinking Like a Geologist Can Help Save the World*. Princeton, NJ: Princeton University Press.

Blaauw, A, C. S. Gum, J. L. Pawsey, and G. Westerhout. 1960. "The New IAU System of Galactic Coordinates (1958 Revision)." *Monthly Notices of the Royal Astronomical Society* 121 (2): 123–31. https://doi.org/10.1093/mnras/121.2.123.

Block, Ned, and Robert Stalnaker. 1999. "Conceptual Analysis, Dualism, and the Explanatory Gap." *Philosophical Review* 108 (1): 1–46.

Bogen, James, and James Woodward. 1988. "Saving the Phenomena." *Philosophical Review* 97 (3): 303–52.

Borges, Jorge Luis. 1999. "On Exactitude in Science." In *Collected Fictions*, translated by Andrew Hurley, 325. London: Penguin. Originally published as "Del rigor en la ciencia," 1946.

Bosker, Bianca. 2018. "The Nastiest Feud in Science." *Atlantic*, September. https://www.theatlantic.com/magazine/archive/2018/09/dinosaur-extinction-debate/565769/.

Bouzat, Juan L. 2014. "Darwin's Diagram of Divergence of Taxa as a Causal Model for the *Origin of Species*." *Quarterly Review of Biology* 89 (1): 21–38.

Boyd, Richard. 1992. "Constructivism, Realism, and Philosophical Method." In *Inference, Explanation, and Other Frustrations: Essays in the Philosophy of Science*, edited by John Earman, 131–98. Berkeley: University of California Press.

———. 1999. "Kinds as the 'Workmanship of Men': Realism, Constructivism, and Natural Kinds." In *Rationalität, Realismus, Revision: Proceedings of the*

*Third International Congress, Gesellschaft für Analytische Philosophie,* edited by Julian Nida-Rümelin, 52–89. Berlin: De Gruyter.

Brassel, Kurt E., and Robert Weibel. 1988. "A Review and Conceptual Framework of Automated Map Generalization." *International Journal of Geographical Information Systems* 2 (3): 229–44.

Bratman, Michael. 1999. *Faces of Intention: Selected Essays on Intention and Agency.* Cambridge: Cambridge University Press.

Bredekamp, Horst. 2019. *Darwin's Corals: A New Model of Evolution and the Tradition of Natural History.* Berlin: De Gruyter. Originally published as *Darwins Korallen: Frühe Evolutionsmodelle und die Tradition der Naturgeschichte,* 2005.

Bridges, Calvin B. 1935. "Salivary Chromosome Maps: With a Key to the Banding of the Chromosomes of *Drosophila melanogaster.*" *Journal of Heredity* 26 (2): 60–64.

Brigandt, Ingo. 2010. "Beyond Reduction and Pluralism: Toward an Epistemology of Explanatory Integration in Biology." *Erkenntnis* 73 (3): 295–311.

Bronner, Ben. 2015. "Maps and Absent Symbols." *Australasian Journal of Philosophy* 93 (1): 43–59.

Brotton, Jerry. 2012. *A History of the World in Twelve Maps.* New York: Viking.

———. 2014. *Great Maps: The Worlds Masterpieces Explored and Explained.* New York: Dorling Kindersley.

Brown, Theodore L. 2003. *Making Truth: Metaphor in Science.* Urbana: University of Illinois Press.

Bueno, Otávio, and Mark Colyvan. 2011. "An Inferential Conception of the Application of Mathematics." *Noûs* 45 (2): 345–74.

Bugayevskiy, Lev M., and John P. Snyder. 1995. *Map Projections: A Reference Manual.* Boca Raton, FL: CRC Press.

Buisseret, David. 1992. Introduction to *Monarchs, Ministers, and Maps: The Emergence of Cartography as a Tool of Government in Early Modern Europe,* 1–4. Edited by David Buisseret. Chicago: University of Chicago Press.

Burke, Ariane. 2012. "Spatial Abilities, Cognition and the Pattern of Neanderthal and Modern Human Dispersals." *Quaternary International* 247 (January 9): 230–35. https://doi.org/10.1016/j.quaint.2010.10.029.

Buzan, Barry, and Ole Wæver. 2003. *Regions and Powers: The Structure of International Security.* Cambridge: Cambridge University Press.

Byrne, Joseph P. 2004. *The Black Death.* Westport, CT: Greenwood Press.

Callender, Craig, and Jonathan Cohen. 2006. "There Is No Special Problem about Scientific Representation." *Theoria: An International Journal for Theory, History and Foundations of Science* 21 (1): 67–85.

Callon, Michel. 1998. "The Embeddedness of Economic Markets in Economics." Introduction to *The Laws of the Market,* edited by Callon, 1–57. Oxford: Blackwell Publishers/*Sociological Review.*

Calvino, Italo. 1974. *Invisible Cities.* Orlando, FL: Harcourt. Originally published as *Le città invisibili,* 1972.

Camerini, Jane. 1993. "Evolution, Biogeography, and Maps: An Early History of Wallace's Line." *Isis* 84 (4): 700–727.

Camp, Elisabeth. 2007. "Thinking with Maps." *Philosophical Perspectives* 21 (1): 145–82.

Campbell, Jeremy. 1989. *The Improbable Machine: What the Upheavals in Artificial Intelligence Research Reveal about How the Mind Really Works*. New York: Simon & Schuster.

Campbell, Tony. 1987. "Portolan Charts from the Late Thirteenth Century to 1500." In *The History of Cartography: Cartography in Prehistoric, Ancient, and Medieval Europe and the Mediterranean*, vol. 1, edited by John B. Harley and David Woodward, 371–463. Chicago: University of Chicago Press.

Caporael, Linnda R., James R. Griesemer, and William C. Wimsatt. 2014. *Developing Scaffolds in Evolution, Culture, and Cognition*. Cambridge, MA: MIT Press.

Carnap, Rudolf. 1963. "Replies and Systematic Expositions." In *The Philosophy of Rudolf Carnap*, edited by Paul Arthur Schilpp, 859–1013. La Salle, IL: Open Court.

———. (1967) 2003. *The Logical Structure of the World and Pseudoproblems in Philosophy*. Translated by Rolf A. George. Peru, IL: Carus Publishing. Originally published as *Der logische Aufbau der Welt*, 1928.

Carneiro, Maurício, and Daniel L. Hartl. 2010. "Adaptive Landscapes and Protein Evolution." *Proceedings of the National Academy of Sciences (USA)* 107 (suppl. 1): 1747–51. https://doi.org/10.1073/pnas.0906192106.

Carroll, Lewis. (1893) 2010. *Sylvie and Bruno Concluded*. Charleston, SC: Nabu Press.

Carter, Rita. 2010. *Mapping the Mind*. Rev. ed. Berkeley: University of California Press.

Cartwright, Nancy. 1983. *How the Laws of Physics Lie*. New York: Oxford University Press.

———. 1989. *Nature's Capacities and Their Measurement*. New York: Oxford University Press.

———. 1999. *The Dappled World: A Study of the Boundaries of Science*. Cambridge: Cambridge University Press.

———. 2007. *Hunting Causes and Using Them: Approaches in Philosophy and Economics*. Cambridge: Cambridge University Press.

———. 2010. "Models: Parables v Fables." In *Beyond Mimesis and Convention: Representation in Art and Science*, edited by Roman Frigg and Matthew C. Hunter, 19–31. Dordrecht: Springer.

Cartwright, Nancy, Towfic Shomar, and Mauricio Suárez. 1995. "The Tool Box of Science: Tools for the Building of Models with a Superconductivity Example." In *Theories and Models in Scientific Processes*. Vol. 44 of Poznan Studies in the Philosophy of the Sciences and the Humanities, edited by William Herfel, Wladyslaw Krajewski, Ilkka Niiniluoto, and Ryszard Wojcicki, 137–49. Amsterdam: Rodopi.

Casares, Pelayo. 2007. "A Corrected Haldane's Map Function to Calculate Genetic Distances from Recombination Data." *Genetica* 129 (3): 333–38.

Casey, Edward. 1997. *The Fate of Place: A Philosophical History*. Berkeley: University of California Press.

Cassirer, Ernst. 1957. *The Philosophy of Symbolic Forms: The Phenomenology of Knowledge*. New Haven, CT: Yale University Press.

Castelvecchi, Davide. 2011a. "The Geoid: Why a Map of Earth's Gravity Yields a Potato-Shaped Planet." *Scientific American*, April 1. https://blogs.scientific american.com/observations/the-geoid-why-a-map-of-earths-gravity-yields-a -potato-shaped-planet/.

———. 2011b. "Under a Blood Red Sky." *Scientific American*, July 6. https://blogs .scientificamerican.com/degrees-of-freedom/httpblogsscientificamericancom degrees-of-freedom20110706under-a-blood-red-sky/.

Cavalli-Sforza, Luigi Luca, Italo Barrai, and Anthony W. F. Edwards. 1964. "Analysis of Human Evolution under Random Genetic Drift." *Cold Spring Harbor Symposia on Quantitative Biology* 29, 9–20. Reprinted in Winther 2018a.

Cavalli-Sforza, Luigi Luca, and Anthony W. F. Edwards. 1967. "Phylogenetic Analysis: Models and Estimation Procedures." *Evolution* 21 (3): 550–70. Reprinted in Winther 2018a.

Cavalli-Sforza, Luigi Luca, Paolo Menozzi, and Alberto Piazza. 1994. *The History and Geography of Human Genes*. Princeton, NJ: Princeton University Press.

Chakravartty, Anjan. 2007. *A Metaphysics for Scientific Realism: Knowing the Unobservable*. Cambridge: Cambridge University Press.

———. 2016. "Scientific Realism." In *Stanford Encyclopedia of Philosophy*, Winter 2016 ed., edited by Edward N. Zalta. https://plato.stanford.edu/archives/win 2016/entries/scientific-realism/.

Chalmers, David, and Frank Jackson. 2001. "Conceptual Analysis and Reductive Explanation." *Philosophical Review* 110 (3): 315–61.

Chang, Hasok. 2004. *Inventing Temperature: Measurement and Scientific Progress*. Oxford: Oxford University Press.

———. 2011. "The Philosophical Grammar of Scientific Practice." *International Studies in the Philosophy of Science* 25 (3): 205–21.

Chrisman, Nicholas R. 2017. "Calculating on a Round Planet." *International Journal of Geographical Information Science* 31 (4): 637–57.

Cohen, Jonathan. 1988. "The Naming of America: Fragments We've Shored against Ourselves." *American Voice* 13:56–72.

Coles, Peter. 1999. "Last Scattering Surface." *The Routledge Critical Dictionary of the New Cosmology*, edited by Peter Coles, 245–46. New York: Routledge. Adapted at https://ned.ipac.caltech.edu/level5/Glossary/Essay_lss.html.

Collins, Francis S., and Anna D. Barker. 2007. "Mapping the Cancer Genome." *Scientific American*, March, 50–57.

Colón, Cristóbal. (1498) 1892. "Tercer viaje del almirante D. Cristóbal Colón" In *Relaciones y cartas de Cristóbal Colón*, 268–93. Madrid: Librería de la Viuda de Hernando y Ca.

Columbus, Christopher. 1493. Christopher Columbus to King Ferdinand, February 15. A nonoriginal version was published in *Select Documents Illustrating the Four Voyages of Columbus*, vol. 1, edited by Lionel Cecil Jane, 2–19. London: Hakluyt Society, 1930. Available at https://www.ensayistas.org/antologia/XV/colon/.

Combe, George. 1847. *The Constitution of Man Considered in Relation to External Objects*. 8th ed. Edinburgh: Maclachlan, Stewart.

Conley, Tom. 1996. *The Self-Made Map: Cartographic Writing in Early Modern France*. Minneapolis: University of Minnesota Press.

Cook, Karen S. 2005. "Arthur H. Robinson (1915–2004)." *Imago Mundi* 57 (2): 195–97.

Cornfield, Jerome, William Haenszel, E. Cuyler Hammond, Abraham M. Lilienfeld, Michael B. Shimkin, and Ernst L. Wynder. (1959) 2009. "Smoking and Lung Cancer: Recent Evidence and a Discussion of Some Questions." *International Journal of Epidemiology* 38 (5): 1175–91. https://doi.org/10.1093/ije/dyp289. Originally published in the *Journal of the National Cancer Institute* 22 (1): 173–203.

Costandi, Mo. 2008. "Wilder Penfield: Neural Cartographer." *Neurophilosophy* (blog), August 27. https://neurophilosophy.wordpress.com/2008/08/27/wilder_penfield_neural_cartographer/.

Cottrell, W. Fred. 1939. "Of Time and the Railroader." *American Sociological Review* 4 (2): 190–98.

Couclelis, Helen. 1992. "People Manipulate Objects (but Cultivate Fields): Beyond the Raster-Vector Debate in GIS." In *Theories and Methods of Spatial-Temporal Reasoning in Geographic Space*, edited by A. U. Frank, I. Campari, and U. Formentini, 65–77. Berlin: Springer.

Council of the League of Nations. 1922. "The Palestine Mandate." Available at http://avalon.law.yale.edu/20th_century/palmanda.asp.

Courtois, Hélène M., Daniel Pomarède, R. Brent Tully, Yehuda Hoffman, and Denis Courtois. 2013. "Cosmography of the Local Universe." *Astronomical Journal* 146:69. https://doi.org/10.1088/0004-6256/146/3/69.

Cox, Francis E. G. 2010. "History of the Discovery of the Malaria Parasites and Their Vectors." *Parasites and Vectors* 3:5. https://doi.org/10.1186/1756-3305-3-5.

Crane, Nicholas. 2002. *Mercator: The Man Who Mapped the Planet*. London: Orion Books.

Craver, C. 2007. *Explaining the Brain: Mechanisms and the Mosaic Unity of Neuroscience*. New York: Oxford University Press.

Creighton, Harriet B., and Barbara McClintock. 1931. "A Correlation of Cytological and Genetical Crossing-Over in *Zea mays*." *Proceedings of the National Academy of Sciences (USA)* 17 (8): 492–97.

Crombie, Alistair C. 1994. *Styles of Scientific Thinking in the European Tradition*, vols. 1–3. London: Duckworth.

———. 1996. "Commitments and Styles of European Scientific Thinking." *Theoria* 11 (1): 65–76.

Crone, Gerald Roe. 1978. *Maps and Their Makers: An Introduction to the History of Cartography*. 5th ed. Folkestone, Kent: Dawson.

Curry, Leslie. 1962. "Climatic Changes as a Random Series." *Annals of the Association of American Geographers* 52 (1): 21–31.

Dalton, Craig M. 2013. "Sovereigns, Spooks, and Hackers: An Early History of Google Geo Services." *Cartographica* 48 (4): 261–74.

Darwin, Charles R. 1842. *The Structure and Distribution of Coral Reefs*. London: Smith, Elder. Available at http://darwin-online.org.uk/content/frameset?itemID=F271&viewtype=text&pageseq=1.

———. (1859) 1964. *On the Origin of Species by Means of Natural Selection, or the Preservation of Favoured Races in the Struggle for Life*. Cambridge, MA: Harvard University Press.

Daston, Lorraine. 1995. *Classical Probability in the Enlightenment*. Princeton, NJ: Princeton University Press.

Daston, Lorraine J., and Peter Galison. 2007. *Objectivity*. Brooklyn: Zone Books.

Davidson, Eric. 2006. *The Regulatory Genome: Gene Regulatory Networks in Development*. San Diego, CA: Academic Press/Elsevier.

Dehaene, Stanislas, and Elizabeth M. Brannon, eds. 2011. *Space, Time and Number in the Brain: Searching for the Foundations of Mathematical Thought*. Amsterdam: Elsevier.

de Lapparent, Valérie, Margaret J. Geller, and John P. Huchra. 1986. "A Slice of the Universe." *Astrophysical Journal* 302: L1–L5. https://doi.org/10.1086/184625.

Derman, Emanuel. 2011. *Models. Behaving. Badly: Why Confusing Illusion with Reality Can Lead to Disaster, on Wall Street and in Life*. New York: Free Press.

De Toffoli, Silvia. 2017. "'Chasing' the Diagram—the Use of Visualizations in Algebraic Reasoning." *Review of Symbolic Logic* 10 (1): 158–86.

De Toffoli, Silvia, and Valeria Giardino. 2014. "Forms and Roles of Diagrams in Knot Theory." *Erkenntnis* 79 (4): 829–42.

de Vaucouleurs, Gérard, Antoinette de Vaucouleurs, Harold G. Corwin Jr., Ronald J. Buta, P. Georges Paturel, and Pascal Fouqué, eds. 1991. *Third Reference Catalogue of Bright Galaxies*. New York: Springer.

Dewey, John. 1976. "The Child and the Curriculum." *The Middle Works, 1899–1924, Vol. 2: 1902–1903*. Edited by J. A. Boydston, 271–91. Carbondale: Southern Illinois University Press.

———. 1982. *Reconstruction in Philosophy*. In *The Middle Works of John Dewey, 1899–1924*, edited by Jo Ann Boydston, 12:77–201. Carbondale: Southern Illinois University Press.

Diaconis, Persi, and Brian Skyrms. 2017. *Ten Great Ideas about Chance*. Princeton, NJ: Princeton University Press.

Dilke, Oswald A. W. 1985. *Greek and Roman Maps*. London: Thames and Hudson.

———. 1987. "Cartography in the Ancient World: A Conclusion." In *The History of Cartography: Cartography in Prehistoric, Ancient, and Medieval Europe and the Mediterranean*, vol. 1, edited by John B. Harley and David Woodward, 276–79. Chicago: University of Chicago Press.

Dobzhansky, Theodosius. (1937) 1951. *Genetics and the Origin of Species*. 3rd ed. New York: Columbia University Press.

Doel, Ronald E., Tanya J. Levin, and Mason K. Marker. 2006. "Extending Modern Cartography to the Ocean Depths: Military Patronage, Cold War Priorities, and the Heezen–Tharp Mapping Project, 1952–1959." *Journal of Historical Geography* 32 (3): 605–26. https://doi.org/10.1016/j.jhg.2005.10.011.

Doll, Richard. 2002. "Proof of Causality: Deduction from Epidemiological Observation." *Perspectives in Biology and Medicine* 45 (4): 499–515. https://doi.org/10.1353/pbm.2002.0067.

Doll, Richard, and A. Bradford Hill. 1950. "Smoking and Carcinoma of the Lung: Preliminary Report." *British Medical Journal* 2 (4682): 739–48. http://www.jstor.org/stable/25358512.

Donnelly, Jack. 2000. *Realism and International Relations*. Cambridge: Cambridge University Press.

Doolittle, W. Ford. 1999. "Phylogenetic Classification and the Universal Tree." *Science* 284 (5423): 2124–28. https://doi.org/10.1126/science.284.5423.2124.

Doolittle, W. Ford, and Eric Bapteste. 2007. "Pattern Pluralism and the Tree of Life Hypothesis." *Proceedings of the National Academy of Sciences (USA)* 104 (7): 2043–49. https://doi.org/10.1073/pnas.0610699104.

Douglas, David H., and Thomas K. Peucker. 1973. "Algorithms for the Reduction of the Number of Points Required to Represent a Digitized Line or Its Caricature." *Canadian Cartographer* 10 (2): 112–22.

*Drosophila* 12 Genomes Consortium. 2007. "Evolution of Genes and Genomes on the *Drosophila* Phylogeny." *Nature* 450 (7167): 203–18. https://www.nature.com/articles/nature06341.

Dunn, Leslie Clarence. 1965. *A Short History of Genetics: The Development of Some of the Main Lines of Thought 1864–1939*. New York: McGraw-Hill.

Dunnington, G. Waldo. 1955. *Carl Friedrich Gauss: Titan of Science*. New York: Hafner.

Dupré, John. 1993. *The Disorder of Things: Metaphysical Foundations of the Disunity of Science*. Cambridge, MA: Harvard University Press.

———. 2004. "Understanding Contemporary Genomics." *Perspectives on Science* 12 (3): 320–38.

Dym, Jordana, and Karl Offen, eds. 2011. *Mapping Latin America: A Cartographic Reader*. Chicago: University of Chicago Press.

Eagleman, David. 2011. *Incognito: The Secret Lives of the Brain*. New York: Random House.

Eames, Charles, and Roy Eames. 1977. *Powers of Ten and the Relative Size of Things in the Universe*. Video, 9 minutes. Available at http://www.eamesoffice.com/the-work/powers-of-ten/.

Ebrey, Patricia B. 2010. *The Cambridge Illustrated History of China*. 2nd ed. Cambridge: Cambridge University Press.

Eckert, Max. 1921. *Die Kartenwissenschaft: forschungen und grundlagen zu einer kartographie als wissenschaft*. Berlin: Walter de Gruyter.

Eco, Umberto. 1994. "On the Impossibility of Drawing a Map of the Empire on a

Scale of 1 to 1." In *How to Travel with a Salmon and Other Essays*, 95–106. New York: Harcourt. Originally published as *Il Secondo Diario Minimo*, 1992.

Edge, Michael D. 2019. *Statistical Thinking from Scratch*. New York: Oxford University Press.

Edwards, Anthony W. F. 2005. "R. A. Fisher: *Statistical Methods for Research Workers*, First Edition (1925)." In *Landmark Writings in Western Mathematics, 1640–1940*, edited by I. Grattan-Guinness, 856–70. Amsterdam: Elsevier.

Ekbia, Hamid. 2008. *Artificial Dreams: The Quest for Non-biological Intelligence*. Cambridge: Cambridge University Press.

Élisée Reclus, Jacques. 1890. *The Earth and Its Inhabitants: North America*. Vol. 1 of British North America. Edited by A. H. Keane. New York: Appleton.

Elwick, James. 2012. "Layered History: Styles of Reasoning as Stratified Conditions of Possibility." *Studies in History and Philosophy of Science, Part A* 43 (4): 619–27.

Estany, Anna, and Sergio F. M. Martínez. 2013. "'Scaffolding' and 'Affordance' as Integrative Concepts in the Cognitive Sciences." *Philosophical Psychology* 27 (1): 98–111.

Fauconnier, Gilles. 1997. *Mappings in Thought and Language*. Cambridge: Cambridge University Press.

Fauconnier, Gilles, and Mark Turner. 2002. *The Way We Think: Conceptual Blending and the Mind's Hidden Complexities*. New York: Basic Books.

Felsenstein, Joseph. 2004. *Inferring Phylogenies*. Sunderland, MA: Sinauer Associates.

Felt, Hali. 2012. *Soundings: The Story of the Remarkable Woman Who Mapped the Ocean Floor*. New York: Picador. Kindle.

Fenna, Donald. 2007. *Cartographic Science: A Compendium of Map Projections, with Derivations*. London: CRC Press.

Ferreirós, José. 2012. "Matemáticas y pensamiento: en torno a imágenes, modelos, abstracción y figuración." In *Rondas en Sais: Ensayos sobre matemáticas y cultura contemporánea*, edited by Fernando Zalamea, 15–38. Bogotá: Universidad Nacional de Colombia.

Few, Stephen. 2006. "Beautiful Evidence: A Journey through the Mind of Edward Tufte." BeyeNetwork, August 8. http://www.b-eye-network.com/view/3226.

Feyerabend, Paul. (1975) 2010. *Against Method*. 4th ed. With an introduction by Ian Hacking. London: Verso.

Fine, Arthur. 1986. *The Shaky Game: Einstein, Realism, and the Quantum Theory*. Chicago: University of Chicago Press.

Fisher, Ronald A. 1925. *Statistical Methods for Research Workers*. Edinburgh: Oliver and Boyd.

———. (1935) 1971. *The Design of Experiments*. New York: Hafner.

———. 1959. *Smoking: The Cancer Controversy: Some Attempts to Assess the Evidence*. Edinburgh: Oliver and Boyd. Available at http://www.med.mcgill.ca/epidemiology/hanley/c609/material/FisherOnSmokingAndCancer.pdf.

Fleischmann, R. D., M. D. Adams, O. White, R. A. Clayton, E. F. Kirkness, A. R.

Kerlavage, C. J. Bult, et al. 1995. "Whole-Genome Random Sequencing and Assembly of *Haemophilus influenza Rd.*" *Science* 269 (5223): 496–512. https://doi.org/10.1126/science.7542800.

Fontana, Walter. 2002. "Modelling 'Evo-Devo' with RNA." *Bioessays* 24 (12): 1164–77.

Foucault, Michel. 1972. *The Archaeology of Knowledge.* Translated by A. Sheridan Smith. New York: Harper & Row. Originally published as *L'Archéologie du savoir*, 1969.

Fraczek, Witold. 2003. "Mean Sea Level, GPS, and the Geoid." *ArcUser*, July–September. Environmental Systems Research Institute. http://www.esri.com/news/arcuser/0703/geoid1of3.html.

Frankel, Christian. 2015. "The Multiple-Markets Problem." *Journal of Cultural Economy* 8 (4): 538–46.

Frase, Peter. 2016. *Four Futures: Life after Capitalism.* London: Verso.

Freedman, David. A. 2009. *Statistical Models: Theory and Practice.* 2nd ed. New York: Cambridge University Press.

Friedman, Michael. 1983. *Foundations of Space-Time Theory: Relativistic Physics and Philosophy of Science.* Princeton, NJ: Princeton University Press.

———. 1999. *Reconsidering Logical Positivism.* Cambridge: Cambridge University Press.

———. 2001. *Dynamics of Reason.* Stanford, CA: CSLI Publications.

———. 2008. "Einstein, Kant, and the A Priori." *Royal Institute of Philosophy Supplement* 63:95–112.

Frigg, Roman, and Stephan Hartmann. 2017. "Models in Science." In *Stanford Encyclopedia of Philosophy*, Spring 2017 ed., edited by Edward N. Zalta. https://plato.stanford.edu/archives/spr2017/entries/models-science/.

Frischkorn, Kyle. 2017. "Why the First Complete Map of the Ocean Floor Is Stirring Controversial Waters." *Smithsonian*, July 13. https://www.smithsonianmag.com/science-nature/first-complete-map-ocean-floor-stirring-controversial-waters-180963993.

Frodeman, Robert. 1995. "Geological Reasoning: Geology as an Interpretive and Historical Science." *GSA Bulletin* 107 (8): 960–68.

Gaddis, John Lewis. 2002. *The Landscape of History: How Historians Map the Past.* Oxford: Oxford University Press.

Galison, Peter. 1987. *How Experiments End.* Chicago: University of Chicago Press.

———. 1997. *Image and Logic: A Material Culture of Microphysics.* Chicago: University of Chicago Press.

———. 2003. *Einstein's Clocks and Poincaré's Maps: Empires of Time.* New York: W. W. Norton.

Gall, F. J. 1798. Dr. F. J. Gall to Joseph F. von Retzer, October 1. In *The History of Phrenology on the Web*, by John van Wyhe. Available at http://www.historyofphrenology.org.uk/texts/retzer.htm.

———. 1835. *On the Functions of the Brain and of Each of Its Parts: With Observations of the Possibility of Determining the Instincts, Propensities, and Talents, or the Moral and Intellectual Dispositions of Men and Animals, by the Configuration of the Brain and Head*. Boston: Marsh, Capen & Lyon. Available at https://archive.org/details/onfunctionsofbra0506gall/page/n5.

Gall, James. 1856. "On Improved Monographic Projections of the World." In *Report of the Twenty-fifth Meeting of the British Association for the Advancement of Science (Glasgow)*, 148.

Galton, David. 2009. "Did Darwin Read Mendel?" *QJM: An International Journal of Medicine* 102 (8): 587–89.

Gammeltoft-Hansen, Thomas, and Ninna Nyberg Sørensen. 2013. *The Migration Industry and the Commercialization of International Migration*. London: Routledge.

Gannett, Lisa. 2014. "The Human Genome Project." In *Stanford Encyclopedia of Philosophy*, Winter 2014 ed., edited by Edward N. Zalta. http://plato.stanford.edu/archives/win2014/entries/human-genome/.

Garelli, Glenda, Federica Sossi, and Martina Tazzioli, eds. 2013. *Spaces in Migration: Postcards of a Revolution*. London: Pavement Books. Available at https://www.academia.edu/11933576/.

Gasparini, Graziano, and Luise Margolies. 1980. *Inca Architecture*. Translated by Patricia J. Lyon. Bloomington: Indiana University Press.

Gauss, Carl Friedrich. 1873. "Allgemeine Auflösung der Aufgabe die Theile einer gegebenen Fläche auf einer andern gegebenen Fläche so abzubilden, dass die Abbildung dem Abgebildeten in den kleinsten Theilen ähnlich wird." In *Carl Friedrich Gauss Werke*, vol. 4, 189–216. Göttingen Royal Society of Sciences.

———. (1902) 2005. *General Investigations of Curved Surfaces*. Edited by Peter Pesic. Mineola, NY: Dover.

Gavrilets, Sergey. 2004. *Fitness Landscapes and the "Origin of Species."* Princeton, NJ: Princeton University Press.

Geller, Margaret J., and John P. Huchra. 1989. "Mapping the Universe." *Science* 246 (4932): 897–903.

Gelles, Paul H. 1995. "Equilibrium and Extraction: Dual Organization in the Andes." *American Ethnologist* 22 (4): 710–42.

Gentner, Dedre. 2003. "Analogical Reasoning, Psychology of." In *Encyclopedia of Cognitive Science*, edited by Lynn Nadel, 106–12. London: Nature Publishing Group.

Gentner, Dedre, and Michael Jeziorski. 1993. "The Shift from Metaphor to Analogy in Western Science." In *Metaphor and Thought*, 2nd ed., edited by Andrew Ortony, 447–80. Cambridge: Cambridge University Press.

Gerstein, Mark B., Can Bruce, Joel S. Rozowsky, Deyou Zheng, Jiang Du, Jan O. Korbel, Olof Emanuelsson, et al. 2007. "What Is a Gene, Post-ENCODE? History and Updated Definition." *Genome Research* 17:669–81. https://doi.org/10.1101/gr.6339607.

Gessell, Bryce, and Felipe De Brigard. 2018. "The Discontinuity of Levels in Cognitive Science." *Teorema* 37 (3): 151–65.

Giaquinto, Marcus. 2007. *Visual Thinking in Mathematics: An Epistemological Study*. New York: Oxford University Press.
Gibson, James J. 1979. *The Ecological Approach to Visual Perception*. Boston, MA: Houghton Mifflin.
Giere, Ronald N. 1999. *Science without Laws*. Chicago: University of Chicago Press.
———. 2004. "How Models Are Used to Represent Reality." *Philosophy of Science* 71 (5): 742–52.
———. 2006. *Scientific Perspectivism*. Chicago: University of Chicago Press.
Gieryn, Thomas F. 1999. *Cultural Boundaries of Science: Credibility on the Line*. Chicago: University of Chicago Press.
Gilovich, Thomas. 1991. *How We Know What Isn't So: The Fallibility of Human Reason in Everyday Life*. New York: Free Press.
Givón, Talmy. 1986. "Prototypes: Between Plato and Wittgenstein." In *Noun Classes and Categorization*, edited by Colette G. Craig, 77–102. Amsterdam: John Benjamins B.V.
Goddard Space Flight Center. n.d. "CMB Measured Intensity vs Frequency." National Aeronautics and Space Administration. Last updated April 20, 2012. https://asd.gsfc.nasa.gov/archive/arcade/cmb_intensity.html.
Godfrey-Smith, Peter. 1996. *Complexity and the Function of Mind in Nature*. Cambridge: Cambridge University Press.
———. 2003. *Theory and Reality: An Introduction to the Philosophy of Science*. Chicago: University of Chicago Press.
———. 2008. "Recurrent Transient Underdetermination and the Glass Half Full." *Philosophical Studies* 137 (1): 141–48. https://doi.org/10.1007/s11098-007-9172-2.
Goodchild, Michael F. 2004. "The Validity and Usefulness of Laws in Geographic Information Science and Geography." *Annals of the Association of American Geographers* 94 (2): 300–303.
Goode, John Paul. 1923. *Goode's School Atlas*. 1st ed. Chicago: Rand McNally.
Goodman, Nelson. 1963. "The Significance of *Der logische Aufbau der Welt*." In *The Philosophy of Rudolf Carnap*, edited by Paul Arthur Schilpp, 545–58. La Salle, IL: Open Court.
———. 1978. *Ways of Worldmaking*. Indianapolis: Hackett.
Gould, Stephen Jay. (1981) 1996. *The Mismeasure of Man*. New York: W. W. Norton.
———. 1988. *Time's Arrow, Time's Cycle: Myth and Metaphor in the Discovery of Geological Time*. Cambridge, MA: Harvard University Press.
———. 1997. "Self-Help for a Hedgehog Stuck on a Molehill." *Evolution* 51 (3): 1020–23.
Goyder, David George. 1857. *My Battle for Life: The Autobiography of a Phrenologist*. London: Simpkin, Marshall.
GPS.gov. n.d. "GPS Accuracy." National Coordination Office for Space-Based

Positioning, Navigation, and Timing. Last updated December 5, 2017. https://www.gps.gov/systems/gps/performance/accuracy.

Graziano, Michael. 2006. "The Organization of Behavioral Repertoire in Motor Cortex." *Annual Review of Neuroscience* 29:105–34.

Green, R. T., and M. C. Courtis. 1966. "Information Theory and Figure Perception: The Metaphor That Failed." *Acta Psychologica* 25 (1): 12–35.

Greenhood, D. 1964. *Mapping*. Chicago: University of Chicago Press.

Griesemer, James. 1990. "Modeling in the Museum: On the Role of Remnant Models in the Work of Joseph Grinnell." *Biology and Philosophy* 5 (1): 3–36.

Griffiths, Paul, and Karola Stotz. 2013. *Genetics and Philosophy: An Introduction*. Cambridge: Cambridge University Press.

Gusella, James F., Nancy S. Wexler, P. Michael Conneally, Susan L. Naylor, Mary Anne Anderson, Rudolph E. Tanzi, Paul C. Watkins, et al. 1983. "A Polymorphic DNA Marker Genetically Linked to Huntington's Disease." *Nature* 306 (5940): 234–38. https://doi.org/10.1038/306234a0.

Haack, Susan. 2014. "Do Not Block the Way of Inquiry." *Transactions of the Charles S. Peirce Society* 50 (3): 319–39.

Hacking, Ian. 1975. *The Emergence of Probability*. Cambridge: Cambridge University Press.

———. 1983. *Representing and Intervening: Introductory Topics in the Philosophy of Natural Science*. Cambridge: Cambridge University Press.

———. 1988. "Telepathy: Origins of Randomization in Experimental Design." *Isis* 79 (3): 427–51.

———. 1990. *The Taming of Chance*. Cambridge: Cambridge University Press.

———. 1993. "Working in a New World: The Taxonomic Solution." In *World Changes: Thomas Kuhn and the Nature of Science*, edited by Paul Horwich, 275–310. Cambridge, MA: MIT Press.

———. 1996. "The Disunities of Science." In *The Disunity of Science: Boundaries, Contexts and Power*, edited by Peter Galison and David Stump, 37–74. Stanford: Stanford University Press.

———. 1999. *The Social Construction of What?* Cambridge, MA: Harvard University Press.

———. 2002. *Historical Ontology*. Cambridge, MA: Harvard University Press.

———. 2007a. "Kinds of People: Moving Targets." *Proceedings of the British Academy* 151:285–317.

———. 2007b. "Natural Kinds: Rosy Dawn, Scholastic Twilight." *Royal Institute of Philosophy Supplements* 61:203–39.

———. 2009. *Scientific Reason*. Taipei: National Taiwan University Press.

———. 2012. "Introductory Essay." In *The Structure of Scientific Revolutions*, 50th anniv. ed. (4th ed.), by Thomas Kuhn, vii–xxxvii. Chicago: University of Chicago Press.

Haining, R. 2003. *Spatial Data Analysis*. Cambridge: Cambridge University Press.

Hall, Stephen S. 1992. *Mapping the Next Millennium: The Discovery of New Geographies*. New York: Vintage Books.

Hammer, Marie, and John A. Wallwork. 1979. *A Review of the World Distribution of Oribatid Mites (Acari: Cryptostigmata) in Relation to Continental Drift*. Copenhagen: The Royal Danish Academy of Sciences and Letters.

Haraway, Donna. 1988. "Situated Knowledges: The Science Question in Feminism and the Privilege of Partial Perspective." *Feminist Studies* 14 (3): 575–99.

Hardaker, Terry. 2002. "Cartographic Introduction." In *Compact Peters World Atlas*, edited by Mapmakers for the 21st Century, 6–7. Union, NJ: Hammond.

Harding, Sandra. 1986. *The Science Question in Feminism*. Ithaca, NY: Cornell University Press.

Hardison, Ross C. 2012. "Evolution of Hemoglobin and Its Genes." *Cold Spring Harbor Perspectives in Medicine* 2 (12): a011627. https://doi.org/10.1101/csh perspect.a011627.

Harley, John B. 1962. *Christopher Greenwood, County Map-Maker and His Worcestershire Map of 1822*. Worcester, UK: Worcestershire Historical Society.

———. 1987. "The Map and the Development of the History of Cartography." In *The History of Cartography: Cartography in Prehistoric, Ancient, and Medieval Europe and the Mediterranean*, vol. 1, edited by John B. Harley and David Woodward, 1–42. Chicago: University of Chicago Press.

———. 1989. "Deconstructing the Map." *Cartographica* 26 (2): 1–20.

———. 2001a. "Maps, Knowledge, and Power." In *The New Nature of Maps: Essays in the History of Cartography*, edited by Paul Laxton, 51–81. Baltimore: Johns Hopkins University Press.

———. 2001b. "Can There Be a Cartographic Ethics?" In *The New Nature of Maps: Essays in the History of Cartography*, edited by Paul Laxton, 197–207. Baltimore: Johns Hopkins University Press.

———. 2001c. *The New Nature of Maps: Essays in the History of Cartography*, edited by Paul Laxton. Baltimore: Johns Hopkins University Press.

Harmon, Katharine. 2010. *The Map as Art: Contemporary Artists Explore Cartography*. New York: Princeton Architectural Press.

Harrie, Lars, and Robert Weibel. 2007. "Modelling the Overall Process of Generalisation." In *Generalisation of Geographic Information: Cartographic Modelling and Applications*, edited by William A. Mackaness, Anne Ruas, and L. Tiina Sarjakoski, 67–87. Oxford: Elsevier.

Harris, Elizabeth. 1985. "The Waldseemüller World Map: A Typographic Appraisal." *Imago Mundi* 37:30–53.

Harrison, Richard Edes. 1941. "The World Divided" (map). *Fortune*, July 7.

———. 1944a. "Perspective Maps: Harrison Atlas Gives Fresh New Look to Old World." *Life*, February 28, 56–61.

———. 1944b. *Look at the World: The FORTUNE Atlas for World Strategy*. Text by the editors of *Fortune*. New York: A. A. Knopf.

Hartl, Daniel L., and Andrew C. Clark. 1989. *Principles of Population Genetics*. 2nd ed. Sunderland, MA: Sinauer Associates.

Hashimoto, Takuya, Nicolas Laporte, Ken Mawatari, Richard S. Ellis, Akio K. Inoue, Erik Zackrisson, Guido Roberts-Borsani, et al. 2018. "The Onset of Star

Formation 250 Million Years after the Big Bang." *Nature* 557 (7705): 392–95. https://doi.org/10.1038/s41586-018-0117-z.

Hébert, John. 2011. "America." In *Mapping Latin America: A Cartographic Reader*, edited by Jordana Dym and Karl Offen, 29–32. Chicago: University of Chicago Press.

Henrikson, Alan K. 1979. "All the World's a Map." *Wilson Quarterly* 3 (2): 164–77.

Hermans, Felienne. 2013. "Excel Turing Machine." Blog post, September 19. http://www.felienne.com/archives/2974.

Hesse, Mary. 1966. *Models and Analogies in Science*. South Bend, IN: University of Notre Dame Press.

———. 1974. *The Structure of Scientific Inference*. Berkeley: University of California Press.

Hessler, John W. 2006. "Warping Waldseemüller: A Phenomenological and Computational Study of the 1507 World Map." *Cartographica* 41:101–14.

———. 2008. *The Naming of America: Martin Waldseemüller's 1507 World Map and the Cosmographiae Introductio*. London: Giles.

Hill, Austin Bradford. 1965. "The Environment and Disease: Association or Causation?" *Proceedings of the Royal Society of Medicine* 58:295–300.

Hoffman, Robert R. 1980. "Metaphor in Science." In *Cognition and Figurative Language*, edited by Richard P. Honeck, 393–423. Hillsdale, New Jersey: Lawrence Erlbaum Associates.

Hofstadter, Douglas. 2006. "Analogy as the Core of Cognition." Stanford Presidential Lecture in the Humanities and Arts, Cubberley Auditorium, Stanford University. Video, 1:08:36. https://www.youtube.com/watch?v=n8m7lFQ3njk.

Hofstadter, Douglas, and Emmanuel Sander. 2013. *Surfaces and Essences: Analogy as the Fuel and Fire of Thinking*. New York: Basic Books.

Hogg, David W. 1999. "Distance Measures in Cosmology." Unpublished manuscript, last revised December 16, 2000. https://arxiv.org/pdf/astro-ph/9905116.pdf.

Holdich, Thomas H. 1901. "How Are We to Get Maps of Africa?" *Geographical Journal* 18 (6): 590–601.

Holton, Gerald. 1977. "Sociobiology: The New Synthesis?" *Newsletter on Science, Technology, and Human Values* 21 (October): 28–43.

Holyoak, Keith J., and Paul Thagard. 1995. *Mental Leaps: Analogy in Creative Thought*. Cambridge, MA: MIT Press.

Hooten, Christopher. 2016. "Prince Dead: When Prince Changed His Name to a Symbol, Warner Bros Had to Mail Floppy Disks with a Custom Font." *Independent*, April 21. https://www.independent.co.uk/arts-entertainment/music/news/prince-dead-when-prince-changed-his-name-to-a-symbol-warner-bros-had-to-mass-mail-floppy-disks-with-a6995196.html.

Horgan, John. 2016. "Are Brains Bayesian?" *Scientific American*. Blog post, January 6. https://blogs.scientificamerican.com/cross-check/are-brains-bayesian/.

Horst, Steven. 2007. *Beyond Reduction: Philosophy of Mind and Post-Reductionist Philosophy of Science*. New York: Oxford University Press.
Houle, David, Christophe Pélabon, Günter P. Wagner, and Thomas F. Hansen. 2011. "Measurement and Meaning in Biology." *Quarterly Review of Biology* 86 (1): 3–34.
Houser, Nathan, and Christian Kloesel, eds. 1992. *The Essential Peirce: Selected Philosophical Writings*, vol. 1. Bloomington: Indiana University Press.
———. 1998. *The Essential Peirce: Selected Philosophical Writings*, vol. 2. Bloomington: Indiana University Press.
Howse, Derek. 1980. *Greenwich Time and the Discovery of Longitude*. Oxford: Oxford University Press.
Hubel, David H., and Torsten N. Wiesel. 2004. *Brain and Visual Perception: The Story of a 25-Year Collaboration*. New York: Oxford University Press.
Hughes, R. I. G. 1997. "Models and Representation." *Philosophy of Science* (Proceedings) 64 (2): S325–36.
Hull, David. 1975. "Central Subjects and Historical Narratives." *History and Theory* 14 (3): 253–74.
Humane Borders / Fronteras Compasivas and Pima County Office of the Medical Examiner. n.d. "Custom Map of Migrant Mortality." Arizona OpenGIS Initiative for Deceased Migrants. Accessed January 31, 2019. http://www.humaneborders.info/app/map.asp.
Imhof, Eduard. 2013. *Cartographic Relief Presentation*. Translated by H. J. Steward. Redlands, CA: ESRI Press. Originally published as *Kartographische Geländedarstellung*, 1965.
Ingold, Timothy. 2000. *The Perception of the Environment: Essays on Livelihood, Dwelling and Skill*. London: Routledge.
———. 2007. *Lines: A Brief History*. New York: Routledge.
Ismael, Jennan. 1999. "Science and the Phenomenal." *Philosophy of Science* 66 (3): 351–69.
———. 2007. *The Situated Self*. Oxford: Oxford University Press.
Ito, Takashi, Kosuke Tashiro, Shigeru Muta, Ritsuko Ozawa, Tomoko Chiba, Mayumi Nishizawa, Kiyoshi Yamamoto, et al. 2000. "Toward a Protein–Protein Interaction Map of the Budding Yeast: A Comprehensive System to Examine Two-Hybrid Interactions in All Possible Combinations between the Yeast Proteins." 2000. *Proceedings of the National Academy of Sciences (USA)* 97 (3): 1143–47. https://doi.org/10.1073/pnas.97.3.1143.
Jacob, Christian. 2006. *The Sovereign Map: Theoretical Approaches in Cartography throughout History*. Translated by Tom Conley. Chicago: University of Chicago Press. Originally published as *L'empire des cartes: Approche théorique de la cartographie à travers l'histoire*, 1992.
Jacobs, Frank. 2014. "Africa, Uncolonized: A Detailed Look at an Alternate Continent." *Big Think*, November 11. https://bigthink.com/strange-maps/africa-uncolonized.
James, William. (1911) 1977. "Percept and Concept—The Abuse of Concepts." In

*The Writings of William James*, edited by John McDermott, 243–52. Chicago: University of Chicago Press.

Janik, Allan, and Stephen Toulmin. 1973. *Wittgenstein's Vienna*. New York: Simon & Schuster.

Jewish People's Council. 1948. "The Declaration of the Establishment of the State of Israel." *Official Gazette* 1 (May 14): 1. Available at https://www.knesset.gov.il/docs/eng/megilat_eng.htm.

Johnson, Aylmer. 2004. *Plane and Geodetic Surveying: The Management of Control Networks*. London: Spon Press.

Jones, Edward G. 2008. "Cortical Maps and Modern Phrenology." *Brain* 131 (8): 2227–33.

Jones, M. D. H., and Ann Henderson-Sellers. 1990. "History of the Greenhouse Effect." *Progress in Physical Geography: Earth and Environment* 14 (1): 1–18. https://doi.org/10.1177%2F030913339001400101.

Jones, Morgan D. 1998. *The Thinker's Toolkit: Fourteen Powerful Techniques for Problem Solving*. New York: Three Rivers Press.

Jung, Tobias. n.d. *Compare Map Projections*. Last updated August 6, 2018. https://map-projections.net/index.php.

Jungck, John R. 1985. "Mendel, Mendeleev and Me: Teaching's Impact upon Research." *American Biology Teacher* 47 (4). 197–201.

Kahneman, Daniel. 2002. "Maps of Bounded Rationality: A Perspective on Intuitive Judgment and Choice." In *Les Prix Nobel: The Nobel Prizes 2002*, edited by Tore Frängsmyr, 449–89. Stockholm: Almquist & Wiksell International.

———. 2011. *Thinking, Fast and Slow*. New York: Farrar, Straus, and Giroux.

Kant, Immanuel. 1987. *The Critique of Judgment*. Translated by Werner S. Pluhar. Indianapolis: Hackett.

———. 1992. "Concerning the Ultimate Ground of the Differentiation of Directions in Space." In *Theoretical Philosophy, 1755–1770: The Cambridge Edition of the Works of Immanuel Kant*, edited and translated by David Walford, in collaboration with Ralf Meerbote, 361–72. Cambridge: Cambridge University Press. Originally published as "Von dem ersten Grunde des Unterschiedes der Gegenden im Raume," 1768.

Kaplan, Abraham. 1964. *The Conduct of Inquiry*. San Francisco: Chandler Publishing.

Kaplan, Jonathan Michael, and Rasmus Grønfeldt Winther. 2013. "Prisoners of Abstraction? Genetic Diversity, Differentiation, and Heterozygosity, and the Very Concept of 'Race.'" *Biological Theory* 7:401–12. https://doi.org/10.1007/s13752-012-0048-0.

Keay, John. 2000. *The Great Arc: The Dramatic Tale of How India Was Mapped and Everest Was Named*. London: Harper Collins.

Keller, Evelyn Fox. 1983. *A Feeling for the Organism: The Life and Work of Barbara McClintock*. New York: W. H. Freeman.

———. 2002a. *The Century of the Gene*. Cambridge: Cambridge University Press.

———. 2002b. *Making Sense of Life: Explaining Biological Development with Models, Metaphors, and Machines*. Cambridge, MA: Harvard University Press.

Keynes, John Maynard. 1921. *A Treatise on Probability*. London: Macmillan.

Khalidi, Muhammad A. 1998. "Natural Kinds and Crosscutting Categories." *Journal of Philosophy* 95 (1): 33–50.

Khan Academy. n.d. "Homeotic Genes." Accessed August 24, 2018. https://www.khanacademy.org/science/biology/developmental-biology/signaling-and-transcription-factors-in-development/a/homeotic-genes.

Kimerling, A. Jon, Aileen R. Buckley, Phillip C. Muehrcke, and Juliana O. Muehrcke. 2009. *Map Use: Reading and Analysis*. 6th ed. Redlands, CA: ESRI Press.

Kimmerling, Baruch. 2004. "Benny Morris's Shocking Interview." *Logos* 3.1 (Winter). http://www.logosjournal.com/kimmerling.pdf.

Kitcher, Philip. 1993. *The Advancement of Science*. New York: Oxford University Press.

———. 2001. *Science, Truth, and Democracy*. New York: Oxford University Press.

———. 2002. "Reply to Helen Longino." *Philosophy of Science* 69 (4): 569–72.

Kitchin, Rob, and Martin Dodge. 2007. "Rethinking Maps." *Progress in Human Geography* 31 (3): 331–44.

Knuuttila, Tarja. 2011. "Modelling and Representing: An Artefactual Approach to Model-Based Representation." *Studies in History and Philosophy of Science, Part A* 42 (2): 262–71.

Kohler, Robert E. 1994. *Lords of the Fly: Drosophila Genetics and the Experimental Life*. Chicago: University of Chicago Press.

Kohn, Kurt W. 1999. "Molecular Interaction Map of the Mammalian Cell Cycle Control and DNA Repair Systems." *Molecular Biology of the Cell* 10 (8): 2703–34.

Koláčný, A. 1969. "Cartographic Information—a Fundamental Concept and Term in Modern Cartography." *Cartographic Journal* 6 (1): 47–49.

Korzybski, Alfred. 1933. "A Non-Aristotelian System and Its Necessity for Rigour in Mathematics and Physics." In *Science and Sanity*, 747–61. New York: Institute of General Semantics.

Kosambi, D. D. 1944. "The Estimation of Map Distances from Recombination Values." *Annals of Eugenics* 12 (1): 172–75.

Kraak, Menno-Jan. 2014. *Mapping Time: Illustrated by Minard's Map of Napoleon's Russian Campaign of 1812*. Redlands, CA: ESRI Press.

Kripke, Saul. 1972. *Naming and Necessity*. Cambridge, MA: Harvard University Press.

Krygier, John, and Denis Wood. 2011. *Making Maps: A Visual Guide to Map Design for GIS*. 2nd ed. New York: Guilford Press.

Kuhn, Thomas S. 1970. *The Structure of Scientific Revolutions*. 2nd ed. Chicago: University of Chicago Press.

———. 1974. "Second Thoughts on Paradigms." In *The Structure of Scientific Theories*, edited by Frederick Suppe, 459–82. Urbana: University of Illinois Press.

———. 1977. "The Essential Tension: Tradition and Innovation in Scientific

Research." In *The Essential Tension: Selected Studies in Scientific Tradition and Change*, 225–39. Chicago: University of Chicago Press.

———. 2000a. "Commensurability, Comparability, Communicability." In *The Road since Structure: Philosophical Essays, 1970–1993, with an Autobiographical Interview*, by Kuhn, edited by James Conant and John Haugeland, 33–57. Chicago: University of Chicago Press.

———. 2000b. "Possible Worlds in History of Science." In *The Road since Structure: Philosophical Essays, 1970–1993, with an Autobiographical Interview*, by Kuhn, edited by James Conant and John Haugeland, 58–89. Chicago: University of Chicago Press.

———. 2012. *The Structure of Scientific Revolutions*. 50th anniv. ed. (4th ed.). Chicago: University of Chicago Press.

Lacoste, Yves. 1976. *La géographie, ça sert, d'abord, à faire la guerre*. Paris: Maspero.

Lahsen, Myanna. 2005. "Seductive Simulations? Uncertainty Distribution around Climate Models." *Social Studies of Science* 35 (6): 895–922.

Lapaine, Miljenko, and Ana Kuveždić Divjak. 2017. "Famous People and Map Projections." In *Choosing a Map Projection*, edited by Miljenko Lapaine and E. Lynn Usery, 259–326. Cham, Switzerland: Springer.

Latour, Bruno. 1990. "Drawing Things Together." In *Representation in Scientific Practice*, edited by Michael Lynch and Steven Woolgar, 19–68. Cambridge, MA: MIT Press.

Laudan, Larry. 1990. *Science and Relativism: Some Key Controversies in the Philosophy of Science*. Chicago: University of Chicago Press.

Laudan, Rachel. 1987. *From Mineralogy to Geology: The Foundations of a Science, 1650–1830*. Chicago: University of Chicago Press.

Leonelli, Sabina. 2016. *Data-Centric Biology: A Philosophical Study*. Chicago: University of Chicago Press.

Leslie, Camilo A. 2016. "Territoriality, Map-Mindedness, and the Politics of Place." *Theory and Society* 45 (2): 169–201.

Levins, Richard. 1966. "The Strategy of Model Building in Population Biology." *American Scientist* 54 (4): 421–31.

Levins, Richard, and Richard Lewontin. 1980. "Dialectics and Reductionism in Ecology." *Synthese* 43 (1): 47–78.

———. 1985. *The Dialectical Biologist*. Cambridge, MA: Harvard University Press.

Lewis, Charlton T. and Charles Short. 1879. *A Latin Dictionary*. New York: Harper & Brothers. Available online: http://www.perseus.tufts.edu/hopper/resolveform?redirect=true&lang=Latin.

Lewontin, Richard C. 1974. *The Genetic Basis of Evolutionary Change*. New York: Columbia University Press.

———. 1991. *Biology as Ideology*. New York: Harper Perennial.

Lewontin, R. C., and M. J. D. White. 1960. "Interaction between Inversion Polymorphisms of Two Chromosome Pairs in the Grasshopper, *Moraba scurra*." *Evolution* 14 (1): 116–29.

Liebenberg, Louis. 1990. *The Art of Tracking: The Origin of Science*. Claremont, South Africa: David Philip.

*Life*. 1943. "R. Buckminster Fuller's Dymaxion World." March 1, 41–55.

Lightman, Alan. 1991. *Ancient Light: Our Changing View of the Universe*. Cambridge, MA: Harvard University Press.

Lloyd, Elisabeth A. 1987. "Confirmation of Ecological and Evolutionary Models." *Biology and Philosophy* 2 (3): 277–93.

———. (1988) 1994. *The Structure and Confirmation of Evolutionary Theory*. Princeton, NJ: Princeton University Press.

———. 1995. "Objectivity and the Double Standard for Feminist Epistemologies." *Synthese* 104 (3): 351–81.

———. 1996. "Science and Anti-Science: Objectivity and Its Real Enemies." In *Feminism, Science, and the Philosophy of Science*, edited by Lynn Hankinson Nelson and Jack Nelson, 217–59. Boston: Kluwer Academic Publishers.

———. 2010. "Confirmation and Robustness of Climate Models." *Philosophy of Science* 77 (5): 971–84.

———. 2012. "The Role of 'Complex' Empiricism in the Debates about Satellite Data and Climate Models." *Studies in History and Philosophy of Science* 43 (2): 390–401.

Logothetis, Nikos K. 2008. "What We Can Do and What We Cannot Do with fMRI." *Nature* 453 (June 12): 869–78. https://doi.org/10.1038/nature06976.

Longino, Helen E. 1990. *Science as Social Knowledge: Values and Objectivity in Scientific Inquiry*. Princeton, NJ: Princeton University Press.

———. 2002a. *The Fate of Knowledge*. Princeton, NJ: Princeton University Press.

———. 2002b. "Science and the Common Good: Thoughts on Philip Kitcher's *Science, Truth, and Democracy*." *Philosophy of Science* 69 (4): 560–68.

———. 2013. *Studying Human Behavior: How Scientists Investigate Aggression and Sexuality*. Chicago: University of Chicago Press.

———. Forthcoming. "Scaling Up; Scaling Down: What's Missing." *Synthese*.

Longley, Paul A., Michael F. Goodchild, David J. Maguire, and David W. Rhind. 2011. *Geographic Information Systems and Science*. 3rd ed. Hoboken, NJ: John Wiley & Sons.

López Beltrán, C. 1987. "La Explicación Evolucionista y el Uso de Modelos." Master's thesis, Posgrado en Filosofía de la Ciencia, Universidad Autónoma Metropolitana, Iztapalapa. http://www.filosoficas.unam.mx/~lbeltran/tesis 1987.html.

Lukk, Margus, Misha Kapushesky, Janne Nikkilä, Helen Parkinson, Angela Goncalves, Wolfgang Huber, Esko Ukkonen, and Alvis Brazma. 2010. "A Global Map of Human Gene Expression." *Nature Biotechnology* 28 (4): 322–24. https://doi.org/10.1038/nbt0410-322.

Lützen, Jesper. 2003. "Between Rigor and Applications: Developments in the Concept of Function in Mathematical Analysis." In *Cambridge History of Science: The Modern Physical and Mathematical Sciences*, edited by Mary Jo Nye, 5:468–87. Cambridge: Cambridge University Press.

Lyons, Sherrie L. 2009. *Species, Serpents, Spirits, and Skulls: Science at the Margins in the Victorian Age*. Albany: State University of New York Press.

MacEachren, Alan M. 2004. *How Maps Work: Representation, Visualization, and Design*. New York: Guilford Press.

Mac Lane, Saunders. 1986. *Mathematics: Form and Function*. New York: Springer.

Maderspacher, Florian. 2006. "The Captivating Coral—The Origins of Early Evolutionary Imagery." *Current Biology* 16 (13): R476–R478.

Maling, D. H. 1992. *Coordinate Systems and Map Projections*. New York: Pergamon Press.

Malmgren, Anna-Sara. 2018. "Varieties of Inference?" *Philosophical Issues* 28 (1): 221–54.

Mandal, Ananya. n.d. "What Is Gene Expression?" *News Medical*. Last updated August 23, 2018. https://www.news-medical.net/life-sciences/What-is-Gene-Expression.aspx.

Mann, Charles C. 2011. *1493: Uncovering the New World Columbus Created*. New York: Alfred A. Knopf.

Maroto Camino, Mercedes. 2005. *Producing the Pacific: Maps and Narratives of Spanish Exploration, 1567–1606*. Amsterdam: Rodopi.

Martínez, Sergio F. M. 2001. "Historia y combinatoria de las representaciones científicas." Comentarios a la propuesta de Ibarra y Mormann. *Crítica* 33 (99): 75–95.

Martínez, Sergio F. M., and Xiang Huang. 2011. "Epistemic Groundings of Abstraction and Their Cognitive Dimension." *Philosophy of Science* 78 (3): 490–511.

Masaryk University Mendel Museum. n.d. "The Mathematics of Inheritance." Exhibition text. Brno: Czech Republic. Accessed January 31, 2019. Available at https://web.archive.org/web/20130131134854/http://www.mendel-museum.com/eng/1online/room2.htm.

Massey, Doreen. 2005. *For Space*. London: SAGE.

Maxwell, James Clerk. 1890. "On Faraday's Lines of Force." In *The Scientific Papers of James Clerk Maxwell*, edited by W. D. Niven, 155–229. Cambridge: Cambridge University Press.

McDermott, John J., ed. 1977. *The Writings of William James: A Comprehensive Edition*. Chicago: University of Chicago Press.

McGranahan, Lucas. 2017. *Darwinism and Pragmatism: William James on Evolution and Self-Transformation*. London: Routledge.

McGuffie, Kendal, and Ann Henderson-Sellers. 2005. *A Climate Modelling Primer*. 3rd ed. Chichester, UK: John Wiley & Sons.

McLean, Ian. 2008. "'The Circumference Is Everywhere and the Centre Nowhere': Modernity and the Diasporic Discovery of Columbus as Told by Tzvetan Todorov." *Third Text*, June 18, 5–10. https://doi.org/10.1080/09528829208576378.

McMaster, Robert B., and K. Stuart Shea. 1992. *Generalization in Digital Cartography*. Washington, DC: Association of American Geographers.

McQuillan, Colin. 2010. "Philosophical Archaeology in Kant, Foucault, and Agamben." *Parrhesia* 10:39–49.

Mendel, Gregor. 1996. "Experiments in Plant Hybridization (1865)." Translated by William Bateson, 1901, with edits by Roger Blumberg, 1996. Electronic Scholarly Publishing Project. http://www.esp.org/foundations/genetics/classical/gm-65.pdf.

Menozzi, P., A. Piazza, and L. Cavalli-Sforza. 1978. "Synthetic Maps of Human Gene Frequencies in Europeans." *Science* 201 (4358): 786–92.

Menzel, Christopher. 2017. "Possible Worlds." In *Stanford Encyclopedia of Philosophy*, Winter 2017 ed., edited by Edward N. Zalta. https://plato.stanford.edu/archives/win2017/entries/possible-worlds/.

Mermin, David. 2009. "What's Bad about This Habit?" *Physics Today* 62 (5): 8–9.

Miller, Arthur I. 1972. "The Myth of Gauss' Experiment on the Euclidean Nature of Physical Space." *Isis* 63 (3): 345–48.

Miller, Greg. 2018. "A 500-Year-Old Map Used by Columbus Reveals Its Secrets." *National Geographic*, October 8. https://www.nationalgeographic.com/culture/2018/10/columbus-map-discovery-secrets-new-world/.

Miller, Osborn M. 1942. "Notes on Cylindrical World Map Projections." *Geographical Review* 32 (3): 424–30. http://doi.org/10.2307/210384.

Millikan, Ruth G. 1984. *Language, Thought, and Other Biological Categories*. Cambridge, MA: MIT Press.

Mills, Charles W. 2010. "Does Race Exist?" Video, 7:30. https://www.youtube.com/watch?v=epAv6Q6da_o.

Mitchell, Mary Ames. 2015. "Marco Polo and Paulo Toscanelli." *Crossing the Ocean Sea: Little-Known Trivia, Legends, and Mysteries about Exploring the Atlantic*. http://www.crossingtheoceansea.com/OceanSeaPages/OS-49-PoloToscanelli.html.

Mitchell, Sandra D. 2003. *Biological Complexity and Integrative Pluralism*. Cambridge: Cambridge University Press.

Monmonier, Mark. 1995. *Drawing the Line: Tales of Maps and Cartocontroversy*. New York: Henry Holt.

———. 1996. *How to Lie with Maps*. 2nd ed. Chicago: University of Chicago Press.

———. 2004. *Rhumb Lines and Map Wars: A Social History of the Mercator Projection*. Chicago: University of Chicago Press.

Montemayor, Carlos. 2013. *Minding Time: A Philosophical and Theoretical Approach to the Psychology of Time*. Leiden, NL: Brill.

Montemayor, Carlos, and Rasmus Grønfeldt Winther. 2015. "Review of Space, Time, and Number in the Brain." *Mathematical Intelligencer* 37 (2): 93–98.

Moretti, Franco. 2005. *Graphs, Maps, Trees: Abstract Models for Literary History*. New York: Verso.

Morgan, Edmund S. 2009. *American Heroes: Profiles of Men and Women Who Shaped Early America*. New York: W. W. Norton.

Morgan, Mary. 2006. "Economic Man as Model Man: Ideal Types, Idealization and Caricatures." *Journal of the History of Economic Thought* 28 (1): 1–27.

———. 2012. *The World in the Model: How Economists Work and Think*. New York: Cambridge University Press.

Morgan, Mary, and Margaret Morrison, eds. 1999. *Models as Mediators: Perspectives on Natural and Social Science*. Cambridge: Cambridge University Press.

Morgan, Thomas H. 1939. "Personal Recollections of Calvin B. Bridges." *Journal of Heredity* 30 (9): 354–58.

Morison, Samuel Eliot. 1942. *Admiral of the Ocean Sea: A Life of Christopher Columbus*. Vol. 1. Boston: Little, Brown.

Morton, Newton E. 1991. "Parameters of the Human Genome." *Proceedings of the National Academy of Sciences (USA)* 88 (17): 7474–76.

Moszkowski, Alexander. (1921) 2014. *Einstein the Searcher: His Work Explained from Dialogues with Einstein*. Translated by Henry L. Brose. New York: E. P. Dutton.

Muehrcke, Phillip C. 1969. "Visual Pattern Analysis: A Look at Maps." PhD diss., University of Michigan.

———. 1972. *Thematic Cartography*. Resource paper no. 19. Commission on College Geography. Washington, DC: Association of American Geographers.

———. 1974a. "Map Reading and Abuse." *Journal of Geography* 73 (5): 11–23.

———. 1974b. "Beyond Abstract Map Symbols." *Journal of Geography* 73 (8): 35–52.

———. 1978. *Map Use: Reading, Analysis and Interpretation*. 1st ed. Madison, WI: JP Publications.

Muehrcke, Phillip C., and Juliana O. Muehrcke. 1998. *Map Use: Reading, Analysis and Interpretation*. 4th ed. Redlands, CA: ESRI Press.

Mundy, Barbara E. 2011. "Litigating Land." In *Mapping Latin America: A Cartographic Reader*, edited by Jordana Dym and Karl Offen, 56–60. Chicago: University of Chicago Press.

Myers, Greg. 1988. "Every Picture Tells a Story: Illustrations in E. O. Wilson's *Sociobiology*." *Human Studies* 11 (2–3): 235–69.

Nagel, Ernest. 1961. *The Structure of Science: Problems in the Logic of Scientific Explanation*. New York: Harcourt, Brace & World.

National Aeronautics and Space Administration. n.d.a. "Wilkinson Microwave Anisotropy Probe." Last updated December 22, 2017. https://map.gsfc.nasa.gov/.

———. n.d.b. "Will the Universe Expand Forever?" Universe 101: Our Universe. Last updated January 24, 2014. https://map.gsfc.nasa.gov/universe/uni_shape.html.

National Ocean Service. n.d. "What Is the Geoid?" National Oceanic and Atmospheric Administration. Accessed August 16, 2018. https://oceanservice.noaa.gov/facts/geoid.html.

Nature Video. 2014. *Laniakea: Our Home Supercluster*. September 3. Video, 4:10. Macmillan Publishers UK. https://www.youtube.com/watch?v=rENyyRwxpHo.

Nave, Carl R. 2017. "Ideal Gas Law." *HyperPhysics*. Accessed October 24, 2018. http://hyperphysics.phy-astr.gsu.edu/hbase/Kinetic/idegas.html.

Nersessian, Nancy. 2002. "Maxwell and 'The Method of Physical Analogy':

Model-Based Reasoning, Generic Abstraction, and Conceptual Change." In *Reading Natural Philosophy*, edited by David B. Malament, 129–66. Chicago: Open Court.

Newsom, David D. 2001. *The Imperial Mantle: The United States, Decolonization, and the Third World*. Bloomington: Indiana University Press.

Nietzsche, Friedrich. 2008. *On the Genealogy of Morals*. Translated by Douglas Smith. Oxford: Oxford University Press. Originally published as *Zur Genealogie der Moral: Eine Streitschrift*, 1887.

Noë, Alva. 2010. *Out of Our Heads: Why You Are Not Your Brain, and Other Lessons from the Biology of Consciousness*. New York: Hill and Wang.

Nolte, David D. 2010. "The Tangled Tale of Phase Space." *Physics Today*, April, 33–38.

Norment, Lynn. 1997. "The Artist Formerly Known as Prince Has a New Wife, New Baby, and a New Attitude." *Ebony*, January.

Northcott, Robert. 2008. "Can ANOVA Measure Causal Strength?" *Quarterly Review of Biology* 83 (1): 47–55.

Norton, John D. 2016. *Analogy*. Unpublished manuscript. http://www.pitt.edu/~jdnorton/papers/material_theory/4.%20Analogy.pdf.

Nussbaum, Martha. 1995. "Objectification." *Philosophy and Public Affairs* 24 (4): 249–91.

Nüsslein-Volhardt, Christiane. 1995. "The Identification of Genes Controlling Development in Flies and Fishes." Nobel Lecture, December 8. Video and transcript. Nobel Foundation. https://www.nobelprize.org/prizes/medicine/1995/nusslein-volhard/lecture/.

Odenwald, Sten, and Rich Fienberg. 1993. "Galaxy Redshifts Reconsidered." *Sky and Telescope*, February. Available at https://archive.org/details/Sky_and_Telescope_1993-02-cbr/page/n65.

O'Gorman, Edmundo. (1958) 2006. *La invención de América*. 4th ed. Mexico City: Fondo de Cultura Económica.

Ohlsson, Stellan, and Erno Lehtinen. 1997. "Abstraction and the Acquisition of Complex Ideas." *International Journal of Educational Research* 27 (1): 37–48.

Okasha, Samir. 2002. *Philosophy of Science: A Very Short Introduction*. New York: Oxford University Press.

Olby, Robert. 1979. "Mendel No Mendelian?" *History of Science* 17 (1): 53–72.

Olsson, Gunnar. 1965. "Distance and Human Interaction: A Migration Study." *Geografiska Annaler: Series B, Human Geography* 47 (1): 3–43.

O'Neill, Craig, A. Lenardic, M. Weller, L. Moresi, S. Quenette, and S. Zhang. 2016. "A Window for Plate Tectonics in Terrestrial Planet Evolution." *Physics of the Earth and Planetary Interiors* 255:80–92. https://doi.org/10.1016/j.pepi.2016.04.002.

O'Neill, Onora. 1987. "Abstraction, Idealization, and Ideology in Ethics." In *Moral Philosophy and Contemporary Problems*, edited by J. D. G. Evans, 55–69. Cambridge: Cambridge University Press.

Oppenheimer, J. Robert. 1956. "Analogy in Science." *American Psychologist* 11 (3): 127–35.
Oreskes, Naomi. 1999. *The Rejection of Continental Drift: Theory and Method in American Earth Science*. New York: Oxford University Press.
———. 2004. "The Scientific Consensus on Climate Change." *Science* 306 (5702): 1686.
Oreskes, Naomi, and Eric Conway. 2011. *Merchants of Doubt: How a Handful of Scientists Obscured the Truth on Issues from Tobacco Smoke to Global Warming*. New York: Bloomsbury Press.
Oyama, Susan. (1985) 2000. *The Ontogeny of Information: Developmental Systems and Evolution*. 2nd ed. Durham, NC: Duke University Press.
Painter, Theophilus S. 1933. "A New Method for the Study of Chromosome Rearrangements and the Plotting of Chromosome Maps." *Science* 78 (2034): 585–86.
Pais, Abraham. 1982. *Subtle Is the Lord: The Science and Life of Albert Einstein*. Oxford: Oxford University Press.
Palumbo, Michael. 1991. *The Palestinian Catastrophe: The 1948 Expulsion of a People from Their Homeland*. New York: Olive Branch Press.
Papadopoulos, Athanase. 2016. "Euler and Chebyshev: From the Sphere to the Plane and Backwards." *Proceedings in Cybernetics* 22 (2): 55–69.
———. 2017. "Nicolas-Auguste Tissot: A Link between Cartography and Quasi-conformal Theory." *Archive for History of Exact Sciences* 71 (4): 319–36.
Pariser, Eli. 2011. *The Filter Bubble: How the New Personalized Web Is Changing What We Read and How We Think*. New York: Penguin Books.
PawełMM. 2010. "Reproduction of the Globe of Martin Behaim, 1492." Work derived from *Nouveau Larousse illustré* (1898). https://commons.wikimedia.org/wiki/File:MartinBehaim1492.png.
Pearl, Judea, and Dana Mackenzie. 2018. *The Book of Why: The New Science of Cause and Effect*. New York: Basic Books.
Peirce, Charles S. 1989. *Writings of Charles S. Peirce: A Chronological Edition, Vol. 4: 1879–1884*. Edited by Christian J. W. Kloesel. Bloomington: Indiana University Press.
———. 1992a. "Grounds of Validity of the Laws of Logic." In Houser and Kloesel, *The Essential Peirce*, 1:56–82. Originally published in *Journal of Speculative Philosophy* 2 (1869): 193–208.
———. 1992b. "Deduction, Induction, and Hypothesis." In Houser and Kloesel, *The Essential Peirce*, 1:186–99. Originally published in *Popular Science Monthly* 13 (August 1878), 470–82.
———. 1992c. "How to Make Our Ideas Clear." In Houser and Kloesel 1992, 124–41. Originally published in *Popular Science Monthly* 12 (January 1878): 286–302.
———. 1998a. "What Is a Sign?" In Houser and Kloesel 1998, 4–10. Likely composed in 1894.
———. 1998b. "The First Rule of Logic." In Houser and Kloesel 1998, 42–56.

Penfield, Wilder, and Herbert Jasper. 1954. *Epilepsy and the Functional Anatomy of the Human Brain*. Boston: Little, Brown.

Penfield, Wilder, and Theodore Rasmussen. 1950. *The Cerebral Cortex of Man*. New York: MacMillan.

Pereda, Carlos. 1999. *Crítica de la razón arrogante*. Mexico City: Taurus.

Perini, Laura. 2005. "Explanation in Two Dimensions: Diagrams and Biological Explanation." *Biology and Philosophy* 20 (2–3): 257–69.

Peters, Arno. 1983. *The New Cartography*. New York: Friendship Press.

———. 1993. *Compact Peters World Atlas*. Union, NJ: Hammond.

Pevsner, Jonathan. 2015. *Bioinformatics and Functional Genomics*. 3rd ed. Chichester, UK: John Wiley.

Phan, Doantam, Ling Xiao, Ron Yeh, Pat Hanrahan, and Terry Winograd. 2005. "Flow Map Layout." http://graphics.stanford.edu/papers/flow_map_layout.

Phillips, William D. Jr., and Carla Rahn Phillips. 1992. *The Worlds of Christopher Columbus*. Cambridge: Cambridge University Press.

Pickering, Andrew. 1984. *Constructing Quarks: A Sociological History of Particle Physics*. Chicago: University of Chicago Press.

Pincock, Chris. 2007. "A Role for Mathematics in the Physical Sciences." *Noûs* 41 (2): 253–75.

Plutynski, Anya. 2008. "The Rise and Fall of the Adaptive Landscape." *Biology and Philosophy* 23 (5): 605–23.

Poldrack, Russell A. 2010. "Mapping Mental Function to Brain Structure: How Can Cognitive Neuroimaging Succeed?" *Perspectives on Psychological Science* 5 (6): 753–61. https://doi.org/10.1177/1745691610388777.

———. 2018. *The New Mind Readers: What Neuroimaging Can and Cannot Reveal about Our Thoughts*. Princeton, NJ: Princeton University Press.

Poldrack, Russell A., Yaroslav O. Halchenko, and Stephen José Hanson. 2009. "Decoding the Large-Scale Structure of Brain Function by Classifying Mental States Across Individuals." *Psychological Science* 20 (11): 1364–72. https://doi.org/10.1111/j.1467-9280.2009.02460.x.

Pollack, George H. 2005. "Revitalizing Science in a Risk-Averse Culture: Reflections on the Syndrome and Prescriptions for Its Cure." *Cellular and Molecular Biology* 51 (8): 815–20.

Provine, William B. 1986. *Sewall Wright and Evolutionary Biology*. Chicago: University of Chicago Press.

———. 2001. *The Origins of Theoretical Population Genetics*. Chicago: University of Chicago Press.

Psillos, Stathis. 1999. *Scientific Realism: How Science Tracks the Truth*. London: Routledge.

———. 2000. "Abduction: Between Conceptual Richness and Computational Complexity." In *Abduction and Induction: Essays on Their Relation and Integration*, edited by Peter A. Flach and Antonis C. Kakas, 59–74. Dordrecht: Kluwer.

Putnam, Hilary. 1973. "Meaning and Reference." *Journal of Philosophy* 70 (19): 699–711.

———. 1975. *Mathematics, Matter, and Method*. Cambridge: Cambridge University Press,
———. 1984. "What Is Realism?" In *Scientific Realism*, edited by Jarrett Leplin, 140–53. Berkeley: University of California Press.
Raaflaub, Kurt A., and Richard J. A. Talbert. 2010. *Geography and Ethnography: Perceptions of the World in Pre-modern Societies*. Chichester, UK: John Wiley.
Radder, Hans. 2006. *The World Observed / The World Conceived*. Pittsburgh: University of Pittsburgh Press.
Raisz, Erwin. 1938. *General Cartography*. New York: McGraw-Hill.
———. 1962. *Principles of Cartography*. New York: McGraw Hill.
Ramer, Urs. 1972. "An Iterative Procedure for the Polygonal Approximation of Plane Curves." *Computer Graphics and Image Processing* 1 (3): 244–56.
Ramsey, Frank P. 1990. "General Propositions and Causality." In *F. P. Ramsey: Philosophical Papers*, edited by David H. Mellor, 145–63. Cambridge: Cambridge University Press.
Rankin, William. 2016. *After the Map: Cartography, Navigation, and the Transformation of Territory in the Twentieth Century*. Chicago: University of Chicago Press.
Ratajski, Lech. 1972. "Cartology." *Geographia Polonica* 21:63–78.
Ravenstein, Ernest G. 1885. "The Laws of Migration." *Journal of the Statistical Society of London* 48:167–235.
———. 1908. "Composite: Globe Gores 1–4. Martin Behaim's Erdapfel, 1492. (Facsimile of Behaim's Globe)." London: George Philip, Son & Nephew. Available at https://www.davidrumsey.com/luna/servlet/detail/RUMSEY~8~1~291869~90063414:Composite--Globe-Gores-1---4--Marti.
Rescher, Nicholas. 2001. *Philosophical Reasoning: A Study in the Methodology of Philosophizing*. Malden, UK: Blackwell.
Rescorla, Michael. 2017. "From Ockham to Turing—and Back Again." In *Philosophical Explorations of the Legacy of Alan Turing: Turing 100*, edited by Juliet Floyd and Alisa Bokulich, 279–304. Cham, Switzerland: Springer.
Rhodes, Margaret. 2016. "The Fascinating Origin Story of Prince's Iconic Symbol." *Wired*, April 22. https://www.wired.com/2016/04/designers-came-princes-love-symbol-one-night/.
Richardus, Peter, and Ron K. Adler. 1972. *Map Projections: For Geodesists, Cartographers and Geographers*. New York: American Elsevier.
Riemann, Bernhard. (1851) 1867. *Grundlagen für eine allgemeine Theorie der Functionen einer veränderlichen complexen Grösse*. Inaugural dissertation, 2nd ed. (unchanged). Göttingen: Adalbert Rente.
Robinson, Arthur H. 1953. *Elements of Cartography*. New York: John Wiley & Sons.
———. 1985. "Arno Peters and His New Cartography." *American Cartographer* 12 (2): 103–11.
———. 1986. *Which Map Is Best? Projections for World Maps*. Special publication of the ACA no. 1. American Cartographic Association, Committee on

Map Projections. Falls Church, VA: American Congress on Surveying and Mapping.

———. 1988. *Choosing a World Map: Attributes, Distortions, Classes, Aspects*. Special publication of the ACA no. 2. American Cartographic Association, Committee on Map Projections. Falls Church, VA: American Congress on Surveying and Mapping.

———. 1990. "Rectangular World Maps—No!" *Professional Geographer* 42 (1): 101–4.

Robinson, Arthur H., Joel L. Morrison, Phillip C. Muehrcke, A. Jon Kimerling, and Stephen C. Guptill. 1995. *Elements of Cartography*. 6th ed. New York: John Wiley & Sons.

Robinson, Arthur H., and Barbara B. Petchenik. 1976. *The Nature of Maps: Essays toward Understanding Maps and Mapping*. Chicago: University of Chicago Press.

Robinson, Arthur H., and Randall D. Sale. 1969. *Elements of Cartography*. 3rd ed. New York: John Wiley & Sons.

Robinson, Arthur H., and John P. Snyder, eds. 1991. *Matching the Map Projection to the Need*. Special publication of the ACA no. 3. American Cartographic Association, Committee on Map Projections. Bethesda, MD: American Congress on Surveying and Mapping.

Rocke, Alan J. 2010. *Image and Reality: Kekulé, Kopp, and the Scientific Imagination*. Chicago: University of Chicago Press.

Romm, Joseph. 2016. *Climate Change: What Everyone Needs to Know*. New York: Oxford University Press.

Rose, Nikolas, and Joelle M. Abi-Rached. 2013. *Neuro: The New Brain Sciences and the Management of the Mind*. Princeton, NJ: Princeton University Press.

Rosenberg, Daniel, and Anthony Grafton. 2012. *Cartographies of Time: A History of the Timeline*. New York: Princeton Architectural Press.

Rosenberg, Eugene and Ilana Zilber-Rosenberg. 2018. "The Hologenome Concept of Evolution after 10 Years." *Microbiome* 6:78. https://doi.org/10.1186/s40168-018-0457-9.

Roston, Eric, and Blacki Migliozzi. 2015. "What's Really Warming the World?" *Bloomberg Business Week*, June 24. https://www.bloomberg.com/graphics/2015-whats-warming-the-world/.

Rothstein, David. 2015. "What Is the Difference between the 'Doppler' Redshift and the 'Gravitational' or 'Cosmological' Redshift? (Advanced)." *Ask an Astronomer*, June 27. http://curious.astro.cornell.edu/physics/104-the-universe/cosmology-and-the-big-bang/expansion-of-the-universe/610-what-is-the-difference-between-the-doppler-redshift-and-the-gravitational-or-cosmological-redshift-advanced.

Rouse, Joseph. 2002. *How Scientific Practices Matter: Reclaiming Philosophical Naturalism*. Chicago: University of Chicago Press.

———. 2003. "Kuhn's Philosophy of Scientific Practice." In *Thomas Kuhn*, edited by Thomas Nickles, 101–21. Cambridge: Cambridge University Press.

Roush, Sherrilyn. 2005. *Tracking Truth: Knowledge, Evidence, and Science.* Oxford: Oxford University Press.

Russell, Bertrand. (1912) 1997. *The Problems of Philosophy.* New York: Oxford University Press.

Ryckman, Thomas. 2017. *Einstein.* New York: Routledge.

Ryden, Barbara. 2006. "Introduction to Cosmology." Unpublished manuscript. http://carina.fcaglp.unlp.edu.ar/extragalactica/Bibliografia/Ryden_Intro Cosmo.pdf.

———. 2016. *Introduction to Cosmology.* 2nd ed. Cambridge: Cambridge University Press.

Ryle, Gilbert. 1949. *The Concept of Mind.* Chicago: University of Chicago Press.

———. 1971. "Abstractions." In *Collected Papers, Vol. 2: Collected Essays, 1929– 1968*, 435–45. London: Hutchinson.

Saarinen, Tom F. 1999. "The Euro-Centric Nature of Mental Maps of the World." 1 (2): 136–78.

Sardar, Ziauddin. 1992. "Lies, Damn Lies and Columbus: The Dynamics of Constructed Ignorance." *Third Text*, June 19, 47–56. https://doi.org/10.1080/09528829208576384.

Schulten, Susan. 1998. "Richard Edes Harrison and the Challenge to American Cartography." *Imago Mundi* 50.174–88.

Schuurman, Nadine. 1999. "Reconciling Social Constructivism and Realism in GIS." *ACME: An International E-Journal for Critical Geographies* 1 (1): 73–90.

———. 2004. *GIS: A Short Introduction.* Malden, MA: Blackwell Publishing.

Schwartz, Seymour I. 2007. *Putting "America" on the Map: The Story of the Most Important Graphic Document in the History of the United States.* New York: Prometheus.

Scotese, Christopher R. n.d. "During the Early Carboniferous Pangea Begins to Form." Paleomap Project. Accessed January 31, 2019. http://www.scotese.com/newpage4.htm.

Sellars, Wilfrid. 1962. "Philosophy and the Scientific Image of Man." In *Frontiers of Science and Philosophy*, edited by Robert Colodny, 35–78. Pittsburgh, PA: University of Pittsburgh Press.

———. 1981. "Mental Events." *Philosophical Studies* 39 (4): 325–45.

Sen, Amartya. 1977. "Rational Fools." *Philosophy and Public Affairs* 6 (4): 317–44.

Shannon, Claude E. 1956. "The Bandwagon." *IRE Transactions on Information Theory* 2 (1): 3.

Shen, Elaine H., Caroline O. Overly, and Allan R. Jones. 2012. "The Allen Human Brain Atlas: Comprehensive Gene Expression Mapping of the Human Brain." *Trends in Neurosciences* 35 (12): 711–14. https://doi.org/10.1016/j.tins.2012.09.005.

Shin, Sun-Joo, Oliver Lemon, John Mumma. 2014. "Diagrams." In *Stanford Encyclopedia of Philosophy*, Winter 2014 ed., edited by Edward N. Zalta. http://plato.stanford.edu/archives/win2014/entries/diagrams/.

Shine, James M., Matthew J. Aburn, Michael Breakspear, and Russell A. Poldrack. 2018. "The Modulation of Neural Gain Facilitates a Transition between Functional Segregation and Integration in the Brain." *eLife* 7:e31130. https://doi.org/10.7554/eLife.31130.

Shiraev, Eric B., and Vladislav M. Zubok. 2016. *International Relations*. Oxford: Oxford University Press.

Shirley, Rodney W. 1983. *The Mapping of the World: Early Printed World Maps 1472–1700*. London: Holland Press.

Shoshan, Malkit. 2012. *Atlas of the Conflict: Israel-Palestine*. Portland, OR: Publication Studio.

Simon & Schuster Books. 2012. "Ursula Le Guin and the Map of Earthsea." Video, 49 seconds. https://www.youtube.com/watch?v=BGHbYByBnZ4.

Simpson, George G. (1944) 1984. *Tempo and Mode in Evolution*. New York: Columbia University Press.

Sismondo, Sergio. 1998. "The Mapping Metaphor in Philosophy of Science." *Cogito* 12 (1): 41–50.

———. 2004. "Maps and Mapping Practices: A Deflationary Approach." In *From Molecular Genetics to Genomics: The Mapping Cultures of Twentieth-Century Genetics*, edited by Jean-Paul Gaudillière and Hans-Jörg Rheinberger, 203–9. London: Routledge.

Sismondo, Sergio, and Nicholas Chrisman. 2001. "Deflationary Metaphysics and the Natures of Maps." *Philosophy of Science* 68 (3): S38–49.

Sklar, Lawrence. 1981. "Do Unborn Hypotheses Have Rights?" *Pacific Philosophical Quarterly* 62 (1): 17–29. https://doi.org/10.1111/j.1468-0114.1981.tb00039.x.

Slocum, Terry A., Robert B. McMaster, Fritz C. Kessler, and Hugh H. Howard. 2005. *Thematic Cartography and Geographic Visualization*. 2nd ed. Upper Saddle River, NJ: Pearson Prentice Hall.

Smith, Barry. 2008. "Ontology (Science)." In *Formal Ontology in Information Systems*, edited by Carola Eschenbach and Michael Grüninger, 21–35. IOS Press: Amsterdam. https://doi.org/10.3233/978-1-58603-923-3-21.

Smith, Brian Cantwell. 1996. *On the Origin of Objects*. Cambridge, MA: MIT Press.

———. 2015. "The Couch or the Bottle: Levels of Abstraction and the Anxious Mind." Lecture, February 26. Video, 1:17:30. https://www.youtube.com/watch?v=C0g4OwtkoCQ.

Smith, Jonathan Z. 1978. "Map Is Not Territory." In *Map Is Not Territory: Studies in the History of Religions*, 289–309. Leiden: E. J. Brill.

Snyder, John P. 1987. *Map Projections—A Working Manual*. US Geological Survey Professional Paper 1395. Washington, DC: US Government Printing Office.

———. 1993. *Flattening the Earth: Two Thousand Years of Map Projections*. Chicago: University of Chicago Press.

Snyder, Peter J., and Harry A. Whitaker. 2013. "Neurological Heuristics and Artistic Whimsy: The Cerebral Cartography of Wilder Penfield." *Journal of the History of the Neurosciences* 22 (3): 277–91.

Sobel, Dava. 1995. *Longitude*. London: Walker Books.
Solnit. Rebecca. 2010. *Infinite City: A San Francisco Atlas*. Berkeley: University of California Press.
Solomon, Miriam. 2001. *Social Empiricism*. Cambridge, MA: MIT Press.
Spanier, Bonnie B. 1995. *Im/partial Science: Gender Ideology in Molecular Biology*. Bloomington: Indiana University Press.
Sprackling, Michael. 1991. *Thermal Physics*. London: Macmillan.
Stanford, Kyle. 2006. *Exceeding Our Grasp: Science, History, and the Problem of Unconceived Alternatives*. New York: Oxford University Press.
Stanley, Matthew L., Bryce Gessell, and Felipe De Brigard. 2019. "Network Modularity as a Foundation for Neural Reuse." *Philosophy of Science* 86 (1): 23–46.
Stark, Alexander, Michael F. Lin, Pouya Kheradpour, Jakob S. Pedersen, Leopold Parts, Joseph W. Carlson, Madeline A. Crosby, et al. 2007. "Discovery of Functional Elements in 12 *Drosophila* Genomes Using Evolutionary Signatures." *Nature* 450 (7167): 219–32. https://www.nature.com/articles/nature06340.
Steinhardt, Paul J. 2011. "The Inflation Debate." *Scientific American*, April, 37–43.
Stich, Stephen. 1990. *The Fragmentation of Reason*. Cambridge: MIT Press.
Stockton, Nick. 2014. "Get to Know a Projection: Azimuthal Orthographic." *Wired*, November 20. https://www.wired.com/2014/11/get-to-know-a-projection-azimuthal-orthographic/.
Stokes, George Gabriel. (1883) 2009. "On the Variation of Gravity at the Surface of the Earth." In *Mathematical and Physical Papers: George Gabriel Stokes*, 131–71. Cambridge: Cambridge University Press.
Stolley, Paul D. 1991. "When Genius Errs: R. A. Fisher and the Lung Cancer Controversy." *American Journal of Epidemiology* 133 (5): 416–25. https://doi.org/10.1093/oxfordjournals.aje.a115904.
Strauss, Michael A. 2016. "The Expansion of the Universe." In *Welcome to the Universe*, by Neil deGrasse Tyson, J. Richard Gott, and Michael A. Strauss, 207–21. Princeton, NJ: Princeton University Press.
Strebe, Daan. 2012. "Why Mercator for the Web? Isn't the Mercator Bad?" Mapthematics Forum, March 15, 2012. http://www.mapthematics.com/forums/viewtopic.php?f=8&t=251.
Sturtevant, Alfred Henry. 1913. "The Linear Arrangement of Six Sex-Linked Factors in *Drosophila*, as Shown by Their Mode of Association." *Journal of Experimental Zoology* 14 (1): 43–59.
Suárez, Mauricio. 2003. "Scientific Representation: Against Similarity and Isomorphism." *International Studies in the Philosophy of Science* 17 (3): 225–44.
———. 2015. "Deflationary Representation, Inference, and Practice." *Studies in History and Philosophy of Science* 49:36–47.
Suppe, Frederick. 1989. *The Semantic Conception of Theories and Scientific Realism*. Urbana: University of Illinois Press.
Suppes, Patrick. 2002. *Representation and Invariance of Scientific Structures*. Stanford, CA: CSLI Publications.

———. 2012. "Models and Simulations in Brain Experiments." In *Models, Simulations, and Representations*, edited by Paul Humphreys and Cyrille Imbert, 188–206. New York: Routledge.

Suppes, Patrick, David. H. Krantz, R. Duncan Luce, and Amos Tversky. 1989. *Foundations of Measurement, Volume II: Geometrical, Threshold, and Probabilistic Representations*. New York: Academic Press.

Sutton, Walter S. 1903. "The Chromosomes in Heredity." *Biological Bulletin* 4 (5): 231–50.

Svensson, Erik I., and Ryan Calsbeek. 2012. "The Past, the Present, and the Future of the Adaptive Landscape." In *The Adaptive Landscape in Evolutionary Biology*, edited by Svensson and Calsbeek, 299–308. Oxford: Oxford University Press.

Swinburne University of Technology. n.d. "Cosmological Redshift." *COSMOS: The SAO Encyclopedia of Astronomy*. Accessed November 29, 2018. https://astronomy.swin.edu.au/cosmos/C/Cosmological+Redshift.

Swiss Society of Cartography. 1977. *Cartographic Generalization: Topographic Maps*. 2nd ed. Zürich: SGK-Publikationen.

Taubes, Gary. 1997. "Beyond the Soapsuds Universe." *Discover*, August 1. http://discovermagazine.com/1997/aug/beyondthesoapsud1197.

Tan, Shu Ling, Vera Storm, Dominique A. Reinwand, Julian Wienert, Hein de Vries, and Sonia Lippke. 2018. "Understanding the Positive Associations of Sleep, Physical Activity, Fruit and Vegetable Intake as Predictors of Quality of Life and Subjective Health across Age Groups: A Theory Based, Cross-Sectional Web-Based Study." *Frontiers in Psychology* 9, article 977, 1–13. https://doi.org/10.3389/fpsyg.2018.00977.

Tazzioli, Martina. 2015. "Which Europe? Migrants' Uneven Geographies and Counter-Mapping at the Limits of Representation." *Movements: Journal für kritische Migrations und Grenzregimeforschung* 1 (2): 1–20. https://movements-journal.org/issues/02.kaempfe/04.tazzioli-europe-migrants-geographies-counter-mapping-representation.html.

Teller, Paul. 2008. "Of Course Idealizations Are Incommensurable!" In *Rethinking Scientific Change and Theory Comparison: Stabilities, Ruptures, Incommensurabilities?*, edited by Léna Soler, Howard Sankey, and Paul Hoyningen-Huene, 247–64. Dordrecht, Netherlands: Springer. https://doi.org/10.1007/978-1-4020-6279-7_18.

Tenenbaum, Joshua B., Charles Kemp, Thomas L. Griffiths, and Noah D. Goodman. 2011. "How to Grow a Mind: Statistics, Structure, and Abstraction." *Science* 331 (6022): 1279–85. https://doi.org/10.1126/science.1192788.

Tharp, Marie. 1999. "Connect the Dots: Mapping the Seafloor and Discovering the Mid-ocean Ridge." In *Lamont-Doherty Earth Observatory: Twelve Perspectives on the First Fifty Years, 1949–1999*, edited by Laurence Lippsett. Palisades, NY: Lamont-Doherty Earth Observatory of Columbia University. Excerpted on the website of Woods Hole Oceanographic Institution, April 1, 1999. https://www.whoi.edu/news-insights/content/marie-tharp/.

Tharp, Marie, and Henry Frankel. 1986. "Mappers of the Deep." *Natural History*, October, 48–62.

Think Astronomy. n.d. "The Galactic Coordinate System." *Where Is M13? User Manual*, version 2.3. Accessed October 2, 2018. http://www.thinkastronomy.com/M13/Manual/common/galactic_coords.html.

Thomas, Julian. 2004. *Archaeology and Modernity*. Milton Park, UK: Routledge.

Thompson, Paul. 1989. *The Structure of Biological Theories*. Albany: State University of New York Press.

———. 2007. "Formalisations of Evolutionary Biology." in *Philosophy of Biology*, edited by Mohan Matthen and Christopher Stephens, 485–523. Amsterdam: Elsevier. https://doi.org/10.1016/B978-044451543-8/50023-X.

Tissot, Auguste. 1881. *Mémoire sur la représentation des surfaces et les projections des cartes géographiques*. Paris: Gauthier-Villars.

Tobler, Waldo R. 1966. "Medieval Distortions: The Projections of Ancient Maps." *Annals of the Association of American Geographers* 56 (2): 351–60.

———. 1981. "A Model of Geographical Movement." *Geographical Analysis* 13 (1): 1–20.

———. 1987. "Experiments in Migration Mapping by Computer." *American Cartographer* 14 (2): 155–63.

Tong, Rosemarie, and Nancy Williams. 2014. "Feminist Ethics." In *Stanford Encyclopedia of Philosophy*, Fall 2014 ed., edited by Edward N. Zalta. http://plato.stanford.edu/archives/fall2014/entries/feminism-ethics/.

Töpfer, F., and W. Pillewizer, with notes by D. H. Maling. 1966. "The Principles of Selection, a Means of Cartographic Generalisation." *Cartographic Journal* 3 (1): 10–16.

Toulmin, Stephen. (1953) 1960. *The Philosophy of Science: An Introduction*. Reprint, New York: Harper & Row.

Transport for London. n.d. "Harry Beck's Tube Map." Accessed March 20, 2018. https://tfl.gov.uk/corporate/about-tfl/culture-and-heritage/art-and-design/harry-becks-tube-map.

Tully, R. Brent, Hélène Courtois, Yehuda Hoffman, and Daniel Pomarède. 2014. "The Laniakea Supercluster of Galaxies." *Nature* 513 (7516): 71–73.

Turconi, Cristina. 1997. "The Map of Bedolina, Valcamonica Rock Art." *Tracce Online Rock Art Bulletin*, September 18. http://www.rupestre.net/tracce/?p=2422.

Turing, Alan M. 1937. "On Computable Numbers, with an Application to the Entscheidungsproblem." *Proceedings of the London Mathematical Society*, ser. 2, 42 (1): 230–65.

———. 1938. "On Computable Numbers, with an Application to the Entscheidungsproblem. A Correction." *Proceedings of the London Mathematical Society*, ser. 2, 43 (1): 544–46.

———. 1950. "Computing Machinery and Intelligence." *Mind* 59 (236): 433–60.

Turkheimer, Eric. 2000. "Three Laws of Behavior Genetics and What They

Mean." *Current Directions in Psychological Science* 9 (5): 160–64. https://doi.org/10.1111%2F1467-8721.00084.

Turkheimer, Eric, and Mary Waldron. 2000. "Statistical Analysis, Experimental Method, and Causal Inference in Developmental Behavioral Genetics." *Human Development* 43 (1): 51–52. https://doi.org/10.1159/000022656.

Turnbull, David. 1993. *Maps Are Territories: Science Is an Atlas*. Chicago: University of Chicago Press.

———. 2004. "Genetic Mapping: Approaches to the Spatial Topography of Genetics." In *Classical Genetic Research and Its Legacy: The Mapping Cultures of Twentieth-Century Genetics*, edited by Hans-Jörg Rheinberger and Jean-Paul Gaudillière, 207–19. London: Routledge.

Tversky, Amos. 1977. "Features of Similarity." *Psychological Review* 84 (4): 327–52.

Twain, Mark. 1894. *Tom Sawyer Abroad*. Available at http://www.gutenberg.org/files/91/91-h/91-h.htm.

Tyner, Judith A. 2005. "Elements of Cartography: Tracing Fifty Years of Academic Cartography." *Cartographic Perspectives* 51 (1): 4–13.

United Nations Committee on the Exercise of the Inalienable Rights of the Palestinian People. 1979. *The Origins and Evolution of the Palestine Problem: 1917–1988*. June 30. https://unispal.un.org/DPA/DPR/unispal.nsf/9a798adbf322aff38525617b006d88d7/d442111e70e417e3802564740045a309.

United Nations General Assembly. 1947. *Resolution 181 (II): Future Government of Palestine*. A/RES/181(II). November 29. https://unispal.un.org/DPA/DPR/unispal.nsf/0/7F0AF2BD897689B785256C330061D253.

Utrilla, P., C. Mazo, M. C. Sopena, M. Martínez-Bea, and R. Domingo. 2009. "A Palaeolithic Map from 13,660 calBP: Engraved Stone Blocks from the Late Magdalenian in Abauntz Cave (Navarra, Spain)." *Journal of Human Evolution* 57 (2): 99–111. https://doi.org/10.1016/j.jhevol.2009.05.005.

Van de Guchte, Maarten, Hervé M. Blottière, and Joël Doré. 2018. "Humans as Holobionts: Implications for Prevention and Therapy." *Microbiome* 6:81. https://doi.org/10.1186/s40168-018-0466-8.

Van Duzer, Chet. 2012. "Waldseemüller's World Maps of 1507 and 1516: Sources and Development of His Cartographical Thought." *Portolan*, Winter, 8–20.

———. 2019. *Henricus Martellus's World Map at Yale (c. 1491): Multispectral Imaging, Sources, and Influence*. Cham, Switzerland: Springer.

van Fraassen, Bas C. 1980. *The Scientific Image*. Oxford: Oxford University Press.

———. 1988. "The Invisible City." In *Le città del mondo e il futuro delle metropoli: Oltre la città, la metropoli*. Catalogue of the XVII Milan Triennale Exhibition, 155–56, 158.

———. 1989. *Laws and Symmetry*. New York: Oxford University Press.

———. 2002. *The Empirical Stance*. New Haven, CT: Yale University Press.

———. 2008. *Scientific Representation: Paradoxes of Perspective*. New York: Oxford University Press.

van Hoogstraten, Samuel. 1678. *Inleyding tot de Hooge Schoole der Schilderkonst*. Rotterdam.

Van Sant, Tom. 2013. *The Geosphere Project*. Video, 13:03. https://vimeo.com/50738196.

Van Syoc, Robert J., and Rasmus Grønfeldt Winther. 1999. "Sponge-Inhabiting Barnacles of the Americas: A New Species of *Acasta* (Cirripedia, Archaeobalanidae), First Record from the Eastern Pacific, Including Discussion of the Evolution of Cirral Morphology." *Crustaceana* 72 (5): 467–86.

Venter, J. Craig, Mark D. Adams, Eugene W. Myers, Peter W. Li, Richard J. Mural, Granger G. Sutton, Hamilton O. Smith, et al. 2001. "The Sequence of the Human Genome." *Science* 291 (5507): 1304–51. https://doi.org/10.1126/science.1058040.

Vicedo, Marga. 1995. "Scientific Styles: Towards Some Common Ground in the History, Philosophy, and Sociology of Science." *Perspectives on Science* 3 (2): 231–54.

Vinod, K. K. 2011. "Kosambi and the Genetic Mapping Function." *Resonance* 16 (6): 540–50. https://doi.org/10.1007/S12045-011-0060-X.

von Neumann, John. 1932. *Mathematische Grundlagen der Quantenmechanik*. Berlin: Julius Springer.

———. (1955) 1983. *Mathematical Foundations of Quantum Mechanics*. Translated by Robert T. Beyer. Princeton, NJ: Princeton University Press.

Vox. 2016. "Why All World Maps Are Wrong." December 2. Video, 6 minutes. https://www.youtube.com/watch?v=kIID5FDi2JQ.

Vujakovic, Peter. 2002. "Whatever Happened to the 'New Cartography'? The World Map and Development Mis-education." *Journal of Geography in Higher Education* 26 (3): 369–80.

Wachtel, Nathan. 1973. "Pensamiento salvaje y aculturación: el espacio y el tiempo en Felipe Guaman Poma de Ayala y el Inca Garcilaso de la Vega." In *Sociedad e ideología: ensayos de historia y antropología andinas*, edited by Wachtel, 165–228. Lima: Instituto de Estudios Peruanos.

Wade, M. J. 1992. "Sewall Wright: Gene Interaction and the Shifting Balance Theory." In *Oxford Surveys of Evolutionary Biology VI*, edited by J. Antonovics and Douglas Futuyma, 35–62. New York: Oxford University Press.

———. 2012. "Wright's Adaptive Landscape: Testing the Predictions of His Shifting Balance Theory." In *The Adaptive Landscape in Evolutionary Biology*, edited by Erik Svensson and Ryan Calsbeek, 58–73. Oxford: Oxford University Press.

Wagner, David. 2011. "Glimpses of Unsurveyable Maps." In *Image and Imaging in Philosophy, Science and the Arts*, vol. 2, edited by Richard Heinrich, Elisabeth Nemeth, Wolfram Pichler, and David Wagner, 365–76. Frankfurt: Ontos Verlag.

Walsh, Denis. 2015. *Organisms, Agency, and Evolution*. Cambridge: Cambridge University Press.

Walsh, James. 2013. "Remapping the Border: Geospatial Technologies and Border Activism." *Environment and Planning D: Society and Space* 31 (6): 969–87.

Wegener, Alfred. 1966. *The Origin of Continents and Oceans*. Translated by John Biram. New York: Dover. Originally published as *Die Entstehung der Kontinente und Ozeane*, 1929.

Weibel, Robert. 1991. "Amplified Intelligence and Rule-Based Systems." In *Map Generalization: Making Rules for Knowledge Representation*, edited by Barbara P. Buttenfield and Robert B. McMaster, 172–86. Essex, UK: Longman Scientific & Technical.

Weinberg, Steven. 1972. *Gravitation and Cosmology: Principles and Applications of the General Theory of Relativity*. New York: John Wiley & Sons.

———. 2008. *Cosmology*. Oxford: Oxford University Press.

Weisberg, Michael. 2007. "Who Is a Modeler?" *British Journal for the Philosophy of Science* 58 (2): 207–33.

———. 2013. *Simulation and Similarity: Using Models to Understand the World*. New York: Oxford University Press.

Wendt, Alexander. 1992. "Anarchy Is What States Make of It: The Social Construction of Power Politics." *International Organization* 46 (2): 391–425.

West-Eberhard, Mary Jane. 2003. *Developmental Plasticity and Evolution*. New York: Oxford University Press.

Wexler, Alice. 2010. "Stigma, History, and Huntington's Disease." *Lancet* 376 (9734): 18–19. https://doi.org/10.1016/S0140-6736(10)60957-9.

Wey Gómez, Nicolás. 2008. *The Tropics of Empire: Why Columbus Sailed South to the Indies*. Cambridge, MA: MIT Press.

Weyl, Hermann. 1940. "The Mathematical Way of Thinking." *Science* 92 (2394): 437–46.

———. 1950. *The Theory of Groups and Quantum Mechanics*. 2nd ed. Translated by H. P. Robertson. New York: Dover. Originally published as *Gruppentheorie und Quantenmechanik*, 1928.

———. 1952. *Space—Time—Matter*. 4th ed. Translated by Henry L. Brose. New York: Dover. Originally published as *Raum—Zeit—Materie*, 1918.

Wikipedia. 2017. "List of Map Projections." Accessed August 22, 2018. https://en.wikipedia.org/wiki/List_of_map_projections.

Wilford, John N. 2000. *The Mapmakers*. Revised 2nd ed. New York: Vintage.

Williamson, Timothy. 2008. *The Philosophy of Philosophy*. Malden, MA: Blackwell.

Wilson, Edward O. 1999. *Consilience: The Unity of Knowledge*. New York: Vintage.

Wilson, George, and Samuel Shpall. 2016. "Action." In *Stanford Encyclopedia of Philosophy*, Winter 2016 ed., edited by Edward N. Zalta. https://plato.stanford.edu/archives/win2016/entries/action/.

Wilson, Mark. 2006. *Wandering Significance: An Essay on Conceptual Behavior*. New York: Oxford University Press.

Wilson, Robert A. 2005. *Genes and the Agents of Life: The Individual in the Fragile Sciences: Biology*. Cambridge: Cambridge University Press.

Wimsatt, William C. 2007. *Re-engineering Philosophy for Limited Beings: Piecewise Approximations to Reality*. Cambridge, MA: Harvard University Press.

Winsor, Justin. 1886. *Narrative and Critical History of America*. Vol. 2. Boston: Houghton, Mifflin.

Winther, Rasmus Grønfeldt. 2000. "Darwin on Variation and Heredity." *Journal of the History of Biology* 33 (3): 425–55.

———. 2001. "August Weismann on Germ-Plasm Variation." *Journal of the History of Biology* 34 (3): 517–55.

———. 2003. "Formal Biology and Compositional Biology as Two Kinds of Biological Theorizing." PhD diss., History and Philosophy of Science Department, Indiana University. http://philpapers.org/archive/WINFBA.1.pdf.

———. 2006a. "Fisherian and Wrightian Perspectives in Evolutionary Genetics and Model-Mediated Imposition of Theoretical Assumptions." *Journal of Theoretical Biology* 240 (2): 218–32.

———. 2006b. "On the Dangers of Making Scientific Models Ontologically Independent: Taking Richard Levins' Warnings Seriously." *Biology and Philosophy* 21 (5): 703–24.

———. 2006c. "Parts and Theories in Compositional Biology." *Biology and Philosophy* 21 (4): 471–99.

———. 2008. "Systemic Darwinism." *Proceedings of the National Academy of Sciences (USA)* 105 (33): 11833–38.

———. 2009a. "Character Analysis in Cladistics. Abstraction, Reification, and the Search for Objectivity." *Acta Biotheoretica* 57 (1–2): 129–62.

———. 2009b. "A Dialogue: Review of Kyle Stanford's *Exceeding Our Grasp: Science, History, and the Problem of Unconceived Alternatives* (Oxford UP, 2006)." *Metascience* 18:370–79.

———. 2009c. "Prediction in Selectionist Evolutionary Theory." *Philosophy of Science* 76 (5): 889–901.

———. 2009d. "Schaffner's Model of Theory Reduction: Critique and Reconstruction." *Philosophy of Science* 76 (2): 119–42.

———. 2010. *Networks: Ontologies, Modeling, and Ethics*. Presentation. http://www.rgwinther.com/Publications/WintherRG2010NetworksPres.pdf.

———. 2011a. "Consciousness Modeled: Reification and Promising Pluralism." *Pensamiento* 67 (254): 617–30.

———. 2011b. "Part-Whole Science." *Synthese* 178 (3): 397–427. https://www.researchgate.net/publication/220608082_Part-whole_science.

———. 2012a. "Interweaving Categories: Styles, Paradigms, and Models." *Studies in History and Philosophy of Science, Part A* 43 (4): 628–39.

———. 2012b. "Mathematical Modeling in Biology: Philosophy and Pragmatics." *Frontiers in Plant Evolution and Development* 3, article 102, 1–3. http://www.frontiersin.org/Plant_Evolution_and_Development/10.3389/fpls.2012.00102/full.

———. 2014a. "Determinism and Total Explanation in the Biological and Behavioral Sciences." *Encyclopedia of Life Sciences*, July. http://www.els.net/WileyCDA/ElsArticle/refId-a0024143.html.

———. 2014b. "The Genetic Reification of 'Race'? A Story of Two Mathematical Methods." *Critical Philosophy of Race* 2 (2): 204–23.

———. 2014c. "James and Dewey on Abstraction." *Pluralist* 9 (2): 1–28.

———. 2014d. "World Navels." *Cartouche of the Canadian Cartographic Association* 89:15–21.

———. 2015a. "Evo-Devo as a Trading Zone." In *Conceptual Change in Biology: Scientific and Philosophical Perspectives on Evolution and Development*, edited by Alan Love, 459–82. Berlin: Springer.

———. 2015b. "Mapping Kinds in GIS and Cartography." In *Natural Kinds After the Practice-Turn*, edited by Catherine Kendig, 197–216. London: Routledge.

———. 2016. "The Structure of Scientific Theories." In *Stanford Encyclopedia of Philosophy*, Winter 2016 ed., edited by Edward N. Zalta. https://plato.stanford.edu/archives/win2016/entries/structure-scientific-theories/.

———. 2018a. *Phylogenetic Inference, Selection Theory, and History of Science: Selected Papers of A. W. F. Edwards with Commentaries*. Cambridge: Cambridge University Press.

———. 2018b. "Race and Biology." In *The Routledge Companion to the Philosophy of Race*, edited by Paul C. Taylor, Linda Martin Alcoff, and Luvell Anderson, 305–20. New York: Routledge.

———. 2019a. "Cutting the Cord: A Corrective for World Navels in Cartography and Science." *Cartographic Journal*. https://doi.org/10.1080/00087041.2018.1534043.

———. 2019b. "Mapping the Deep Blue Oceans." In *The Philosophy of GIS*, edited by Timothy Tambassi, 99–123. Cham, Switzerland: Springer.

Winther, Rasmus Grønfeldt, Ryan Giordano, Michael D. Edge, and Rasmus Nielsen. 2015. "The Mind, the Lab, and the Field: Three Kinds of Populations in Scientific Practice." *Studies in History and Philosophy of Science, Part C* 52:12–21. https://doi.org/10.1016/j.shpsc.2015.01.009.

Winther, Rasmus Grønfeldt, and Jonathan M. Kaplan. 2013. "Ontologies and Politics of Bio-Genomic 'Race.'" *Theoria: A Journal of Social and Political Theory (South Africa)* 60 (136): 54–80.

Winther, Rasmus Grønfeldt, Michael J. Wade, and Christopher C. Dimond. 2013. "Pluralism in Evolutionary Controversies: Styles and Averaging Strategies in Hierarchical Selection Theories." *Biology and Philosophy* 28 (6): 957–79.

Wittgenstein, Ludwig. 1922. *Tractatus Logico-Philosophicus*. Translated by Frank P. Ramsey and Charles Kay Ogden. Introduction by Bertrand Russell. London: Kegan Paul, Trench, Trubner. Originally published as *Logisch-Philosophische Abhandlung*, 1921.

———. (1979) 2001. *Wittgenstein's Lectures, Cambridge, 1932–1935: From the Notes of Alice Ambrose and Margaret Macdonald*. Edited by Alice Ambrose. Amherst, NY: Prometheus.

———. 2009. *Philosophical Investigations*. Translated by Gertrude E. M.

Anscombe, Peter M. S. Hacker, and Joachim Schulte; rev. ed. (4th ed.) by Hacker and Schulte. Chichester, UK: Wiley-Blackwell. Originally published as *Philosophische Untersuchungen*, 1953.

Wolter, John A. 1975. "Cartography—an Emerging Discipline." *Canadian Cartographer* 12 (2): 210–16.

Wood, Denis. 1992a. "How Maps Work." *Cartographica* 29 (3–4): 66–74.

———. 1992b. *The Power of Maps*. With John Fels. New York: Guilford Press.

———. 2010. *Rethinking the Power of Maps*. With John Fels and John Krygier. New York: Guilford Press.

———. 2012a. "The Anthropology of Cartography." In *Mapping Cultures: Place, Practice, Performance*, edited by Les Roberts, 280–303. Basingstoke, UK: Palgrave Macmillan.

———. 2012b. Review of *Mapping Latin America: A Cartographic Reader*, edited by Jordana Dym and Karl Offen. *Cartographica* 47 (2): 136–38.

Wood, Denis, and John Fels. 1986. "Designs on Signs/Myth and Meaning in Maps." *Cartographica* 23 (3): 54–103.

———. 2008. *The Natures of Maps: Cartographic Constructions of the Natural World*. Chicago: University of Chicago Press.

Wood, Denis, Ward L. Kaiser, and Bob Abramms. 2006. *Seeing through Maps: Many Ways to See the World*. 2nd ed. Amherst, MA: ODT.

Woodward, David, and G. Malcolm Lewis. 1998. "Introduction." In *The History of Cartography: Cartography in the Traditional African, American, Arctic, Australian, and Pacific Societies*, vol. 2, book 3, edited by David Woodward and G. Malcolm Lewis, 1–10. Chicago: University of Chicago Press.

Woodward, James. 2003. *Making Things Happen: A Theory of Causal Explanation*. New York: Oxford University Press.

Wright, John K. 1942. "Map Makers Are Human: Comments on the Subjective in Maps." *Geographical Review* 32 (1): 527–44.

Wright, Sewall. 1920. "The Relative Importance of Heredity and Environment in Determining the Piebald Pattern of Guinea-Pigs." *Proceedings of the National Academy of Sciences (USA)* 6 (6): 320–32. https://www.ncbi.nlm.nih.gov/pmc/articles/PMC1084532/.

———. 1932. "The Roles of Mutation, Inbreeding, Crossbreeding, and Selection in Evolution." *Proceedings of the Sixth International Congress of Genetics* 1:356–66. http://www.esp.org/books/6th-congress/facsimile/contents/6th-cong-p356-wright.pdf.

———. 1977. *Evolution and the Genetics of Populations, Volume 3: Experimental Results and Evolutionary Deductions*. Chicago: University of Chicago Press.

———. 1988. "Surfaces of Selective Value Revisited." *American Naturalist* 131 (1): 115–23.

Wynn, Thomas. 1989. *The Evolution of Spatial Competence*. Champaign: University of Illinois Press.

Yang, Qihe, John P. Snyder, and Waldo R. Tobler. 2000. *Map Projection Transformation: Principles and Applications*. London: Taylor & Francis.

Yu, Zhuoyun, compiler. 1984. *Palaces of the Forbidden City*. Translated by Ng Mau-sang, Chan Sinwai, and Puwen Lee. Consultant editor, Graham Hutt. New York: Viking.

Ziman, John M. 2000. *Real Science: What It Is, and What It Means*. Cambridge: Cambridge University Press.

Zook, Matthew A., and Mark Graham. 2007. "The Creative Reconstruction of the Internet: Google and the Privatization of Cyberspace and DigiPlace." *Geoforum* 38 (6): 1322–43. https://doi.org/10.1016/j.geoforum.2007.05.004.

Zuber, Mike A. 2011. "The Armchair Discovery of the Unknown Southern Continent: Gerardus Mercator, Philosophical Pretensions and a Competitive Trade." *Early Science and Medicine* 16 (6): 505–41.

# Index

Page numbers followed by *t* or *f* refer to tables or figures, respectively.

*Abbildung*, 170, 172, 173
abduction, 34–35
abstraction process, phases, 81. *See also* calibration of units and coordinates; data collection and management; generalization protocols
abstractions: concepts as, 44–64; and map making, 61; map/world link, 60; reversal, 82; stages of, 61. *See also specific abstraction stages*
abstractive-averaging assumptions, 180, 208t, 209; arrowized assumptions, 181; and cognitive neuroscience, 189–90; path diagrams, 204; and phrenology, 189–90; testing of, 201
ACA. *See* American Cartographic Association
adaptive landscape maps, 215, 231–34; critiques of, 233n37; evolutionary opportunities, 233; organismal complexity, evolution of, 231
adaptive value, 232
*Advancement of Science, The* (Kitcher), defending realism, 247
affordance, 197–98
AFKAP. *See* Artist Formerly Known as Prince
aim and value assumptions, 108, 109t
Alekbu-Lan map, 138n59
algorithmic reduction, 77n47

alleles, 212, 216; of genes, 217–18; in state-space maps, 231, 232f
"America," source of name, 16–18, 22
American Cartographic Association (ACA), 106
AmerInd: maps, 20n59; territories, 14
analogies: and abduction, 35; characterization, 32; definition, 30, 32, 35n19; example (Darwin), 32–33; failure of, 34; and induction, 35; of interest, 33; and metaphors, difference, 31n6; philosopher/cartographer, 41–42; spatial maps, 36f, 37, 38, 40n24, 141, 225; "structure-mapping" theory, 35n19; types of, 31–34; uncertainty of, 34–35. *See also* map analogies; *and specific analogies*
analogous maps. *See* general (analogous) maps
analogous-to-maps, representations, 36, 37, 39
analysis of variance (ANOVA), 68–69, 226
Anderson, M.: dynamic interactive-brain model, 197; functional fingerprints, 197–99, 198f; neural-reuse perspective, 199
Anglo-American analytic philosophy, 40–43

ANOVA. *See* analysis of variance
antirealism, 247–48
anti-*Turing Phrenunculus* (Anderson), 199
ArcGIS maps: color use in, 126; contour lines in, 122, 123; worldview, 137
ArcGIS software, 121
Arrhenius, S., climate model building, 139
arrowized assumptions: archaeology of, 183; countermapping, 185; in migration maps, 180, 181–82; simplified, 181–82; temporary migrants type illustrating, 183
arrowized maps, 182–85
Artist Formerly Known as Prince (AFKAP), 119n2
assumption archaeology, 52–56, 110–12, 252; decompositional assumptions, brain maps, 190–91; de-reification, tool for, 54; and Foucault, 53; genetic maps, 214; and integration platforms, 134; in map of Copenhagen, 242f; and pluralistic ontologizing, 134; and Renaissance cartographic worldview, 179; and representations, 53
*Aufbau* (Carnap), 47

"banging mirror," 23
baptism event, 16
Bateson, G., 43, 134
Baudrillard, J., simulacrum and reality, 43–44
Beck, H., 79, 96
Behaim, M., globe of, 179
behavior genetics, laws of, 207
belief, definition (Ramsey), 40–41
beyond-Mercator integration platform, 90, 108–9, 112, 117
Borges, J. L., 1–2, 25; map of the empire, 3, 23
brain: abnormality, explanation of, 190–91; countermapping, 195–99; functions, 187; mapping, fMRI, 194–95; and mind, philosophy of, 188–89, 188n30
brain maps/structure: decompositional assumptions, 180, 190–91; modules, 188
Budassi, P. C., cosmographic map, 155, plate 8
Bureau of Land Management, 67

cadastral maps, 29, 85
calibration of modernity, temporal, 66
calibration of units and coordinates, 61–62, 66–68, 81; calibrating time, 65–66; geodesic surveying, 62, 63–65, 74; plane surveying, 62; of universe, in space and time, 157
Calvino, I., 2, 3
Carboniferous period, continent locations in, 161, 162f
"cargo cult" narrative, 49–50
Carnap, R.: on scientific inquiry, 250; on scientific objectivity, 46–47
carpet map of Eudoxia (Calvino), 2, 3, 12
cartographer's maps, 11n25
cartographic generalization, 9, 69–70; definition, 71–72; protocols, 80
cartographic representations, 4, 12n35, 55; example, 120–21; flat coordinate system, 75; map projections, 71; ontologizing pluralistically, 136; partitioning, 15, 192; perspectivizing, 15, 81, 128; testing of, 82–83; Web Mercator (online), 76, 90, 107
cartography: definition, 5; exaggeration, 70, 79; images, 254–55; mathematics and, 170; rules (Harley), 13–14; social sciences, analogizing protocols, 80
cartology. *See* map thinking
cartopower, 22; analyzing of, 132n40; cartographic context, 129, 130–32,

131f; of geological maps, 165; of theories of disease, 133–34
Cartwright, N., on inconsistency, 115–16
Cassini de Thury, C.-F., map of France, 62–63
Cassirer, E., on space, 54
causal complexity, 204
causal maps: analogy, 36, 38–39, 56; guinea pig development, 205–6, 206f; path diagrams, 200, 204–5; statistical entanglement, 204. *See also* statistical causal maps
causal model sleep quality study, 204–5, 204f
causal relations, litter mates, 206f
causal world structure, and maps, 206–8
Cavalli-Sforza, L. L., 215, 227–29
centimorgans (map units), 218–19
chloroplasts, 145f, 237n56
chromosomes, 214; crossover process, 218; homologous, 218
classification protocol, 78–79
climate change: greenhouse gases, 138–39; merely-seeing-as, 140–41; modeling, 138–41; multiple representations, 138–41; ontologizing, 140–41. *See also* global climate change
CMB map. *See* cosmic microwave background (CMB) map
codons, 222
Cognitive Atlas (Poldrack), 196
cognitive mapping, 25n78, 86–87; and Kuhn's paradigm, 48; ontological layers, 143; and space (Harley), 7
cognitive ontology, 196
cognitive science, contextual objectivity, 187
Columbus, C., voyages, 177–78
*Columbus's Worldview*, 176f, 178–79
Combe, G., 191; phrenology, 187
communication paradigm, 60–61

comoving distance, 156–57
comoving frame, 156–57
comparative genetic maps, 215; evolutionary changes, display, 229–30; partitioning frame of, 230–31
computational maps, 25
*Concept of Mind, The* (Ryle), 41–42
concepts as abstractions (James), 44–46
conceptual analysis, 251
conceptual ontology, 195–96
conformal projection, 44, 45f, 75, 76t
conformal quincuncial projection (Peirce), 44, 45f
Conley, T., on cartographic writing, 43n39
constructivism: Kuhn's, 248n14; map analogy, 245–46; social, 60, 80
constructivists, 80, 245, 247
Contarini-Rosselli map (1506), 16n47
contextualizing, 99, 107; vs. universalizing, 109
contextual objectivity, 23, 89–90; achieving, 99; in cognitive and neuroscience, 187; conformation, 95–96, 96n24; decompositional assumptions, 187–88; definition, 95; elements of, 95–99; essential indexical, 95, 97–99; seeking, 107
continental drift, 159–64
continent locations, Carboniferous period, 161, 162f
conventional (stipulated) similarity, 124–25, 125f, plate 1
cortex, organization, 189f, 193
cortex person (homunculus) maps, 193
cosmic and Hubble flows, distinction, 154
cosmic flows, 154
cosmic microwave background (CMB) map, 148f, 149, plate 7; creation process, 151, 158–59; Euclidian universe, 156; extreme-scale, 152; findings, 151–52; Planck project, 151

cosmic web, 154
*Cosmographiae Introductio* (Waldseemüller and Ringmann), 15, 17
cosmographic map (Budassi), 155, plate 8
cosmography, importance of, 155
cosmological physical theory, 159
cosmological principle, 155–56
cosmological redshift, 153–54
cosmology extreme-scale maps, 149, 152
*cosmovisión*, 22
Couclelis, H., on GIS and traditional cartography, 10
countermapping: affordance, 197–98; arrowized assumptions, challenge, 185; the brain, 195–99; migrant's movements, 181, 186–87, 186f; migration, 185–87; neural reuse, 197; neuroscience, example, 197; and pluralistically ontologizing, 137–38; strategies to genetics, 240
countermaps, 208t, 209; definition, 22; to genotype-phenotype map, 227; statistical, 207
Courtois, H., three-dimensional maps of galactic motion, 154–55, plate 7
Crick, F.: central dogma, 246; three-dimensional metal model, 126
CRISPR, 243
*Critique of Judgment* (Kant), 53n78
cross-variables, explanatory impotence, 202
*Currents of Migration* map, 184f
Curry, L., 91; empiricist map-thinking, 246

*Da Ming Hun Yi Tu* map, 92, 93f
Darwin, C. R., 30, 32; natural selection hypothesis, 234–37
data collection and management, 61, 67–69, 140; in abstraction process, 81; and cosmic-scale maps, 155, 158; hydrographic data, 67, 83; in mapmaking, 68; and scientific experimentation, 68
data structures, 82–83
*Dead Speak Tomorrow, The*, 255f
death maps, 185
decompositional assumptions, 193; brain maps, 180, 190–91; and *The Turing Phrenunculus*, consistency, 191
"Deconstructing the Map" (Harley), 13–14
deep mapping, 5, 6–7, 6n8, 29, 54
Defense Mapping Agency, 67
de Lapparent, V., 154
Department of Parks and Recreation (DPR), California, 121
de-reification, 54
Descartes, R., mind/body split, 41–42
*Design of Experiments, The* (Fisher), 69
developable surfaces, 74, 75t
Dewey, J.: map/explorer relationship, 44, 46; on scientific inquiry, 250
diachronic time, 229
diagrams: and maps, distinction, 115; relations in, 114; representation of gene combinations, 232f
disanalogies, 31, 32, 32n10, 33
disentanglement, statistical, 204
DNA strands, 214
Dobzhansky, T., literal cartographic genetic map, 215, 227–29, 228f
dominant traits, 212
double helix, 222
DPR. *See* Department of Parks and Recreation (DPR), California
dreams: and the brain, 188–89; of cartographic images, 87; and existence, 244; mapping our, 5; of moving, 185; representations influencing, 243; representing our world, 256
*Drosophila melanogaster* (fruit flies): for case studies, 213; gene maps, 216–17, 221f; genetic variation, geographic gradients of, 227, 228f

*Drosophila pseudoobscura*, gene arrangement map, 228f
dynamic maps: computational, 40n24; topographic, 121

Eagleman, D., on the brain, 188
Earth Gravitational Model (EGM) 2008, 65
Earth maps, steady-state, 163
*Earthsea Map* (Le Guin), 2, 3, plate 1
EGM2008. *See* Earth Gravitational Model (EGM) 2008
Einstein, A., 145f; and FLRW, 156; on importance of Gauss, 173n64; influenced by Gauss's map thinking, 173; and Newton's theories, 53
*Elements of Cartography* (Robinson and Sale), 69
ellipsoid, 63–64
*El primer nueva corónica y buen gobierno* (Poma): map in, 18–19; observations about consciousness, 18–20; orientation, 21, 21n63, 21n65; Peru, observations, 20–22
empirical and theoretical (data) models, 83
empiricism, map analogy, 245, 246
end-of-inquiry consistency, 112, 115–16
environmental disease studies, 203–4
Environmental Systems Research Institute (Esri), ArcGIS software, 121
epistemic criteria, 68
equatorial shift, 162–63, 162f
equipotential surface, 65
Esri (Environmental Systems Research Institute), ArcGIS software, 121
ethics and power: in map of Copenhagen, 242f; tracking, 251, 252–53, 254
etiology, diseases, 202–4
Euclidean grid geometric space, 53
Euclidean universe, 156
Eudoxia, carpet map of (Calvino), 2, 3, 12

Euler diagram, 36–37, 36f, 38; and fuzzy map types, 40
European Carboniferous coal beds, formation, 161–63
European Space Agency, Planck project, 151. *See also* cosmic microwave background (CMB) map
evolutionary process, model of, 112n73
*Evolution of Linear Genetic Maps in Four Movements, The*, 221f
excavated representations, 53
existence, objects and processes, 244–45
expansion of space, 156
explanatory frames, 50
explanatory power, 68, 91
exponential doubling, 150
extreme-scale maps: analogy, 36, 37, 38, 54; in cosmology, 148f, 149, 150, 152, plate 7; gene expression mapping, 225

factors (genes). *See* genes
figures of Earth, types of, 65
Fisher, R. A., 69; model of evolutionary processes, 112n73; spurious correlation, smoking and lung cancer, 203
fitness value: equal relative, 232–33, 232n33; relative, 231–32
FLRW metric. *See* Friedmann–Lemaître–Robertson–Walker (FLRW) metric
fMRI (functional magnetic resonance imaging): brain mapping, 194–95; and pernicious reification, 194, 199
*Fortune* wartime atlas (1944), 104, 136
Foucault, M., 52–53
Friedmann, A., 156
Friedmann–Lemaître–Robertson–Walker (FLRW) metric, 155n14; comoving distance and frame, 159
functional fingerprints, 197–99, 198f

functional magnetic resonance imaging. *See* fMRI
functional transformation, dynamic, 174
fuzzy map boundaries, 40, 209

Gaddis, J. L., on historical map analogy, 50–51
galactic coordinate system, 158
galactic motion, three-dimensional maps, 154–55, plate 7
galaxies, patterns of, 154, plate 7
Galen (of Pergamon): humoral theory by, 133, 134, 244; universal model for disease, 133
Gall, F. J., founder of phrenology, 191
Gall–Peters projection, 75f, 76t; equal-area, 103, 111f, 136; Tissot indicatrix, with and without, 110, 111f
gametes, 212t, 216
Gauss, C. F.: curved surfaces, 170–72; influence on Einstein, 173; Nether-Saxon geodetic survey, 170, 171f
Gedde, C., map of Copenhagen (1761), 242f, 254
Geller, M., galaxy patterns, 154, plate 7
genealogy, 33
gene combinations, diagrammatic representation, 232f
gene expression maps, 84–85, 214–15, 224–25
general (analogous) maps, 36f; analogy, 37, 39, 160, 215; mathematical mapping and functions, 174; in mathematics, 149, 169–74
generalization protocols: cartographic, 80; maps, 68, 69, 79. *See also specific protocols and types of selections*
genes, 212; combinations, field of, 232f; co-traveling, 217, 219; definition, 214; homeotic (HOX) mapping, 224; independent, 217, 219; linked, 217–19; orthologous, 230; paralogous, 230

genetic code, 222; genomes, of *Drosophila* species, 230
genetic diseases, mapping, 202–3
genetic information, DNA strand, 221f, 222
genetic maps, types of, 214–15
genome editing, 243
genotype-phenotype maps, 215; Lewontin's map, 210f, 226–27; norms of reaction, 225
genotypes, 214
genotype space, 210f
genotypic state space, 226
geodesic surveying, 62, 63–65; curved surface models, 74
geographic information system (GIS), 6; mapmaking, automated, 68; philosophical implications (Schuurman), 9; and traditional cartography, comparison, 10
*Geography* (Ptolemy), 15
geoid, 11, 64–65, plate 3
geometric projections, 74, 75f
geometric symbols, 125f, 126
GeoSphere map (Van Sant), 79, plate 2; imaging strategies, 23–24; limitations, 24; objectivity, 23
germ theory of disease, 133
Giere, R.: map analogy, uses of, 95, 113, 115; perspectivism, defending, 142–43
GIS. *See* geographic information system
global climate change: and geographic maps, 84n68; and multiple representations, 138–41; temperature anomalies, map of, 118f, plate 5. *See also* climate change
Goodchild, M., on space differentiation, 247
Goodman, N.: on *Aufbau* (Carnap), 47; worldmaking processes, 79n54
Google Earth, 24–25
Google Maps, 24, 25, 107, 121, plate 6;

scale in, 123; seeing as the world, 137; symbolism in, 126
Gould, S. J., on reification of intelligence, 197
GPS locations, calibration, 65
granular pluralism, 112, 113
graticules, 21, 64; arc, 145f; comoving frame, 156–57; cosmological, 158, 174; map, 156, 162f, 164
greenhouse gases, and climate change, 138–39
Gymnasium Vosagense, 15

Habans, R., 185
Hacking, I.: on concept of probability, 7–8; on unity or disunity, sciences, 116
Hardy–Weinberg equilibrium, and assumption archaeology, 52
Harley, J. B., 106; "cartographic rules," 13–14; *History of Cartography*, 7; on map functions, 13–14; on Peters's critique against cartography, 106
Harrison, R. E., 76–77; contextual objectivity, seeking, 106; *Fortune* wartime atlas (1944), 104, 136; perspective maps, 105–6, 105f, 136
Heezen, B., 164
hemoglobin, three-dimensional structure maps, 230
Hercules cluster, 153–54, plate 7
Hesse, M., 31, 32
Himalayas, 159
*History of Cartography* (Harley), 7
Hofstadter, D., analogy definition, 32, 35n19
holobiont, 240
hologenome, 240
homeotic genes. *See* HOX (homeotic) genes
*Homo cartograficus*, 54, 81, 138, 152
homologies, 230
homunculus (cortex person) maps, 193

HOX (homeotic) genes: mapping, 224; tracing mutations, 238
Hubble flow, 153, 156; and cosmic flows, distinction, 154
Huchra, J., galaxy patterns, 154
human consciousness, 145f; mapping impulse in, 7; raising political, 186; understanding emergence of, 174
Humane Borders / Fronteras Compasivas, death maps, 185
Human Genome Project, 223
human migration routes, 227–28
humoral theory of disease, 133, 134, 244
Huntington's disease, 202
hydrographic data, 67, 83
hyperreal, 44

ICA. *See* International Cartographic Association
iceberg analogy, 124; Wegener's, 34, 163
icon, DNA molecules, 126–27
ideal gas law, and state-space maps, 166–69, 168f
imagination: effect on the world, 244–45; and metaphors, 31; projecting from physical realm to philosophical world, 87; representations, influence on, 243; representing our world, 256; Russell on enrichment of intellectual, 253–54
imaging technologies, satellite, 23
imagining "What if . . . ?," 251–52; in map of Copenhagen, 242f
improvements, explanatory, 108
*Incognito* (Eagleman), 188
*Indian Ocean Floor* panorama, 164, plate 9
induction, 34–35
inferences: abduction, 34–35; ampliative, 34–35; analogical, 32, 33–34, 35; deductive, 34–35; phylogenetic, 238

inferential practices, 55
inherent (natural) similarity, 125f; vs. stipulated similarity, 124–25
integration platforms: beyond-Mercator, 90, 108–9, 112, 117; demonstration of limits, 112; map assumptions, 213–14; mapping genetics, 213–15, 239–40, 239t; multiple maps as, 131f, 136–37; narrowing vs. broadening and decentralizing, 110; pernicious reification, blocking, 108; philosophical aspects, 112, 113–16; plurality of perspectives of, 108, 109; robustness of, 116; single maps as, 137
International Cartographic Association (ICA), map definition (1973), 5n6, 11n25
*Invisible Cities* (Calvino), 2, 3
isomorphism: account, 120, 121–24; in cartography, 123; definition, 121
Israeli presence, map of fluctuations, 130, 131f
Israel-Palestine conflict, history, 129–32

James, W., concepts as abstractions, 44–46
Japan, mapping in, 8

Kajita, H. S., 235
Kant, I., 52–53, 97
Kitcher, P.: defending realism, 247; existence of genes, 249; on map use impact, 84; representations, jointly consistent, 115
knowledge: building, 47, 54; production, 22
knowledge maps, types of spatialized, 37, 38n22, 54, 60
Koch, R., germ theory of disease, 133
Korzybski, A., 43, 44, 49, 134
Kripke, S., baptism event, 16

Kuhn, T. S.: constructivism of, 248n14; map analogy, 48; ontological assumptions and values, 252; on paradigms, 48; on paradigm's conceptual ontology, 195–96; puzzle-solving, science as, 47; scientific paradigm and map analogy, 47–48

Lahsen, M., and multiple representations account, 140
Lamarckism, 212
Lambert Conformal Conic projection, 75f, 76t; Tissot indicatrix, with and without, 111f
*Landscape of History, The* (Gaddis), 50
Laniakea, 155, plate 7
Le Guin, U., 2, 3, plate 1
Lewis, G. M., 12n35
linear genetic maps, 214, 215–16; assumptions, 223–24; evolution of, 221f, 222; of nucleotides, 222–23; phenotype linkage, 216–22, 219n9, 223
linear model assumptions, 201–5
linkage groups, gene map distances, 218–19
literal cartographic genetic map, 215, 227–29, 228f
literal maps, 12, 36, 54; analogy, 38; geology in, 149, 159–65; use of space, 114
local group galaxy mapping, 157–58
Logothetis, N., fMRI, 194
London Tube map (Beck), 79, 96
Longino, H.: on conformation, 96; on interactions and causal maps, 208; on statistical causal maps, 207; unified world picture, impossibility of, 116
longitudinal error independence, 201–2
*Look at the World* (Harrison), 104
*Lord of the Rings, The* (Tolkien), 1

*Love Symbol* (Prince), meaning of ⚥, 119
loxodromes (rhumb lines), 100

MacEachren, A. M., 60–61
manifolds, definition, 172
map abstractions, 3, 59, 69, 86, 159; universal, 251
*Mapa mundi de[l] reino de las In[di]as* (Poma), 19f
map analogies: Anglo-American analytic philosophy, 40–43; assumption archaeology of, 52–56; basic, 29, 35; employment of, 40; Euler diagram, 36f; European (continental) analytical philosophy, 40, 43–44; historical, 40, 50–51; across humanities, 51; and knowledge building, 54–55; Kuhnian, 48; lessons from, 51–52; a map of the world, 29; paradigms and revolution, 47–48; philosophy of science, 40, 46–48; pragmatic philosophy, 40, 44–46; religious studies, 40, 49–50, 56; scientific-general, 39; and social sciences, 56, 79–80; typology of, 36–40; uses of, 95, 113, 115. *See also* map thinking
map analogizing, definition, 6
map and territory, distinction, 137
map-based understanding, 7, 9, 14, 29
map generalization, 158; protocols, 79–80
map icons, similarity of, 125, 126
"map is not the territory, the" (MINT), 134, 137, 159
map layers. *See specific layers*
map of Copenhagen, 242f, 254
map of the empire (Borges), 3, 23
*mappae mundi*, 244; orientation, 21, 22
"mapper," 11
mapping: as an activity, 174; chromosomes, 205; and concept of function, 173–74; definition, 5, 40; genetic integration platform, future extensions, 239–40, 239t; history of Western, 8–9; *Homo sapiens*, 208; local group galaxies, 157–58; many-to-many, 190, 198f; in mathematics, definition, 170, 173; modern, 5; ourselves, 207. *See also* Waldseemüller, M., world map
mapping and space, 5, 11, 29, 56; cognitive (Harley), 7; definition, 40; extreme-scaled, 37; and GIS, 9–10; scientific theory/models, 37–38, 39, 54–55; and social structures, 14; space thinking, 7
mapping-genetics integration platform, 239t
map projections: conformal (quincuncial), 44, 45f; development in twentieth century, 104; distortions, 63, 75–77; scale, 70, 71–73; selection, 70–71, 74–77; unfolded, 74, 75f; Tissot indicatrix, with and without, 111f. *See also* map analogies; mapping; maps; *and specific projections*
maps: activities, world changed by, 243; amplifying, 72, 73f; cartographic, 160; and causal world structure, 206–8; characterization, 13–15; in classrooms, 86–87; coal, 162f, 163–64; complexity, omitting, 134–35; digital display, 9; humanists' understanding of, 51; internal, 51; internal characteristics, 135; as "invitations," 14, 24; in literature, 3; meaning of, 14; models for analysis, 41; multimodality, 114; necessary and sufficient conditions, 12–13; of ocean floor, 145f, 164–65, plate 9; partitioning, 15, 81, 192; perspectivizing, 15, 81, 128; popular, 11n25; power and limitations, 3; processual emergence, 13; and scientific theories, similarity, 55; search for

maps (*continued*)
ultimate, 23, 24; selection, "radical law," 72n39; space, 11; studies, definition, 5–6; symbols, 125f, 126; as "theater," 14, 18, 24; thematic, 29; truth in, 95–96; ultimate, search for, 23, 24; understanding of, 52; understanding the world, 85; uses, 85; as "weapons," 14, 24; of the world, student drawn, 58f, 86; world-making functions, 55. *See also* map analogies; map projections; *and specific map types*

maps, definitions: history of, 11n25; modern, 9; *Oxford English Dictionary*, 12; Robinson and Petchenik, 11–12

map thinking, 3, 4–6, 52, 141, 209; about ideology and power, 253; about migration, 180–81; of cartographic images, 254–55; cognitive doublethink, 138; elements of, 5–6; empiricist, 246; of ethical thinking, 252–53; isomorphism, 123; and mathematics, 174; philosophical methodology, 251–52; philosophical projections, plurality, 248; philosophical reflection, 4, 42; and pragmatic philosophy, 44, 47; realism in, 247; in religious studies, 49; requirements, 247; science and philosophy, 29, 243–44; in the sciences, 149, 174–75; on scientific methodology, 249–51; scope of, 14. *See also* map analogies

map units (centimorgans), 218–19

*Map Use* (Muehrcke and Muehrcke), 69

map/world, distinction, 60

map–world–map: feedback, 84n67; gene expression maps, 84–85; global climate change, 84n68

Margulis, L., champion of symbiosis evolutionary process, 145f

Martellus, H., map of, 15n43

mathematical symbols, isomorphism vs. similarity account of, 127

mean sea-level, 64, 65, plate 3; and geoid/ellipsoid/sphere relation, plate 3

Mendel, G.: crossbreeding experiments, 212; inheritance principle, 211–13, 216; traits, 211–12

Mercator projection, 55, 75f, 76t; criticism of, 103; history, 99–102; purpose, 136; scale factor for, 122; Tissot indicatrix, with and without, 110, 111f; transformation equations, 76, 101, 107; transverse, 64n12; universalizing, 91–92; usefulness and failings, 85; in wall maps, 107

Mercator's map (1569), 88f, 89, 90, 244; alternatives to, 102; latitude and longitude presentation, 101; for navigation, 100–101; septentrion inset, 102

merely-seeing-as, 134–35; Carboniferous/contemporary worlds, treated as, 164; and climate change, 140–41; and economists, 142n68; and multiple representation analysis, 128; and ontologizing, doublethink, 138

metaphor, use of, 31

metric layers, 108, 109t, 120, 127

migrants: classification, 183–84; movement countermapping, 181; representations of, 180–81

migration: laws of, 182–83, 184f; by stages, lack of individual agency, 183–85; world routes, 182f

migration maps: arrowing assumptions, 181–82, 189; graphical assumptions, 187

Miller projection, 92

mimetic icons, 125f, 126

MINT. *See* "map is not the territory, the"

mitochondria, 145f

model-based generalization (in GIS), 9
"Modeler's Hippocratic Oath," principles of, 94
model map analogy, 56
*Model Organisms*, 223, 230, plate 10
models: analogizing, maps to, 35, 36–37; canonical idealized, 52; causal, sleep quality, 204–5, 204f; cosmological, 156; curved surfaces, 74; decompositional brain, 188; of Earth, 63–65; as epistemic tools, 142n68; explanatory, 60; isomorphism, 123; mathematical, 80, 112, 127, 142n68, 166, 243; of organismic morphology, 112; pernicious reification, forestalling, 199; and theories, relations, 46; universalizing, 91
Mollweide projection, 76n45, 98f, 148f, plate 7
morphology, 214
motor homunculi, 189f; and pernicious reification, 192–93
Muehrcke, J. O., 134
Muehrcke, P. C., 134, 135
multimodality, 18
multiple representations: and climate change, 138–41; and multiple assumptions, contrast, 108; relation among, 115. *See also* representations
multiple representations account, 120, 127–28; analysis, 128, 137; and climate change modeling, 138–41; limitations, 141–42. *See also* representations
multivariate normal distribution, 201
myoglobin, three-dimensional structure maps, 230

National Geospatial-Intelligence Agency, 24, 65n15, plate 3
natural kinds classifications, 78
negative analogies, 31, 32, 33
Nether-Saxon geodetic survey, 170, 171f
network topologies, 237

neural reuse, 197
neuroplasticity, 198
neuroscience, 187; cognitive ontology, 196; contextual objectivity, 187; fMRI in, 194
neutral analogies, 31, 32; analogy of interest, 33
nongenetic explanations, phenotype, 55
*North Atlantic Physiographic Diagram* (Tharp), 164
nucleotide genetic maps, 216; of free-living organisms, 224; *Homo sapiens*, 223; in mammals, similarity, 230
nucleotides, 222

objective universality, 97–98
objectivity/subjectivity, simultaneous roles, 97
*Observable Universe Logarithmic Map* (Budassi), plate 8
ocean floor maps, 145f, 164–65, plate 9
Olsson, G.: on physical distance replacement, 186; raising political consciousness, 186
one-to-one and 1:1 maps, distinction, 25n78
one-to-one scaled map, 2, 23
"On Exactitude in Science" (Borges), 1–2, 25
*On the Origin of Species* (Darwin), 234
ontogenesis, shift to ontology, 13
ontological assumptions, 108, 109t, 132
ontological layers, 120, 127, 141, 143; of Palestine Plan of Partition, 130, 131f, 132
ontologizing: Carboniferous/contemporary world, treated as, 164; cartographic representation, pluralistically, 109n65, 135–38, 240; in cartographic/scientific contexts, 129–34, 141; and climate change, 140–41; definition, 60, 81; doublethink of, 138; germ theory of

ontologizing (*continued*)
  disease, 133–34; and map use, 61; map/world link, 60; and merely-seeing-as, 128, 138, 140–41; pluralistic, 135–38, 240; and representation testing, 82; social sciences, 142, 142n68; subjective indexicality onto objective universality, 97–98
ontology: cognitive, 195–97, 199; conceptual, 195–96; controlled vocabularies, 195–96; of geography, 179n8; shift to ontogenesis, 13
organismal complexity, evolution of, 231
*Origin of Continents and Oceans, The* (Wegener), 159
orogeny, 159
orthographic azimuthal projection, 99
orthologous genes, 230

Palestine Plan of Partition (UN), ontological layer, 130, 131f, 132
paradigm map analogy, 48
paralogous genes, 230
partitioning, 15; brain regions, 192
partitioning frame, 81, 195, 206; assumptions, genetic maps, 213–14; of comparative genetic maps, 230–31; extreme-scale maps, genes, 225; gene expression maps, 84–85, 214–15, 224–25; of genotype-phenotype map, 227; in linear genetic maps, 223–24; literal genetic maps, parts of, 229; mapping-genetics integration platform, 239t; of phylogenetic analogous maps, 238. *See also* adaptive landscape maps
Pasteur, L., germ theory of disease, 133
path diagrams, 200; abstractive-averaging linear methodological assumptions, derivation from, 204; causal model, sleep quality, 204–5, 204f

Peirce, C. S.: end of inquiry concept, 115; maps and reality, 44, 45f; map signs, 79; modes of influence, 250–51; symbols/icons/indices, distinction, 125–26
Penfield, W.: decompositional assumptions, adoption of, 193; "organs of the brain" model, 192
performance cartography, 12n35
performativity of economics, 85
permissible system states, 166
pernicious reification, 55, 56–57, 82, 89–90; arrowized assumptions and, 181; cognitive basis of, 91n6; components of, 91; definition, 90, 94; dinosaur extinction, example, 92–94; forms of, 134; integration platform, 108; of Mercator projection, consequences, 102–3; and merely-seeing-as, 128; and "Modeler's Hippocratic Oath" principles, 94; narrowing, 91; of scientific representations, 94; and somatosensory homunculi, 192–93; universalizing, 91
personal maps: autobiographical, 98; exercise, 98f
perspective maps (Harrison), 105–6, 105f, 136
perspectives, pluralism of, 41
perspectivism, 142–43
perspectivizing, 15, 81, 128
Pescadero State Beach maps: comparison of, 121, plate 6; symbols in, 126, plate 6
Petchenik, B., 7, 60; on map analogy, 51; on map definition, 11
Peters, A.: cartography, critique against, 106; on Mercator projection, 103, 244; single projection, 103. *See also* Gall–Peters projection
phase diagram, 167, 168f
phenotypes, 214

*Index* [315]

phenotype space, 210f
phenotypic linkage maps, 216, 219n9; mapping logic, 217–22
phenotypic state space, 226
philosopher's maps, 11n25
*Philosophical Investigations* (Wittgenstein), 42–43
phrenology, 187; abstractive-averaging assumptions, 189–90; assumption archaeology, brain structure, 190–91; decompositional assumptions, 193; mind-brain relationship, 192
phylogenetic analogous maps, partitioning frame of, 238
phylogenetic trees, 238
phylogenies, 227, 238–39
pictographic icons, 125f, 126
Pillewizer, W., 72n39
pink Indiana, 3
place, concept of, 11n31, 187, 252n22, 256
Planck, M.: length, 72, 145f; project, 151; time, 145f
plane (flat) surveying, 62
pluralistic ontologizing, 109n65
pluralistic thinking, integration, 196
polar azimuthal equidistant projection, 75f, 76t, 102; Tissot indicatrix, with and without, 111f; *The World Centrifuged*, 77f
Poldrack, R., Cognitive Atlas, 195–97, 199
political and social context, 54, 55, 57
Poma, G.: *cosmovisión*, 22; counter-map, 13, 15, 18–22, 19f, 244
*Pondering Darwin's Forms of Life* (Kajita and Winther), 235, 236f
Popper, K., 250
portolan charts, 100
positive analogies, 31–32, 32n10, 33
*Powers of Ten*, spatial and time scales represented, 73f
practice/theory distinction, 60n1

pregiven world, 60, 119, 132
Prince, Artist Formerly Known as (AFKAP), 119n2
principle of independent assortment (Mendel), and gene mapping, 216
principle of segregation (Mendel), 216
privileging legibility, 50–51
projections: assumptions, evaluation, 108, 109t; composite, 76n45; compromise, 76, 104; conformal, 44, 45f, 75, 76t; distortions, 64n12, 75, 112n72; equal-area, 76, 103, 111f, 136; equidistant, 75, 99; space distortions, 122; Tissot indicatrix, with or without, 110, 111f; transformations, 74; types, 75f, 76t, 99; unconceived, 92
pseudoconic projection, 76t, 99–100
Ptolemy, C., 15
Punnett squares: for crossbreeding, 212t; two-gene, 218t

"radical law" of map selection, 72n39
Ramsey, F. P., "belief," 40–41, 51
Rasmussen, T. B., 192; decompositional assumptions, adoption of, 193
Ravenstein, E. G., 182–84, 184f
realism: defending realism (Kitcher), 247; map analogy, 245, 246–47
recessional velocity, 153, 157n21
recessive traits, 212
recombinant genotype, 218
rectangular maps, views on, 106
regression lines, 201
reification: cartographic, 177; of intelligence, 197; during ontologizing, 82n62. *See also* pernicious reification
representation accounts, multiple. *See* multiple representations account
representational generalization, 158–59
representational pluralism, 112, 113–15

representational practices, 7, 35, 37, 44, 52, 54, 55
representational structures, 82–83
representations: analytical perspective accounts, 120; and assumption archaeology, 53; dynamic, 155; excavated, 53; fitting to realities, 50; influencing dreams, 243; of migrants, 180–81; rebuilding a world, 84; relations, 120–21; similarity of, 124; and targets, similarity, 127n25; testing of, 82–83; use and abuse, 82; and world relation, 119–20. *See also* multiple representations account
*Rethinking the Power of Maps* (Wood), 132
rhumb lines (loxodromes), 100
Riemann, B., 172
Ringmann, M., 15
Robinson, A. H., 7, 11, 51, 60, 69
Russell, B., on enrichment of intellectual imagination, 253–54
Ryle, G., 41–42

Sale, R. D., 69
San Francisco Bay Area, USGS topographic map, 28f, 29, plate 4
*Satellite Map of the Earth* (Van Sant), 13, 15, plate 2
scaffolding, 13, 53
scale factor, 122
scales: in Google Maps, 123; selection protocol, 70, 71–73, 71n35; spatial and temporal, 73f; transformation, 122
science: compositional vs. formal, 38n22; incompleteness of description of, 249
scientific analogies, 30t, 34n13
scientific inquiry, 250
scientific maps, 11n25
scientific methodology, 250–51
scientific models: and contextualizing, 109, 112; evaluation of, 96; map of the world, 46; and simulation, 56

*Scientific Representation* (van Fraassen), 97
scientific theories, a map of the world, 29, 35
similarity account, 120, 124–27
simplification protocol, 77–78
*Simulacra and Simulation* (Baudrillard), 43–44
Smith, J. Z., maps of, 49
Solnit, R., 14
somatosensory homunculi, 189f, 192; and pernicious reification, 192–93
source domain, 31
space: distortion, 122; dual, 162f, 164, 174; importance of, 54–55; literal and geographic, 229; modeling of, 9–10; new, 186; thinking, mapping, 7
*Spaces in Migration* map, 185–86, 186f
Spanier, B.: existence of genes, 249; on gendered language, 245–46
spatial and temporal scales, representation, 73f
spatial calibration, 68; and FLRW metric, 59
spatial information, uses, 54
spatialized cognition/communication, 29. *See also* deep mapping
spatialized knowledge, and map typology, 37, 38n22, 54–55, 60
spatialized worlds, cartographic representations, 15
spatial layers, 122, 141
spatial map analogy, 36f, 37, 38, 40n24, 141, 225
spatial thinking, and space, 54
speciation, 231
standard maps, 22, 208t, 209; misleading tendencies of, 180
Stanford, K.: existence of genes, 249; on history of science, 246
*Star Wars: The Force Awakens*, holographic maps, 1

state-space maps, 36, 38n22, 54, 210f; analogy, 37, 166; gene combinations, diagram, 232f; ideal gas law example, 166–69, 168f; phase diagram, 167, 168f; phenotypic, 226; in physics and physical chemistry, 149, 165–69, 174; relative fitness value, 231–32; time dimension, 40n24; as topographic maps, 169

statistical causal maps, 199–200; linear model assumptions, 180; statistical countermaps and linear assumptions in, 207

*Statistical Methods for Research Workers* (Fisher), 69

statistics: ANOVA, 68–69, 226; independent/dependent factors, correlation, 201; and mapping, 199–200

Stevenson, R. L., 1

stick figure, galaxies, 148f, 154, plate 7

stifling particularity, 50

stipulated vs. inherent similarity, 124–25

strong structure-function mapability, brain structures, 190

"structure-mapping" theory of analogy, 35n19

*Structure of Scientific Revolutions, The* (Kuhn), 47, 248n14

Sturtevant, A. H., gene-mapping logic, 216–18

subjective indexicality, onto objective universality, 97–98

supergalactic coordinate system, 158, plate 7

symbiosis, 145f

symbolic layers, 120, 124, 126, 127, 141–42

symbolization, cartographic, 79

symbols, meaning of, 119

symbol-world analogy, 126

synchronic time, 229

system trajectories, possible, 166

target domain, 31

Tazzioli, M., 186

Tharp, M., 145f, 164–65

theoretical models, and empirical substructures, 83

theory/practice distinction, 60n1

three-dimensional map of local universe, 148f, plate 7

three-dimensional maps: of galactic motion, 154–55, plate 7; of protein structures, 230

time: calibrating, 65–66; map representation of, 40n24; and space, relation, 65–66

Tissot indicatrix, 64n12, 110, 111f

Tolkien, J. R. R., 1

*Tom Sawyer Abroad* (Twain), conflation map and territory, 2–3

Töpfer, F., 72n39

topographic maps, 15, 28f, 29; dynamic, 121; state-space maps as, 169

Toscanelli, P., 178–79

*Tractatus Logico-Philosophicus* (Wittgenstein), 42

traditional cartography and GIS, comparison, 10

transformation equations, 65, 85, 113; Mercator's, 76, 101, 107

transformation functions, generalized, 74

transformation laws, 226

Treasure Island map (Stevenson), 1

Tree of Life (Darwin), 234–37; as an analogical map, 39; analogy example, 32–34; countermapping, 235, 236f; phylogeny, illustrating, 238

triangulation, 62–63

trilateration, 62–63

Tully, R. B., 154–55, plate 7

*Turing Phrenunculus, The*, 188, 189f; decompositional assumptions, consistency with, 191; decompositional brain model, 188

Turkheimer, E., 207
Twain, M., 2, 3, 135

uncorrelated residuals, 201
*Universal Cosmography According to the Tradition of Ptolemy and the Discoveries of Amerigo Vespucci and Others, A* (Waldseemüller), xviii
*Universalis cosmographia* (Waldseemüller), 16–18, 22
universalizing, 90; vs. contextualizing, 109
Universal Transverse Mercator (UTM) coordinate system, 64n12
universe: as energy dominated, 150; history of, 150–51, 152–53; in space and time, maps of, 157
*Universe Particulars*, 145f
USGS topographic map, San Francisco Bay Area, 28f, 29, plate 4
UTM (Universal Transverse Mercator) coordinate system, 64n12

van Fraassen, B. C., 83n64; on maps and scientific theories, 97
van Hoogstraten, S., on ontological layers, 143
Van Sant, T., world map, 13, 15. *See also* GeoSphere map (Van Sant)
variables, explanatory, 201
Venter, C., 224
Venus de Milo (Aphrodite), 119
Vespucci, A., on naming of "America," 17

Waldseemüller, M., world map, 8, 13, 15–18, 15n43; "America," use of, 16, 244; multimodality, 18
Watson, J., 126
Web Mercator projection (online), 76, 90, 107
Wegener, A., 159–64, 162f
West-Eberhard, M. J., vision of evolutionary process, 227

Western mapping, history, 8–9
WGS 84. *See* World Geodetic System 1984
Wilkinson Microwave Anisotropy Probe (WMAP) nine-year map, 148f, 151, plate 7. *See also* cosmic microwave background (CMB) map
Wilson, M., maps as models, 41
Wittgenstein, L.: influence within analytic philosophy, 42–43; on models mapping the world, 42
WMAP nine-year map. *See* Wilkinson Microwave Anisotropy Probe (WMAP) nine-year map
Wood, D., 106; cartographic constructivist suggestion, 245; maps as "weapons," 14, 24; meaning of maps, 14; on ontological assumptions, 132; on symbols, 125
Woodward, D., 12n35
*World Centrifuged, The* (Harrison), 77f
World Geodetic System 1984 (WGS 84), 64; and EGM 2008, 65; ellipsoid, 74
worldmaking process, 79n54
world maps: canonical, 24; evolution, 24; Poma, 18, 19f; scientific models are, 46; scientific theories are (analogy 1), 29, 35; student-drawn, 58f, 86; Van Sant, 13, 15; Waldseemüller, 8, 13, 15–18, 15n43
world migration routes, 182f
world navels, 22; of Earth, 77; as map centers, 110; overtones of, 20n60
*World Ocean Floor Panorama*, 164, plate 9
*World on a Quincuncial Projection, The* (Peirce), 45f
Wright, S., 112n73, 205, 231, 232–33, 232f, 232n33

*Yellow Book* (Wittgenstein), 42

Ziman, J., 113